PREDICTION ANALYSIS

by

JOHN R. WOLBERG

Senior Lecturer
Technion—Israel Institute of Technology
Haifa, Israel

D. VAN NOSTRAND COMPANY, INC.

PRINCETON, NEW JERSEY

TORONTO   LONDON   MELBOURNE

Van Nostrand Regional Offices: *New York, Chicago, San Francisco*

D. Van Nostrand Company, Ltd., *London*

D. Van Nostrand Company, (Canada), Ltd., *Toronto*

D. Van Nostrand Australia Pty. Ltd., *Melbourne*

PRINTED IN THE UNITED STATES OF AMERICA

1083492

To

LAURIE, BETH, and DAVID

"The other type of criticism to which experimental results are exposed is that the experiment itself was ill designed, or, of course, badly executed. If we suppose that the experimenter did what he intended to do, both of these points come down to the question of the *design*, or the *logical structure* of the experiment. This type of criticism is usually made by what I might call a heavyweight *authority*. Prolonged experience, or at least the long possession of a scientific reputation, is almost a pre-requisite for developing successfully this line of attack. Technical details are seldom in evidence. The authoritative assertion "IIis *controls* are *totally* inadequate" must have temporarily discredited many a promising line of work; and such an authoritarian method of judgement must surely continue, human nature being what it is, so long as theoretical notions of the principles of experimental design are lacking—notions just as clear and explicit as we are accustomed to apply to technical details."

Ronald A. Fisher
THE DESIGN OF EXPERIMENTS

סוף מעשה במחשבה תחילה

RABBI SHLOMO HALEVI ELKAVATZ, *Circa* 1500

Preface

In all areas of science and technology there is a continuing need for work of an experimental nature. Often the experiment consists of the determination of a set of data points and then analysis of the data by "fitting" the points to a mathematical model. I refer to this type of experiment as a "quantitative experiment."

Prior to the widespread availability of high-speed digital computers, graphical methods of curve fitting were usually employed for analysis of most experimental data. Although they are time-saving, graphical procedures suffer from the rather arbitrary nature of the results. It is now common practice to analyze data using the method of least squares and for all but the simplest mathematical models it is necessary to perform the analysis with the aid of a digital computer. Besides eliminating the need for arbitrary judgment, the method of least squares yields estimates of the uncertainties associated with the resulting least square parameters. These uncertainties are crucial in the evaluation of the experiment.

As the need for experimental accuracy increases, the cost and complexity of quantitative experiments invariably rise. To satisfy the objectives of such experiments, there is an increasing need for sophisticated methods of experimental planning and design. *Prediction Analysis* is a technique that can be used to facilitate the planning of quantitative experiments.

In this book the method of prediction analysis if formulated and application of the method to a number of important classes of experiments is considered. To make the book comprehensible for the largest possible number of readers, only a knowledge of mathematics through basic calculus has been assumed. All essential background material in the areas of statistics and the method of least squares is included in Chapters 2 and 3.

The book has been organized as a reference and is directed to scientists and engineers interested in work of an experimental nature. Although the primary emphasis is on the planning of experiments, analysis of experimental data using the method of least squares is also discussed in detail. Concepts important to the digital computer programming of prediction and least squares analyses are emphasized throughout the book.

The book does not fit into the framework of any of the courses which now form the basis of a scientific or engineering education. However, university curricula are in a state of change with a greater emphasis being placed upon fundamentals. Undergraduate courses on digital computer programming are already quite common and within the next few years I will not be surprised if experimental planning and data analysis are also included as a part of the basic technological education. To make the book useful as a textbook as well as a reference, problems and computer projects are included at the end of most of the chapters.

The problems and computer projects are also useful for those readers using the book as a reference. The problems offer the reader an opportunity to test his comprehension of the material within the chapters. The simpler computer projects (for example, Project 1 or 2 of Chapter 5) can be used to test the prediction analysis computer programs developed by the readers for their own particular experiments. The more ambitious projects are included for possible student use.

The well known books on experimental design and planning (for example, W. G. Cockran and G. M. Cox, *Experimental Designs*, John Wiley & Sons, Inc.; D. R. Cox, *Planning of Experiments*, John Wiley & Sons, Inc.) are primarily concerned with "comparative experiments." The only other book that I have been able to locate on the planning and analysis of quantitative experiments was originally written in Russian" by physicists and for physicists" but is now available in English. (*Analysis and Planning of Experiments by the Method of Maximum Likelihood*, by N. P. Klepikov and S. N. Sokolov, Pergamon Press Ltd.) The authors are primarily interested in the least squares analyses of experiments in which the fitting functions are linear. The subject of experimental planning is directed toward location of the values of the independent variable which minimize a particular "functional." In their preface, the authors also note that treatment of quantitative experiments is either "completely absent or presented in a very limited form in most books on mathematical statistics and the treatment of experimental data."

Planning of quantitative experiments has been discussed in the statistical literature, notably in the papers by G. E. P. Box and his colleagues. I have not located any prior treatments of the general problem; however, some discussions related to the design of particular classes of experiments resemble the approach presented in this book.

The need for a method such as prediction analysis cannot be disputed. I have found it easiest to convince those scientists and engineers that have been involved with time-consuming or costly experiments of the need for such analyses. For those whose experience has been limited to simple experiments, or who have never been involved with quantitative experiments, I have often detected an attitude of "Why bother?" The answer to this question becomes more obvious as the importance, cost, and time required for a proposed experiment increase. By properly employing the method, the experimental plan can result in considerable savings of time, money, and effort.

Haifa, Israel   JOHN R. WOLBERG
April, 1967

Acknowledgment

I would like to thank J. Helholtz for his critical reading of the manuscript and for his many helpful suggestions. I am indebted to S. Aronson and D. Ilberg for their comments concerning Part I.

The value of the discussions that I had with S. Shalev, W. Rothenstein, and T. Gozani is gratefully acknowledged. They reviewed and contributed to my early papers related to prediction analysis, and their many helpful comments influenced the development of this general presentation.

The staffs of the computation centers of the Technion, the Israel Defence Forces, and New York University (Uptown) were extremely helpful. Their cooperation facilitated the computations leading to the numerical results included throughout the book.

I would like to thank M. Pozdena for typing the manuscript and correcting innumerable spelling mistakes. Finally, my wife deserved special commendation for her patience and help during all phases of the preparation of this book.

Contents

PART II

Part I. Theory

Chapter 1

Introduction

1.1 The Experimental Method

A complete experiment consists of four distinct phases. Chronologically these are:

1. Planning
2. Execution
3. Data Analysis
4. Interpretation

Extraordinary advances have been made in all areas of science and technology to facilitate and improve the "execution phase" of experiments. The potential accuracy of measurements is continually being extended by improvements in all types of measuring instruments. At the same time, the scope of commercially available instruments is being broadened and the time required for completion of a given measurement is being reduced.

The improvements in the "execution phase" of experiments have most certainly been equaled in the "data analysis phase." Digital computers now process in seconds the same amount of data that would require months or years to process with a desk calculator. The digital computer has added a new dimension to data analysis. Techniques of analysis which were invariably rejected because of the complexity of the required calculational procedures are now used on a routine basis.

In the final phase of an experiment, a decision is made concerning the interpretation of the results. This decision-making process is often facilitated by results from similar experiments, and from theoretical analysis of the problem. Typical decisions are to accept or reject the results, to repeat the experiment, or to modify the experiment in light of the results. Once again, advances in science and technology have greatly facilitated this phase of the experiment. Theoretical calculations can now be made to a much greater degree of sophistication, primarily as a result of the digital computer.

As it becomes easier to execute, analyze, and interpret experiments, the accuracy requirements are becoming more stringent. For an experiment in which 5% accuracy was considered acceptable several years ago, current objectives for similar experiments might require 1% accuracy. In addition, the advances in experimental techniques and analysis have not been cheap. The costs of buying equipment and performing experiments have risen considerably.

For these reasons the "planning phase" of an experiment is becoming more important. Considerable amounts of time, money, and effort can be wasted if planning is not approached with the same degree of sophistication as the latter phases of the experiment. Care should be taken to ensure a well-conceived experiment in which the results will be worth the effort expended.

1.2 Planning an Experiment

The first step in the planning phase of an experiment is to specify the objectives. The objectives are often stated in terms of accuracy require-ments. Such factors as cost and time required to prepare and execute the experiment can also be included in a statement of the objectives. These objectives should influence the methods of measuring the variables and the general experimental plan. For example, if the requirements are for results of high precision or accuracy, measuring instruments capable of high precision should most probably be selected and the experiment should be performed in a manner conducive to achieving results of high accuracy.

The next step in the planning phase of an experiment should be a pre-diction of the accuracy of the results as a function of the experimental variables. Such a prediction will provide a basis for answering several important questions:

1. Can the proposed experiment yield results with sufficient accuracy to satisfy the objectives?
2. What choice or choices of the experimental variables (e.g., number of data points, accuracy of the individual data points, etc.) will best satisfy the objectives?

Most quantitative experiments are directed toward determining the values of unknown parameters. If the unknown parameters are to be measured directly, prediction of their accuracies as a function of the ex-perimental variables is a relatively simple task. In many experiments, however, the unknown parameters are not measured directly. A relationship between the dependent and independent variables is used as the basis of an analysis to determine the unknown parameters. For such experiments, prediction of the accuracies as a function of the experimental variables is complicated.

Prediction analysis is a general method for treating problems of this nature. The accuracies to which the unknown parameters can be determined in a proposed experiment are predicted as a function of the experimental variables. Results from a "prediction analysis" of a proposed experiment can then be used as a basis for planning the experiment.

1.3 Definition of Prediction Analysis

Prediction analysis is a general method for predicting the accuracy of results that can be expected from a proposed experiment. Consider an experiment in which values of a dependent variable y are to be related to an independent variable x according to a given function:

$$y = f(x; a_1, a_2, \cdots, a_p),\qquad (1.3.1)$$

where the a_k's are the unknown parameters of the function.

The purpose of the experiment is to determine the unknown parameters a_k. Analysis of the experimental data also yields estimates of the σ_{a_k}'s which are the standard deviations of the values of the a_k's. (See Section 2.4 for a definition of standard deviation.) The magnitude of the σ_{a_k}'s indicates how accurately the values a_k have been determined. A small value of σ_{a_2}, for example, indicates that a_2 has been determined to a high precision of accuracy. The values of σ_{a_k} clearly play an important role in the interpretation of the experiment. For this reason, results are rarely presented without an estimate of the magnitude of these standard deviations.

The purpose of prediction analysis is to predict the values of σ_{a_k} as a function of all the parameters, variables, and standard deviations of the experiment. Consider an experiment in which σ_{x_i} (the standard deviation of the ith measurement of the independent variable x), and σ_{y_i} (the standard deviation of the ith measurement of the dependent variable y) can be related to x and y:

$$\sigma_{x_i} = g(x_i),\qquad (1.3.2)$$

$$\sigma_{y_i} = h(y_i),\qquad (1.3.3)$$

where g and h are two specified functions. Prediction analysis yields predicted values of the σ_{a_k}'s as functions of the number of data points, the values of x, the functions g and h, and the values of the unknown parameters a_k.

For example, consider a "straight-line" experiment:

$$y = a_1 + a_2 x.\qquad (1.3.4)$$

The physical significance of the variables x and y is irrelevant. In Example 4 of Section 4.5, a "straight-line" experiment is discussed in which y is the length of a spring, x is the weight of a load attached to the spring, a_1 is the unextended length of the spring, and a_2 is the spring constant. The same type of analysis could be applied to any experiment in which Equation 1.3.4 is valid. For example, y could be the length of a bar at temperature T, a_1 could be the length at temperature T_0, x could be the temperature difference $(T - T_0)$, and a_2 would then be the temperature coefficient of expansion.

If the values of y can be measured to an accuracy of $100\, K_y$ per cent, then:

$$\sigma_{y_i} = h(y_i) = K_y \cdot y_i, \qquad (1.3.5)$$

and if the standard deviations of the values of x are assumed to be constant:

$$\sigma_{x_i} = g(x_i) = K_x. \qquad (1.3.6)$$

Prediction analysis is then applied to the problem of predicting values of σ_{a_1} and σ_{a_2} as functions of the number of data points n, K_y, K_x, a_1, and a_2, and the values of x for all n data points. The results of the analysis can be used to facilitate the planning of the experiment.

The analysis is not, of course, limited to the functions g and h specified by Equations 1.3.5 and 1.3.6. Any functions that satisfy the realities of the proposed experiment are acceptable. Nor is the analysis limited as to the choice of the function f. To simplify this introductory discussion, in Equation 1.3.1 the dependent variable y is only a function of one independent variable (i.e., the variable x). The method is developed for the more general case in which y can be a function of more than one independent variable.

1.4 Application of Prediction Analysis

The method of prediction analysis can be applied to any proposed experiment in which a set of data points will be "fitted by a curve" in order to determine values of the unknown parameters of the curve. The method is useful for any type of curve, any number of independent variables, and any number of unknown parameters. A significant fraction of all experiments being performed in all areas of science and technology falls within this category.

Part II of the book includes prediction analyses of some of the most important "classes of experiments." An experiment may be classified by the functional form of the curve that is used to fit the data. Regardless of the physical significance of the variables, all experiments, in which the same function is applicable, are assumed to be in the same "class." In Section 1.3, for example, the "straight-line" experiment was discussed. It was emphasized that the prediction analysis applies to any experiment in which the variables y and x are to be related by an equation of the form $y = a_1 + a_2 x$.

1.5 Connection with the Method of Least Squares

Prediction analysis is related to the method of least squares. The method of least squares is a well-known technique for fitting a curve to a set of data points. Prediction analysis is based on the assumption that the method of least squares will be used for analysis of the data once the experiment has been performed.

The method of least squares was developed by Gauss in the early part of the nineteenth century. Although the method of least squares yields the "best" estimates of the unknown parameters,[1] it has been applied extensively only in recent years. The considerable amount of computation required for a solution had been a formidable obstacle for acceptance of the method. Widespread availability of digital computers has completely altered the situation. The calculations, when performed with the aid of a digital computer, no longer present an obstacle to arriving at a solution.

The similarity of prediction and least squares analysis can be exploited by designing similar computer programs which can be used to perform both types of analyses. Block diagrams for such programs are discussed in Section 4.7.

1.6 Scope of the Book

The book includes both the development of the method of prediction analysis and the application of the method to some of the most important classes of experiments. To make the book "self-contained," essential background material has been included. Those aspects of statistics which are important for the development of prediction analysis and development of the method of least squares are discussed in Chapter 2. The method of least squares is treated in Chapter 3, and prediction analysis is developed in Chapter 4.

Modern analysis relies to a considerable extent on the digital computer. For this reason the treatment of both prediction and least squares analysis includes practical aspects related to digital computer solutions. Illustrative examples of the techniques discussed in the book are included.

The book is divided into two parts—Theory (Part I) and Applications (Part II). The applications treated in Part II include some of the most common classes of experiments. The results for these applications can be used directly for the planning of experiments.

[1] See Section 3.1 for an explanation of the word "best".

Chapter 2

Statistical Background

2.1 Introduction

In this chapter the statistical background required for an understanding of prediction analysis is presented. Statistics is a very large and important branch of science and no attempt is made to survey the entire field. Only those topics and concepts related to prediction analysis and least squares theory are treated.

It is worthwhile to precede the discussion of statistics by some general remarks concerning the nature of experiments. Most experiments fall within one of two categories—qualitative or quantitative. For example, consider an experiment for which the objective is to synthesize a particular chemical compound. The experiment is completed by first performing the proposed process and then testing to see if the desired compound is present. Such an experiment is qualitative. The process either works or it does not work. The experiment assumes a quantitative nature when, for example, the rate at which the compound is produced is measured as a function of the temperature at which the chemical reaction takes place.

Prediction of the success or failure of qualitative experiments is usually based upon theory and prior experience. The method of prediction analysis cannot be applied to experiments of this type. It can, however, be applied to quantitative experiments.

Quantitative experiments can be classified according to the mathematical model used to relate the dependent and independent variables. Models can be considered as either *stochastic* or *deterministic*, according to whether they do or do not contain random variables.[1]

"Stochastic" or random effects may enter a model in a fundamental manner. For example, in quantum mechanics, probability distributions

---

[1] P. Whittle, "Prediction and Regulation by Linear Least-Square Methods," English Universities Press, Ltd., 1963.

rather than actual locations of particles are of interest. Often, however, random effects are present in an experiment because of inaccuracies of the measurement techniques or failure to account for all variables affecting the experiment. For example, failure to control the temperature of an experiment might cause random variations in the experimental results.

The opposite of "stochastic" is "deterministic." A deterministic model does not include random variables. For example, prediction of the location of a space vehicle is based upon a deterministic mathematical model.

The ideal model is based upon a theoretical analysis of the problem. For example, prediction of the orbit of a space vehicle is subject to theoretical treatment in which the general laws of motion are applied to the problem. Often, however, the theory is unknown or too complicated for a particular quantitative experiment. For such experiments, an examination of the results might lead the experimenter to a mathematical model that "fits" the data. Although the model might not be theoretically correct, it can be used for such purposes as interpolation, integration, differentiation, etc.

Regardless of whether the model is stochastic or deterministic, or whether it is based upon theory or observation, most quantitative experiments have one unifying aspect. That is, the purpose of the experiment is usually to determine the unknown parameters of the mathematical model. In Chapter 3, the method of least squares is discussed. This method can be applied to the general problem of analyzing experimental data to determine the unknown parameters of the model. The method of prediction analysis is developed in Chapter 4 and can be applied to the general problem of predicting the uncertainty associated with these parameters.

2.2 Statistical Frequency

Many types of observations and measurements exhibit random variations. For such processes, if the observation or measurement is repeated, the resulting values will not necessarily be the same. In statistical terminology, any observed quantity that varies from one repetition of a given process to the next repetition is called a "random" or "stochastic variable."

As an example of a "random variable," consider an experiment in which a coin is tossed ten times and the number of times that heads are observed is tabulated. The number of heads per ten tosses is a "random variable." Only "discrete values" of this variable are possible for each experiment. The variable can therefore be classified as a "discrete random variable." The possible values that this "discrete random variable" can take in a given experiment are $0, 1, 2, \cdots, 10$. The variable does not, however, take each value with the same "frequency." If the experiment were repeated many times, one would expect to observe 5 heads more often than any of the other possible values. One would also expect to observe 0 and 10 heads less often than any of the other possible values. Statistical analysis can be applied to the problem of predicting the "relative frequencies" of all the possible values for this experiment. *For a discrete random variable, the relative frequency of a particular value is the fraction of all experiments in which the particular value can be expected to be observed.*

It is useful to express the concept of "relative frequency of a discrete random variable" in mathematical terms. If the notation x is used for the variable, then the *frequency function* $\phi(x)$ is the relative frequency of the variable x. For the preceding example, the possible values of x are 0, 1, 2, $\cdots$, 10. Summing the relative frequencies gives

$$\phi(0) + \phi(1) + \phi(2) + \cdots + \phi(10) = 1. \tag{2.2.1}$$

(The sum of the frequencies of all possible values of x is, by convention, equal to one.) Equation 2.2.1 can be generalized for any "discrete random variable":

$$\sum_{\text{All values of } x} \phi(x) = 1. \tag{2.2.2}$$

That is, the "frequency function" $\phi(x)$ summed over all possible values of x equals one.

Many observations and measurements are concerned with "continuous" rather than "discrete random variables." A "continuous random variable" can assume any value within a given range. It is not limited to just integer or discrete values. As an example of a "continuous random variable," consider the measured values of the diameters of shafts produced by a lathe. If the measuring instrument is sensitive enough, a variation in diameter from shaft to shaft will be observed. The measured diameters will be "distributed about an average value."

An equation similar to Equation 2.2.2 is applicable for "continuous random variables." Defining $\phi(x)dx$ as the probability (or relative frequency) of observing the variable in the region $x - (dx/2)$ to $x + (dx/2)$ gives

$$\int_{\text{Range of values of } x} \phi(x) \, dx = 1. \tag{2.2.3}$$

That is, the integral of the "frequency function" over the entire range of the continuous random variable is equal to one.

2.3 Measures of Location

The frequency function of a random variable in itself is not considered a *measure of location* of a particular distribution. Measures of location

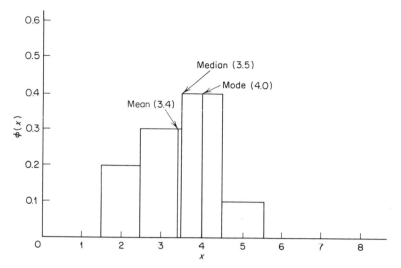

Figure 2.1. A typical *frequency distribution* for a *discrete random variable.*

can, however, be derived from frequency functions. To understand the meaning of the term "measure of location," consider Figures 2.1 and 2.2.

Figure 2.1 is a frequency distribution for a "discrete random variable." Histograms can also be drawn for data from a "continuous random variable" by lumping the observed values into groups. Figure 2.2 is a frequency distribution for a "continuous random variable." To specify the "location" of distributions, several common "measures of location" are used:

1. The *mean*
2. The *median*
3. The *mode*

The "mean" is the average value of x. To determine the "mean value of x" from a set of data, one simply computes the average of all observed values of x:

$$\bar{x} \equiv x_{\text{mean}} = \frac{1}{n} \sum_{i=1}^{n} x_i. \tag{2.3.1}$$

If one has determined a histogram from the data, then $\bar{x}$ can be estimated directly using the values of the frequency function:

$$\bar{x} = \sum_{\text{All values of } x} x_i \phi(x_i). \tag{2.3.2}$$

For a discrete random variable, the same value of $\bar{x}$ is obtained using

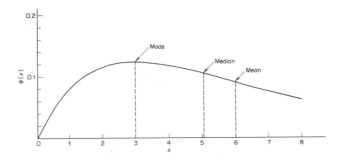

Figure 2.2. A typical *frequency distribution* for a *continuous random variable.*

either Equation 2.3.1 or 2.3.2. It should be noted, however, that if x is a continuous variable, Equation 2.3.1 is the correct equation for computing $\bar{x}$.

If an analytical expression for the frequency function exists, then:

$$\mu = \int_{\text{Range of values of } x} x\phi(x) \, dx, \qquad (2.3.3)$$

where μ is defined as the true mean value of x. The quantity $\bar{x}$ is the value of the mean estimated from the data (i.e., the *sample average*).

The median is defined as the middle measurement for a "random variable." If there is no middle measurement, the interpolated middle value is the median. If an analytical expression exists for the frequency function, then the median is defined as the value of x for which the area under the frequency function to the left equals the area under the frequency function to the right:

$$\int_{\text{Lower limit of } x}^{\text{median}} \phi(x) \, dx = \int_{x_{\text{median}}}^{\text{Upper limit of } x} \phi(x) \, dx = \tfrac{1}{2}. \qquad (2.3.4)$$

The mode is the value of x which corresponds to the maximum frequency. If there is more than one value of x corresponding to the maximum frequency, no completely satisfactory definition of the mode exists.[2] If an analytical expression exists for the frequency distribution, the mode is determined by locating the value of x at which the derivative of x equals zero, and the second derivative is negative:

$$\left(\frac{d\phi(x)}{dx}\right)_{x_{\text{mode}}} = 0. \qquad (2.3.5)$$

The values of the mean, median, and mode can be evaluated for the fre-

[2] P. G. Hoel, "Introduction to Mathematical Statistics," John Wiley & Sons, Inc., New York, 1954.

quency distribution shown in Figure 2.1. From Equation 2.3.2:

$$\bar{x} = 2 \times 0.2 + 3 \times 0.3 + 4 \times 0.4 + 5 \times 0.1 = 3.4. \qquad (2.3.6)$$

The value of x_{median} is determined by noting that one-half the area of the frequency distribution is for $x = 3$ or less, and one-half is for $x = 4$ or more. The interpolated value for the median is, therefore, 3.5. The value of x_{mode} is seen to be equal to 4.

To solve for the values of the mean, median, and mode for the frequency distribution shown in Figure 2.2, an analytical expression for the frequency distribution might be assumed.

We shall try to fit the function

$$\phi(x) = \frac{x}{\sigma^2} e^{-x/\sigma} \qquad (2.3.7)$$

to the curve shown in Figure 2.2 by an appropriate choice of the parameter σ. This function satisfies Equation 2.2.3 if the range of values of x is assumed to be 0 to ∞. Solving for the mean, we obtain:

$$\mu = \int_0^\infty x\phi(x) \, dx = \int_0^\infty \frac{x^2}{\sigma^2} e^{-x/\sigma} \, dx = 2\sigma. \qquad (2.3.8)$$

Solving for the median, Equation 2.3.4 is used:

$$\int_0^{x_{median}} \frac{x}{\sigma^2} e^{-x/\sigma} \, dx = \tfrac{1}{2}, \qquad (2.3.9)$$

which leads to:

$$x_{median} = 1.68\sigma. \qquad (2.3.10)$$

Solving for the mode, we obtain:

$$\frac{d\phi(x)}{dx} = \frac{1}{\sigma^2} \left(-\frac{x}{\sigma} e^{-x/\sigma} + e^{-x/\sigma} \right) = 0, \qquad (2.3.11)$$

$$x_{mode} = \sigma. \qquad (2.3.12)$$

The mode for the "frequency distribution" shown in Figure 2.2 is 3.0. If Equation 2.3.7 is a satisfactory representation of the distribution, then from Equations 2.3.8, 2.3.10, and 2.3.12,

$$x_{mean} = 6.0 \quad \text{and} \quad x_{median} = 5.04.$$

It should be noted that the frequency distributions shown in Figures 2.1 and 2.2 are asymmetrical. *For symmetrical distributions, the mean, the median, and the mode can have the same value.* The most important symmetrical frequency distribution is called the *normal distribution* and is discussed in Section 2.5.

2.4 Measures of Variation

Two distributions can have the same values of the mean, median, and mode, and yet be considerably different. An example of two such distributions is shown in Figure 2.3. Distribution 2 in Figure 2.3 exhibits a greater spread than Distribution 1. A *measure of variation* is used to specify this "spread" in the distribution.

The most common "measure of variation" is called the *standard deviation*. The standard deviation σ_x of a frequency distribution is defined by the following equation:

$$\sigma_x = \left(\int \phi(x)\,(x - \mu)^2\,dx \right)^{1/2}. \tag{2.4.1}$$

The square of the standard deviation (i.e., σ_x^2) is called the *variance*. From Equation 2.4.1 it is seen that the greater the spread or variation in the values of x, the greater the values of standard deviation and variance. The variance is the average value of the square of the deviation of the variable x from its mean value μ.

Often the frequency function for a particular variable is not known. The value of σ_x can, however, be estimated from the measured values of the n data points:

$$s_x = \left(\frac{1}{n - 1} \sum_{i=1}^{n} (x_i - \bar{x})^2 \right)^{1/2}, \tag{2.4.2}$$

where s_x is defined as the *unbiased estimate* of σ_x.

In most introductory statistical textbooks, it is proven that the denominator of Equation 2.4.2 should be $(n - 1)^{1/2}$, and not $n^{1/2}$ as one might expect. Qualitatively, the reason why $(n - 1)^{1/2}$ is used is that the value of $\bar{x}$ in Equation 2.4.2 must be used rather than the true but unknown value μ. To avoid a "biased estimate" of σ_x, $(n - 1)^{1/2}$ rather than $n^{1/2}$ is used.

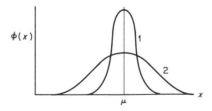

Figure 2.3. Two distributions with the same values of the mean, the median, and the mode.

The concept of an "unbiased estimate" is used throughout the following chapters. An "unbiased estimate" of a parameter is defined as an estimate that, if repeated many times, would yield an average value for the parameter equal to the true value. If, for example, $n^{1/2}$ were used in place of $(n - 1)^{1/2}$ in Equation 2.4.2, the calculated values of s_x would, on the average, be lower than the value of σ_x for the true distribution. Using $(n - 1)^{1/2}$ will yield an average value of s_x equal to σ_x.

The term *uncertainty* is often used in place of *standard deviation*. Experimentalists refer to the "uncertainty of a particular measurement." In this book, the uncertainty is defined as the estimated value of the standard deviation. The estimate does not necessarily have to be based upon an analysis of data. For example, in a particular experiment a length is measured as 21 cm, and the uncertainty is estimated to be 0.1 cm. The value of 0.1 cm might have come from the experimentalists knowledge of the measuring instrument. If the measurement is repeated several times, and if the standard deviation of the measurement is computed, one would expect that the computed value would be close to the quoted uncertainty.

The term *accuracy* is used throughout the book in place of uncertainty or standard deviation only when these quantities are expressed as percentages. A measurement good to 1% accuracy implies that the uncertainty or standard deviation of the measurement is 1% of the measured value.

Another "measure of variation" is the *probable error*. The probability is $\frac{1}{2}$ (i.e., 50%) that a measured value x will fall within a range $\bar{x} \pm$ P.E. The "probable error" (i.e., P.E.) is related to the standard deviation:

$$\text{P.E.} = 0.6745 \cdot \sigma_x. \tag{2.4.3}$$

The value 0.6745 is based upon a normal distribution which is discussed in the following section. In this book, the probable error will not be used as a "measure of variation."

The concept of a "confidence interval" is used in the following chapters. A confidence interval is defined as an interval in which a particular measurement will fall to within a quoted degree of *confidence*. If, for example, the interval $x_l \leq x \leq x_h$ is the 95% confidence interval for x, one would expect that 95% of all measured values of x would be within this range.

2.5 Frequency Distributions

The Binomial Distribution

The *binomial distribution* is used for repetitive experiments in which only the occurrence or nonoccurrence of an event is recorded. The binomial frequency function is:

$$\phi(x) = \frac{n!}{x!(n-x)!} p^x q^{n-x}, \tag{2.5.1}$$

where x is the number of occurrences of an event, p is the probability of an occurrence, q is the probability of a nonoccurrence, n is the total number of trials, and $\phi(x)$ is the relative frequency of x occurrences in n trials. The probability of occurrence plus nonoccurrence must equal one:

$$p + q = 1. \tag{2.5.2}$$

Using Equation 2.5.2, it is next shown that Equation 2.2.2 is satisfied for the binomial frequency function:

$$(q + p)^n = q^n + nq^{n-1}p + \tfrac{1}{2}[n \cdot (n-1)]q^{n-2}p^2 + \cdots + p^n. \tag{2.5.3}$$

Re-expressing Equation 2.5.3:

$$(q + p)^n = \sum_{x=0}^{n} \frac{n!}{x!(n-x)!} p^x q^{n-x}. \tag{2.5.4}$$

Substituting Equation 2.5.2 into the left side of 2.5.4 and 2.5.1 into the right side of 2.5.4 gives

$$1^n = 1 = \sum_{x=0}^{n} \phi(x). \tag{2.5.5}$$

It can be shown[3] that the mean value and standard deviation of the binomial distribution are:

$$\mu = np, \tag{2.5.6}$$

and

$$\sigma_x = (npq)^{1/2} = [np(1-p)]^{1/2} \tag{2.5.7}$$

As an example of the binomial distribution, consider a baseball player with a .300 batting average. If he comes to bat 4 times and is not walked or credited with a sacrifice, what is the probability that he will get 0, 1, 2, 3, and 4 hits? A batting average is defined as the probability that a batter will be successful or, in the idiom of the game, will "get a hit." In other words, $p = 0.300$, and from Equation 2.5.2, $q = 0.700$. The number of trials n equals 4, and x is a variable that can be either 0, 1, 2, 3, or 4. Using Equation 2.5.1, results are determined and then tabulated in Table 2.1. It can be seen, for example, that the probability of getting two or more hits is $0.2646 + 0.0756 + 0.0081 = 0.3483$. In other words, in about 35% of the games that the player has four "official at bats," he can be expected to have 2 or more hits.

[3] P. G. Hoel, "Introduction to Mathematical Statistics," John Wiley & Sons, Inc., New York, 1954, p. 65.

TABLE 2.1. Relative frequency of the number of hits in four "official at bats"
for a baseball player with a 0.300 batting average

Number of Hits	Relative Frequency $\phi(x)$
0	0.2401
1	0.4116
2	0.2646
3	0.0756
4	0.0081

$$1.0000 = \sum_{i=0}^{4} \phi(x)$$

The binomial frequency function can be applied to many real-life problems involving "discrete random variables." It should be noticed, however,
that for large values of n it is difficult to calculate the exact values of
$\phi(x)$ from a binomial distribution. For cases in which n is large and p is
very small, the "Poisson distribution" is a good approximation to the
binomial distribution. For cases in which n is large but p is not small, then
the "normal distribution" is a good approximation to the binomial
distribution.

The Poisson Distribution

The *Poisson distribution* is used for experiments in which the number
of occurrences may vary from 0 to infinity, but the mean or average value
(i.e., μ) of the distribution is known. The Poisson frequency function is:

$$\phi(x) = e^{-\mu}\mu^{x}/x!. \tag{2.5.8}$$

The mean value μ does not necessarily have to be an integer. (If, however,
it is an integer, it can be proven that $\phi(\mu - 1)$ is equal to $\phi(\mu)$.)

The Poisson distribution is a good approximation to the binomial distribution[4] for large values of n and small values of p. If $n \geq 100$ and
$p \leq 0.05$, the approximation is sufficiently accurate for most applications.

It can be shown that the standard deviation (i.e., Equation 2.4.1) of a
Poisson distribution is equal to the square root of the mean of the
distribution:

$$\sigma_x = \mu^{1/2}. \tag{2.5.9}$$

If σ_x is to be estimated from the results of an experiment in which the
Poisson distribution is applicable, then s_x, the unbiased estimate of σ_x, is
simply $(\bar{x})^{1/2}$.

---

[4] Proof that the binomial reduces to the Poisson distribution as $n \rightarrow \infty$, $p \rightarrow 0$,
and $np = \mu$ remains constant, is included in the previously cited textbook by P. G.
Hoel.

TABLE 2.2. Poisson distribution for $\mu = 5.0$.[a]

Counts	Relative Frequency $\phi(x)$
0	0.0067
1	0.0335
2	0.0837
3	0.1395
4	0.1745
5	0.1745
6	0.1455
7	0.1040
8	0.0650
9	0.0362
10	0.0181
	0.9812

[a] Note that the sum of $\phi(x)$ for $x = 0$ to 10 is 0.9812. The sum of $\phi(x)$ for $x = 11$ to ∞ must therefore be $1.00 - 0.9812 = 0.0188$.

An example of an experiment in which "Poisson statistics" (or "counting statistics" as it is referred to in the terminology of the experimentalist) is applicable, is the measurement of the average number of counts per minute recorded by a Gieger counter for a long-lived radioisotope. The number of counts recorded in any given minute could vary from zero to infinity. To determine a mean or average value, one might determine the number of counts in an hour and then divide by sixty to estimate the mean or average number of counts per minute. If 300 counts are recorded in an hour, the average count rate per minute is 5.0. Furthermore, from Equation 2.5.9, the standard deviation can be estimated as $(300)^{1/2}/60$ which equals 0.29 count per minute.

Using Equation 2.5.8, the expected relative frequency for any value of x can be computed. As an example, for $\mu = 5.0$, the values of $\phi(x)$ are tabulated in Table 2.2.

Poisson statistics are applicable to a wide variety of experiments. Experiments, in which the number of occurrences in a given period of time is recorded, usually follow Poisson statistics. Such experiments are sometimes referred to as "counting experiments" and are often performed by converting "an occurrence" into an electrical pulse and then recording the total number of pulses.

The Normal Distribution

The normal (or Gaussian) distribution is applied to continuous random variables in which the distribution is symmetrical about the mean μ, and dies out at the tails. The distributions shown in Figure 2.3 are examples of the normal distribution. The frequency function for this distribution is:

$$\phi(x) = \frac{1}{\sigma_x (2\pi)^{1/2}} \exp \left[-\frac{(x - \mu)^2}{2\sigma_x{}^2} \right], \tag{2.5.10}$$

where μ is the mean value of x and σ_x is the standard deviation. Integration of Equation 2.5.10 shows that Equation 2.2.3 is satisfied. Estimates of μ and σ_x can be made from experimental data using Equations 2.3.1 and 2.4.2, respectively.

The normal distribution can also be used as an approximation to the binomial distribution. The mean value and standard deviation of the binomial distribution are np and $(npq)^{1/2}$, respectively (see Equations 2.5.6 and 2.5.7). The quantity $(x - np)/(npq)^{1/2}$ is "normally distributed" with $\mu = 0.0$ and $\sigma_x = 1.0$ for large values of n and for cases in which p or q are not very small (i.e., when $np > 5.0$ for $p \leq \frac{1}{2}$ or $nq > 5.0$ for $p > \frac{1}{2}$).

For a variable which is normally distributed, 68.3% of all observations will be within the range $\mu \pm \sigma_x$, 95.4% will be within the range $\mu \pm 2\sigma_x$, and 99.7% will be within the range $\mu \pm 3\sigma_x$. The range $\mu \pm 0.6745\sigma_x$ will contain 50% of all the observations. The quantity $0.6745\sigma_x$ is defined as the "probable error" (see Equation 2.4.3).

A variable which is often considered in statistical analysis is:

$$u \equiv \frac{\bar{x} - \mu}{\sigma_x / \sqrt{n}}, \tag{2.5.11}$$

where $\bar{x}$ is the estimated value of μ as determined from n observed values of x (see Equation 2.3.1). The interesting property of the variable u is that it is normally distributed with a mean value of zero and a standard deviation of one. Values of u for a given "confidence interval" are included in Table 3.5.

The χ^2 Distribution

The parameter u was defined by Equation 2.5.11 and has a mean value of zero and a standard deviation of one. The sum of the squares of k independent values of u are denoted by χ^2:

$$\chi^2 = \sum_{i=1}^{k} u_i^2, \tag{2.5.12}$$

where k is called the "number of degrees of freedom" for χ^2. The frequency function of χ^2 is only a function of k. The χ^2 frequency function is[5]:

$$\phi(\chi^2) = [2^{k/2}\Gamma(\tfrac{1}{2}k)]^{-1}(\chi^2)^{k/2-1} \exp(-\tfrac{1}{2}\chi^2), \tag{2.5.13}$$

[5] A. Hald, "Statistical Theory with Engineering Applications," John Wiley & Sons, Inc., New York, 1952, Ch. 10.

where Γ denotes the "gamma function":

$$\Gamma(\tfrac{1}{2}k) = (\tfrac{1}{2}k - 1)! = \begin{cases} (\tfrac{1}{2}k - 1)(\tfrac{1}{2}k - 2)\cdots 3 \cdot 2 \cdot 1 & \text{for } k \text{ even.} \\[2ex] (\tfrac{1}{2}k - 1)(\tfrac{1}{2}k - 2)\cdots \tfrac{3}{2} \cdot \tfrac{1}{2} \cdot \sqrt{\pi} & \text{for } k \text{ odd.} \end{cases}$$

$$(2.5.14)$$

A property of the χ^2 distribution which is used in the development of prediction analysis is that the mean value of χ^2 is equal to the number of degrees of freedom[5]:

$$(\chi^2)_{\text{mean}} = k. \tag{2.5.15}$$

The χ^2 distribution is also useful when applied to the distribution of the estimated standard deviation s_x:

$$s_x^2 = \sigma_x^2 \frac{\chi^2}{k}, \tag{2.5.16}$$

where k is the number of degrees of freedom and equals $n - 1$. Use of this aspect of the χ^2 distribution is considered in greater detail in Section 3.13.

Equations 2.5.13 and 2.5.14 are rarely used when analyzing data. The quantity which is of more general interest is the fractional probability P that χ^2 will exceed a specified value. Values of χ^2 are tabulated as a function of the number of degrees of freedom for values of P from 0.01 to 0.99 in Table 3.6.

The t-Distribution

The t-distribution is used in the following chapter with reference to the "confidence intervals" of experimental results. The variable u is defined by Equation 2.5.11. The parameter σ_x appears in the equation, but this parameter is rarely known. However, an unbiased estimate of σ_x (i.e., s_x) can be determined from the experimental data (see Equation 2.4.2). The variable t is used in place of u if s_x is used in place of σ_x:

$$t \equiv \frac{\bar{x} - \mu}{s_x/\sqrt{n}}. \tag{2.5.17}$$

Substituting Equations 2.5.11 and 2.5.16 into Equation 2.5.17 gives

$$t = \frac{\bar{x} - \mu}{(\sigma_x/\sqrt{n}) \cdot (s_x/\sigma_x)} = \frac{u}{(\chi^2/k)^{1/2}}. \tag{2.5.18}$$

Since χ^2 is only a function of k, then t is only a function of k.

The t-distribution is symmetrical about zero and tends to the u-distribution (i.e., a normal distribution with $\mu = 0$ and $\sigma_x = 1.0$) as $k \to \infty$. The frequency function for the t-distribution is:

$$\phi(t) = \frac{1}{(\pi k)^{1/2}} \frac{\Gamma\frac{1}{2}(k+1)}{\Gamma(\frac{1}{2}k)} \cdot \left(1 + \frac{t^2}{k}\right)^{-(k+1)/2}, \qquad -\infty < t < \infty. \quad (2.5.19)$$

Application of the t-distribution is considered in greater detail in Sections 3.10 and 4.5.

Equation 2.5.19 is rarely used when analyzing data. The quantity which is of more general interest is the fractional probability β that the absolute value of t will exceed a specified value. Values of t are tabulated as a function of the number of degrees of freedom for values of $\beta/2$ from 0.005 to 0.25 in Table 3.5.

2.6 The Uncertainty of a Function of Several Variables

To analyze experiments properly, it is often necessary to estimate the uncertainly of a variable that is actually a combination of several other variables. If we denote the variable as v:

$$v = f(u_1, u_2, \cdots, u_r), \qquad (2.6.1)$$

v is a function of r variables u_j. Defining Δv and Δu_j as the variations from the true values of v and u_j, if these variations are small, we obtain:

$$\Delta v = \frac{\partial f}{\partial u_1} \cdot \Delta u_1 + \frac{\partial f}{\partial u_2} \cdot \Delta u_2 + \cdots + \frac{\partial f}{\partial u_r} \cdot \Delta u_r. \qquad (2.6.2)$$

Squaring Equation 2.6.2 gives:

$$(\Delta v)^2 = \left(\frac{\partial f}{\partial u_1} \cdot \Delta u_1\right)^2 + \left(\frac{\partial f}{\partial u_2} \cdot \Delta u_2\right)^2 + \cdots + \left(\frac{\partial f}{\partial u_r} \cdot \Delta u_r\right)^2$$

$$+ 2\frac{\partial f}{\partial u_1} \cdot \frac{\partial f}{\partial u_2} \cdot \Delta u_1 \cdot \Delta u_2 + \text{other cross-product terms.} \quad (2.6.3)$$

If the measurements are repeated N times, and the average values of the variations are computed:

$$(\Delta v)^2{}_{av} = \left(\frac{\partial f}{\partial u_1} \cdot \Delta u_1\right)^2_{av} + \left(\frac{\partial f}{\partial u_2} \cdot \Delta u_2\right)^2_{av} + \cdots + \left(\frac{\partial f}{\partial u_r} \cdot \Delta u_r\right)^2_{av}$$

$$+ \left(2\frac{\partial f}{\partial u_1} \cdot \frac{\partial f}{\partial u_2} \cdot \Delta u_1 \cdot \Delta u_2\right)_{av} + \text{other cross-product terms.} \quad (2.6.4)$$

If the variables u_j and u_k are independent, they are said to be "uncorrelated" and the average values of the products $\Delta u_j \cdot \Delta u_k$ approach zero as N approaches infinity. Thus, as N approaches infinity:

$$(\Delta v)^2{}_{av} \rightarrow \left(\frac{\partial f}{\partial u_1} \cdot \Delta u_1\right)^2_{av} + \left(\frac{\partial f}{\partial u_2} \cdot \Delta u\right)^2_{av} + \cdots + \left(\frac{\partial f}{\partial u_r} \cdot \Delta u_r\right)^2_{av}, \quad (2.6.5)$$

where av refers to the average values of the squares of the quantities within the parentheses. By definition, the average values of Δv^2 and Δu_j^2 are σ_v^2 and $\sigma_{u_j}^2$ (see Equation 2.4.1); therefore:

$$\sigma_v^2 = \left(\frac{\partial f}{\partial u_1} \cdot \sigma_{u_1}\right)^2 + \left(\frac{\partial f}{\partial u_2} \cdot \sigma_{u_2}\right)^2 + \cdots + \left(\frac{\partial f}{\partial u_r} \cdot \sigma_{u_r}\right)^2$$

$$= \sum_{j=1}^{r} \left(\frac{\partial f}{\partial u_j} \cdot \sigma_{u_j}\right)^2. \quad (2.6.6)$$

Equation 2.6.6 is extremely useful when analyzing experiments. Using this equation, the following well-known relationships can be derived. It should be emphasized that this equation is valid only if the u_j's are uncorrelated.

1. If $v = u_1 + u_2$ or $v = u_1 - u_2$,

$$\sigma_v^2 = \sigma_{u_1}^2 + \sigma_{u_2}^2. \quad (2.6.7)$$

2. If $v = u_1 \cdot u_2$ or u_1/u_2,

$$\left(\frac{\sigma_v}{v}\right)^2 = \left(\frac{\sigma_{u_1}}{u_1}\right)^2 + \left(\frac{\sigma_{u_2}}{u_2}\right)^2. \quad (2.6.8)$$

3. If $v = u_1^2$,

$$\sigma_v^2 = (2u_1 \cdot \sigma_{u_1})^2. \quad (2.6.9)$$

As an example, consider the measurements of the count rates of gamma rays emitted by two radioactive isotopes. The count rate of Specimen 1 is 1600 cpm (counts per minute) and the count rate of Specimen 2 is 900 cpm. Both count rates were determined by recording the number of counts in a one-minute interval. It is required that the sum, difference, product, ratio, and squares of these count rates and their uncertainties be estimated.

For experiments of this type (i.e., counting experiments), the Poisson distribution is applicable. According to the comment below Equation 2.5.9, the unbiased estimates of the uncertainties σ_{u_1} and σ_{u_2} (where u_1 is defined as the count rate of Specimen 1 and u_2 is defined as the count rate of Specimen 2) are determined by taking the square roots of the count rates u_1 and u_2:

$$\sigma_{u_1} = (u_1)^{1/2} = 40 \text{ cpm}, \quad (2.6.10)$$

$$\sigma_{u_2} = (u_2)^{1/2} = 30 \text{ cpm}. \quad (2.6.11)$$

From Equation 2.6.7:

$$\sigma_{u_1+u_2} = \sigma_{u_1-u_2} = (\sigma_{u_1}^2 + \sigma_{u_2}^2)^{1/2} = 50 \text{ cpm.} \qquad (2.6.12)$$

From Equation 2.6.8:

$$\frac{\sigma_{u_1 \cdot u_2}}{u_1 \cdot u_2} = \frac{\sigma_{u_1/u_2}}{u_1/u_2} = \left[\left(\frac{\sigma_{u_1}}{u_1}\right)^2 + \left(\frac{\sigma_{u_2}}{u_2}\right)^2\right]^{1/2} = 0.0417, \qquad (2.6.13)$$

and therefore,

$$\sigma_{u_1 \cdot u_2} = 0.0417 \times 1600 \times 900 = 60,000 \text{ (cpm)}^2, \qquad (2.6.14)$$

$$\sigma_{u_1/u_2} = 0.0417 \times \frac{1600}{900} = 0.0742. \qquad (2.6.15)$$

From Equation 2.6.9:

$$\sigma_{u_1}^2 = 2u_1 \cdot \sigma_{u_1} = 128,000 \text{ (cpm)}^2, \qquad (2.6.16)$$

$$\sigma_{u_2}^2 = 2u_2 \cdot \sigma_{u_2} = 54,000 \text{ (cpm)}^2. \qquad (2.6.17)$$

The sum of the count rates (i.e., $u_1 + u_2$) is equal to 2500 ± 50 cpm. Stated in the terminology of the experimentalist, the sum is known to 2% accuracy. The difference (i.e., $u_1 - u_2$) is equal to 700 ± 50 cpm and is therefore known to about 7% accuracy. The product (i.e., $u_1 \cdot u_2$) and the ratio (i.e., u_1/u_2) are equal to $1.14 \times 10^6 \pm 47,500$ (cpm)^2 and 1.78 ± 0.0742. Both are known to about 4% accuracy. The accuracies of the count rates u_1 and u_2 are 2.5% and 3.33%, respectively. The percent accuracies of the squares of the count rates are twice the percent accuracies of the count rates. These results are listed in Table 2.3.

From the table it can be seen that, on a percentage basis, the sum is determined more accurately than either of the individual count rates. The difference, product, ratio, and squares of the count rates are all determined

TABLE 2.3 Tabulation of results for a counting experiment

Variable	Value	Units	Uncertainty	% Accuracy
u_1	1600	cpm	40	2.50
u_2	900	cpm	30	3.33
u_1+u_2	2500	cpm	50	2.00
u_1-u_2	700	cpm	50	7.14
$u_1 \cdot u_2$	1,440,000	(cpm)^2	60,000	4.17
u_1/u_2	1.778	—	0.0742	4.17
u_1^2	2,560,000	(cpm)^2	128,000	5.00
u_2^2	810,000	(cpm)^2	54,000	6.67

less accurately than either of the count rates. As u_2 approaches u_1 the percent accuracy of the difference approaches infinity. For this reason, if at all possible, it is advisable to avoid an experiment in which an important variable is determined as the small difference of large numbers.

If the experiment was performed by first recording the number of counts in a two-minute rather than one-minute interval and then dividing the number of counts by two, the values of all the quantities would be approximately the same, but the uncertainties would be reduced by approximately a factor of $\sqrt{2}$. The percentage accuracies would also be reduced by a factor of about $\sqrt{2}$.

The preceding example is based upon Equation 2.6.6. This equation is valid only if the u_j's are uncorrelated, and for the case under consideration (i.e., the measurement of count rates of gamma rays emitted by two radioactive isotopes) u_1 and u_2 are indeed uncorrelated. The analysis is more complicated if the u_j's are correlated. The average values of the cross-product terms in Equation 2.6.4 are not zero, and therefore Equation 2.6.6 must be replaced by Equation 2.6.18:

$$\sigma_v^2 = \left(\frac{\partial f}{\partial u_1} \cdot \sigma_{u_1}\right)^2 + \left(\frac{\partial f}{\partial u_2} \cdot \sigma_{u_2}\right)^2 + \cdots + \left(\frac{\partial f}{\partial u_r} \cdot \sigma_{u_r}\right)^2$$

$$+ \left(2 \cdot \frac{\partial f}{\partial u_1} \cdot \frac{\partial f}{\partial u_2} \cdot \sigma_{12}\right) + \text{other cross-product terms}$$

$$= \sum_{j=1}^{r} \sum_{k=1}^{r} \left(\frac{\partial f}{\partial u_j} \cdot \frac{\partial f}{\partial u_k} \sigma_{jk}\right). \tag{2.6.18}$$

If $j \neq k$, σ_{jk} is called the *covariance* of u_j and u_k and is the average value of the product $\Delta u_j \cdot \Delta u_k$. The symbol σ_{jj} is called the "variance" of u_j and is equivalent to the standard deviation of u_j squared:

$$\sigma_{jj} \equiv \sigma_{u_j}^2. \tag{2.6.19}$$

It is important to know if Equation 2.6.6 is sufficient or if Equation 2.6.18 must be used. In the preceding example, the measurements of the two count rates are independent. If, for example, the first measurement is higher than the true count rate for Specimen 1, there is no reason to expect a higher than average measurement for Specimen 2. The measurements are therefore uncorrelated and Equation 2.6.6 is sufficient for the analysis of the experiment.

In Chapter 9, an experiment is analyzed for which the purpose is to determine two unknown parameters a_1 and a_2 and the sum of their squares $a_1^2 + a_2^2$. The values of a_1 and a_2 are determined by a least squares analysis of experimental data. (The method of least squares is discussed in detail in Chapter 3.) The a's are not determined independently and it cannot be assumed that they are uncorrelated. The correct relationship for the uncertainty of the sum of the squares must therefore be derived from Equations 2.6.18 and 2.6.19:

$$v = a_1^2 + a_2^2, \tag{2.6.20}$$

$$\sigma_v^2 = (2 \cdot a_1 \cdot \sigma_{a_1})^2 + (2 \cdot a_2 \cdot \sigma_{a_2})^2 + 8 \cdot a_1 \cdot a_2 \cdot \sigma_{12}. \tag{2.6.21}$$

The variances $\sigma_{a_1}^2$ and $\sigma_{a_2}^2$ and the covariance σ_{12} are estimated from an analysis of the experimental results. The method for determining these quantities is presented in Chapter 3. Prediction of these quantities is discussed in Chapter 4. Equation 2.6.21 can then be used to determine the actual or predicted variance of the sum $a_1^2 + a_2^2$.

Problems

1. What is the mean, median, mode, and estimated standard deviation for the following distribution?

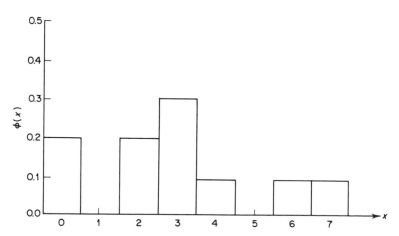

2. What is the mean, median, mode, and standard deviation of the following distribution?

$$\phi(x) = 0.1 + 0.9x.$$

The range of values of x is from 0 to 1.

3. A coin is tossed eight times and the number of heads is recorded. What is the probability that the number of heads is equal to 6 or more? (*Note*—the binomial distribution is applicable to this problem.)

4. The number of cars passing a point on the road is recorded, and the average value is computed to be 4 per minute. What is the probability that 10 cars will pass in a given minute? (*Note*—the Poisson distribution is applicable to this problem.)

5. Derive an equation for σ_v in terms of σ_{u1}, σ_{u2}, and σ_{u3} and the covariances σ_{12}, σ_{13}, and σ_{23}. The variable v is related to the variables u_1, u_2, and u_3 by the following equation:

$$v = \frac{u_1{}^2 \cdot u_2}{u_3}.$$

6. The count rates of two radioactive specimens are 1200 and 500 counts per minute, respectively. If both count rates were determined by recording the number of count in an hour and then dividing by sixty, what are the unbiased estimates of the standard deviations of the count rates? What are the unbiased estimates of the standard deviations of the sum, difference, and ratio of the count rates?

Chapter 3

The Method of Least Squares

3.1 Introduction

The method of least squares is a well-known technique used for the analysis of data. The fundamental principles of least squares theory were developed by Gauss in the early part of the nineteenth century. Prior to the availability of digital computers, however, most scientists and engineers rarely applied the method of least squares to their practical problems of data analysis.

Admittedly least squares analysis of data is tedious if the calculations are to be performed with only the aid of a desk calculator. An excellent "pre-computer age" book by Deming[1] discusses systematic methods for facilitating the analysis; nevertheless, a great deal of time and effort are required for any but the simplest problems.

Widespread availability of digital computers has completely altered the picture. Once a computer program for a particular least squares problem has been completed, the program can be used over and over again with different sets of data. Results of the analysis can be available in a matter of minutes and can therefore provide "feedback" for the next experiment. In fact, at some laboratories, routine experiments are being performed which incorporate "on-line" computers for immediate least squares analysis of the results.

For most quantitative experiments the method of least squares is the "best" analytical technique for extracting information from a set of experimental observations or data points. The method is "best" in the sense that the parameters determined by the least square analysis are "normally distributed" about the true parameters with the least possible

[1] W. E. Deming, "Statistical Adjustment of Data", John Wiley & Sons, Inc., New York, 1943.

"standard deviation."[2] Now that digital computers are available for performing the tedious calculations, the method of least squares is becoming a practical everyday tool for scientists and engineers.

3.2 Curve Fitting

The primary application of the method of least squares is *curve fitting*.[3] A functional form is assumed to express the relationship between the dependent and independent variables. The function includes unknown parameters, and the method of least squares can be used to provide the "best" estimates of these parameters.

Two basic approaches to "curve fitting" are possible:

1. Fitting a curve which passes through every point.
2. Fitting a curve which does not pass through every point but exhibits the general characteristics of the data.

These two possibilities are shown in Figure 3.1.

Fitting a curve which passes through every point is inadvisable if any one of the following three reasons is applicable:

1. *If n, the number of data points, is large.* The number of unknown parameters must equal the number of data points if the curve is to pass through every point. If n is large, the resulting equation will therefore be cumbersome for such purposes as interpolation, extrapolation, differentiation, and integration.

2. *If the uncertainty of the data is large.* The resulting curve for an "exact fit" will exhibit all the vagaries of the data and most probably will oscillate violently about the curve which represents the true function. The resulting equation would be completely useless for the purpose of differentiation.

3. *If, from theoretical considerations, a particular function should be applicable.* For example, if y should be a linear function of x and if 7 data points are available (as shown in Figure 3.1), it is more sensible to fit an equation of the form $y = a_1 + a_2x$ to the data rather than an equation of the form $y = a_1 + a_2x + \cdots + a_7x^6$.

For most experiments, one of the three reasons is applicable, and there-

---

[2] A. Hald, "Statistical Theory with Engineering Applications," John Wiley & Sons, Inc., New York, 1952. (See footnote 5 for additional comments related to this point.)

[3] In this book, only the curve fitting application of least squares theory will be considered. Least squares theory can be applied to other forms of data adjustment. Specific examples are included in the previously cited book by W. E. Deming.

fore fitting a curve which passes through all points is rarely justifiable. If the data are to be fitted by a curve which does not pass through all the points, the method of least squares will provide the best estimate of the unknown parameters of the curve. If the experimenter has access to a digital computer, he can justify least squares analysis for most experiments. In fact, it is becoming the rare experiment for which he can justify not using the method of least squares!

Perhaps several words should be mentioned about the well-known *graphical* method of curve fitting. The experimenter plots his data on a sheet of graph paper and then simply draws what, in his opinion, is the

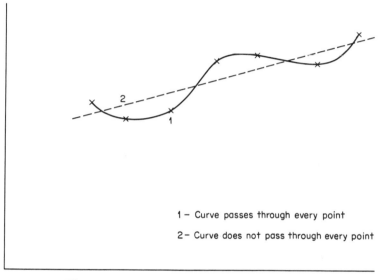

1 — Curve passes through every point

2 — Curve does not pass through every point

Figure 3.1. Two possibilities for curve fitting.

line or curve which best fits the data. Admittedly the resulting curve will vary from man to man if several people plot curves for the same data. The graphical method is, however, quick and will always be useful for preliminary analysis of data. By simply graphing results while an experiment is in progress, the experimenter will often be able to judge whether or not the experiment is progressing satisfactorily. He will be able to spot gross errors in one or several of the data points. The method of least squares is, therefore, not suggested as an alternative to the simple graphical method of curve fitting. It is recommended as a supplementary step which should be used for the *final* analysis of most experimental data.

3.3 General Statement of the Problem

The following notation will be adopted:

$Y_i \equiv$ observed values of the dependent variable.

$\sigma_{y_i} \equiv$ uncertainty (or standard deviation) of Y_i.

$X_{ji} \equiv$ observed values of the jth independent variable.[a]

$\sigma_{x_{ji}} \equiv$ uncertainty (or standard deviation) of X_{ji}.[a]

$y_i \equiv$ calculated (or adjusted) values of the dependent variable.

$x_{ji} \equiv$ calculated (or adjusted) values of the jth independent variable.[a]

$\eta_i \equiv$ true (or mean) values of the dependent variable.

$\xi_{ji} \equiv$ true (or mean) values of the jth independent variable.[a]

$n \equiv$ number of data points $(i = 1, 2, \cdots, n)$.

$m \equiv$ number of independent variables $(j = 1, 2, \cdots, m)$.[a]

$a_k \equiv$ calculated values of the unknown parameters.

$a_{k0} \equiv$ initial estimates of the unknown parameters.

$\alpha_k \equiv$ true values of the unknown parameters.

$\sigma_{a_k} \equiv$ uncertainty (or standard deviation) of a_k.

$s_{a_k} \equiv$ unbiased estimate of σ_{a_k}.

$p \equiv$ number of unknown parameters $(k = 1, 2, \cdots, p; p < n)$.

[a] If $m = 1$, the index j is unnecessary and can therefore be dropped.

The method of least squares is based upon the assumption that the observed values of the variables (i.e., the Y_i's and X_{ji}'s) are distributed about the true (or mean) values[4] (i.e., the η_i's and ξ_{ji}'s).[5] In other words,

[4] The words "true" and "mean" are both used because some measurements can be characterized as having "true" values, whereas other measurements can only be characterized as having "mean" values. As discussed in Section 2.2, "random" variables are distributed about mean values. There is no one true value, but rather a distribution. On the other hand, "deterministic" variables can be considered as having true values. The question of whether a particular variable is "random" or "deterministic" does not affect the method of least squares.

[5] The distributions do not necessarily have to be *normal*. According to P. G. Guest ("Numerical Methods of Curve Fitting," Cambridge University Press, 1961, p. 88), if one is attempting to estimate the α_k's such that the variance has the smallest possible value, whatever the distributions of the observed variables, "the minimum variance postulate leads to estimates which are identical with the least-squares estimates. This is sometimes referred to as the Markoff theorum." The proof of this theorem is included in section 8.3 of the above mentioned text. Alternatively, the values of the α_k's may be estimated on the basis of the *method of maximum likelihood*. This method reduces to the method of least-squares only if the distributions are *normal*. According to P. G. Guest, "in most cases the deviation law will have some symmetrical form not very different from the normal distribution, and it is probable that the least-squares and maximum likelihood estimates will be very nearly equal. Since the former estimates are unbiased and readily calculated, they will almost always be adequate." A discussion of the method of maximum likelihood is included in "Mathematical Methods of Statistics," H. Cramer, Princeton University Press.

if the measurements Y_i and X_{ji} are repeated N times and the average values are computed, these average values are assumed to approach η_i and ξ_{ji} as N approaches infinity. This assumption is valid as long as "systematic errors" do not affect the measurements.

A *systematic error* is an error that consistently causes the measured values to be either larger or smaller than the true values. A systematic error can result from a mistake in the technique used for measuring one of the variables or from a fault in one of the measuring instruments. Systematic errors can also arise if factors which affect the experiment are not taken into consideration. If, for example, the experiment is sensitive to temperature changes, the temperature should be controlled for the duration of the experiment. If the temperature is not controlled, the results of the experiment might contain "systematic errors." Treatment of systematic errors is considered in Section 4.9.

The first step in a least squares analysis is to assume a functional relationship for the dependent and independent variables:

$$\eta_i = f(\xi_{1i}, \xi_{2i}, \cdots, \xi_{mi}; \alpha_1, \alpha_2, \cdots, \alpha_p). \qquad (3.3.1)$$

The function f is usually chosen on the basis of a physical law, and the parameters α_k often have physical significance. It is not necessary, however, that f be based on a physical law. The purpose of the least squares analysis is to use the experimental data to determine the "best estimates" of the parameters α_k. The *best estimates* of α_k are, by definition, the least squares values of α_k and are denoted as a_k. If the experiment is repeated N times, and if the measurements are not subject to systematic errors, then the average values of a_k will approach the true values α_k as N approaches infinity. This statement is, of course, subject to the assumption that the function f adequately represents the relationship between the variables (i.e., Equation 3.3.1 is valid). Techniques for testing the "goodness of the fit" are discussed in Section 3.13. If the function f is invalid or if the data include erroneous values, the tests discussed in Section 3.13 will usually indicate that something has gone wrong.

The least squares analysis is based upon an adjustment of the experimental data in a manner which is described in Section 3.6. The resulting "calculated" or "adjusted" values of the variables are denoted as y_i and x_{ji}. For a case in which $m = 1$ (i.e., only one independent variable), a typical relationship between the observed, calculated, and true values is shown graphically in Figure 3.2.

It is seen from Figure 3.2 that the functional form of the relationship between the dependent and independent variables is the same for the calculated and true variables. The only difference is in the values of the variables and parameters. An equation analogous to Equation 3.3.1 can be written for the calculated variables:

$$y_i = f(x_{1i}, x_{2i}, \cdots, x_{mi}; a_1, a_2, \cdots, a_p). \qquad (3.3.2)$$

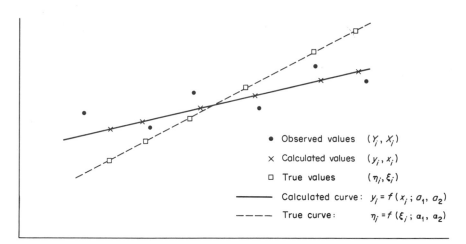

Figure 3.2. A typical situation in curve fitting. This figure shows a typical relationship between the observed, calculated, and true values of the variables.[6]

Several additional definitions are required for a statement of the problem.
Residuals are defined as the differences between the observed and calculated variables:

$$R_{y_i} \equiv Y_i - y_i, \tag{3.3.3}$$

$$R_{x_{ji}} \equiv X_{ji} - x_{ji}. \tag{3.3.4}$$

The *weights* are defined as the reciprocals of the squares of the uncertainties:

$$w_{y_i} \equiv \frac{1}{\sigma_{y_i}^2}, \tag{3.3.5}$$

$$w_{x_{ji}} \equiv \frac{1}{\sigma_{x_{ji}}^2}. \tag{3.3.6}$$

This form of weighting is called *statistical weighting* and is discussed in
greater detail in Section 3.5.

For a case in which $m = 1$ (i.e., only one independent variable), the
"residuals" of the ith point are shown in Figure 3.3.

The slope of the line segment R is equal to R_{y_i}/R_{x_i} which equals
$-(w_x/w_y) \times$ slope of the calculated curve. This relationship can be proved
using Equations 3.6.11 and 3.6.12.

The *weighted sum of the squares of the residuals* is denoted as S:

$$S \equiv \sum_{i=1}^{n} [w_{y_i} R_{y_i}^2 + \sum_{j=1}^{m} w_{x_{ji}} R_{x_{ji}}^2]. \tag{3.3.7}$$

[6] A similar figure appears in W. E. Deming's book, "Statistical Adjustment of
Data," John Wiley & Sons, Inc., New York, 1943.

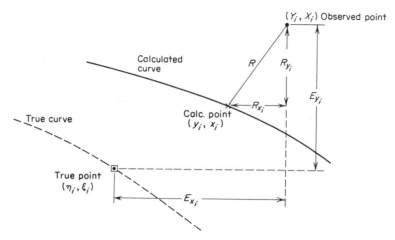

Figure 3.3. A typical relationship between the observed, calculated, and true data points.[6] The "residuals" R_{y_i} and R_{x_i} are shown on the diagram and are determined as a part of the least squares analysis. The "errors" E_{y_i} and E_{x_i} are the errors in the observed values Y_i and X_i. These errors can only be estimated.

The notation can be simplified by using Gauss' brackets { } to indicate a summation over j, and by dropping the index i:

$$S \equiv \sum [w_y R_y^2 + \{w_{x_i} R_{x_i}^2\}]. \tag{3.3.8}$$

The method of least squares is used to determine the values of a_k which minimize S. For most experiments *the values of a_k so determined are the best estimates of the true values* $|\alpha_k$.

Equation 3.3.8 (or 3.3.7) is the fundamental equation of the method of least squares. In elementary discussions of the method, the term $\{w_{x_i} R_{x_i}^2\}$ is often not included in the definition of S. Neglecting this term is permissible only if the uncertainties in the values of the independent variables x_j are negligibly small compared to the uncertainties in the values of y. Another assumption often made is that all the weights (i.e., w_{y_i} and $w_{x_{ii}}$) are equal to one. This assumption is permissible only if it can be assumed that all the uncertainties (i.e., σ_{y_i} and $\sigma_{x_{ii}}$) are approximately the same. The subject of "weighting" is discussed in greater detail in Section 3.5.

3.4 Special Cases of the General Problem

The least squares problem as stated in Section 3.3 is general. The number of independent variables is not limited, the possibility of uncertainty in all the variables is considered, and no limitations on the form of the function f are made. The general solution for the least squares problem is developed in Section 3.6. The formalism developed in Section 3.6 is applicable to the problem in this general form. Often, however, essential simplifications may be made, and a much simpler method of solution may be used.

Case I. Negligible uncertainty in the values of the independent variables.

In many experiments the uncertainty (or standard deviation) of the independent variables is proportionately much less than the uncertainty of the dependent variable. For example, the observed values Y_i might be known to about 2% accuracy (i.e., $\sigma_{y_i} \approx 0.02y_i$), whereas the values of the independent variables X_{ji} are all known to accuracies of better than 0.1% (i.e., $\sigma_{x_{ji}} < 0.001x_{ji}$). For such cases, $\{w_{x_{ji}}R_{x_{ji}}{}^2\}$ is small compared to $w_{y_i}R_{y_i}{}^2$ and very little error arises from neglecting the uncertainties of the independent variables (i.e., assuming $\sigma_{x_{ji}} = 0$). If the $\sigma_{x_{ji}}$'s are assumed to be zero, the residuals $R_{x_{ji}}$ are zero and Equation 3.3.7 reduces to a simpler form:

$$S = \sum_{i=1}^{n} w_{y_i} R_{y_i}{}^2. \tag{3.4.1}$$

Case II. Linear form of Case I.

If the uncertainties of the independent variables can be neglected, and if the function f is "linear" with respect to the unknown parameters a_k, the method of solution is considerably simplified. A direct solution for the values of a_k (i.e., without resorting to iterative techniques) is possible for such cases. Solution for the linear cases is discussed in Section 3.9.

To decide whether or not the function f is linear, one need only examine the function. If f can be expressed as a sum of terms as follows:

$$f = a_1\phi_1(\psi) + a_2\phi_2(\psi) + \cdots + a_p\phi_p(\psi), \tag{3.4.2}$$

where ψ denotes the independent variables:

$$\psi \equiv x_1, x_2, \cdots, x_m, \tag{3.4.3}$$

then the problem is linear. If f cannot be expressed according to Equation 3.4.2, then the problem is nonlinear.

Examples of linear functions are:

$$f = \sum_{k=1}^{p} a_k x^{k-1}, \tag{3.4.4}$$

$$f = \sum_{k=1}^{p} a_k \sin\left(\frac{k\pi x}{L}\right), \tag{3.4.5}$$

$$f = \sum_{k=1}^{p} a_k \frac{x_1{}^2 x_2{}^{2/3}}{\sin(kx_3)}. \tag{3.4.6}$$

Examples of nonlinear functions are:

$$f = a_1 \exp{(-a_2 x)}, \tag{3.4.7}$$

$$f = \sum_{k=1}^{p} a_k^2 \sin\left(\frac{k\pi x_1}{L}\right) \exp{(-k^2 x_2)}, \tag{3.4.8}$$

$$f = \sum_{k=1}^{p} a_k \sin\left(\frac{k\pi x}{a_k}\right). \tag{3.4.9}$$

Equation 3.4.7 could be reduced to a linear form by taking logarithms of both sides of the equation. Logarithms of the values of Y_i would also have to be taken and the least squares process would no longer be based upon minimizing S. A quantity similar to S, the weighted sum of the residuals of the logarithms, would have to be minimized. The resulting values of a_1 and a_2 would be different from the values obtained using Equation 3.4.7. The choice of whether to use Equation 3.4.7 or its logarithmic form would depend upon the purpose of the experiment. How accurately must a_1 and a_2 be known? The availability of computer codes would also enter into the decision. For example, if a general-purpose least squares code was available, there would be no need to linearize the problem. If, however, only a linear least squares code was available, it might be decided that the additional time and money required to write a code capable of treating the nonlinear problem was not worth the improvement in accuracy over a linear treatment of the problem.

Although Equation 3.4.8 is nonlinear, a simple substitution reduces the treatment to a linear problem:

$$f = \sum_{k=1}^{p} b_k \sin\left(\frac{k\pi x_1}{L}\right) \exp{(-k^2 x_2)}, \tag{3.4.10}$$

where

$$b_k = a_k^2. \tag{3.4.11}$$

The same results would be obtained if Equation 3.4.8 were used in conjunction with a nonlinear approach, or Equation 3.4.10 were used in conjunction with a linear approach. The square roots of the values of b_k would equal the values of a_k obtained using Equation 3.4.8, but the saving in computer time using Equation 3.4.10 would be significant.

Equation 3.4.9 is an example of a nonlinear function which cannot be reduced to a linear form. If analysis using this function is required, a nonlinear least squares treatment is unavoidable.

Case III. *A single independent variable.*

For many experiments the function f includes only a single independent variable. Other variables which might significantly affect the measurements (for example, pressure and temperature) should be held constant through-out the experiment. For such cases, Equation 3.3.7 reduces to a simpler form:

$$S = \sum_{i=1}^{n} [w_{y_i} R_{y_i}^2 + w_{x_i} R_{x_i}^2].$$

(3.4.12)

If, in addition, the uncertainty in the values of X_i is negligible, Case III reduces to a subcase of Case I and Equation 3.4.1 is applicable. The simplest cases are those in which there is a single independent variable, the uncertainty in the values of X_i are negligible, and the function is linear (i.e., a subcase of Cases I, II, and III).

3.5 Statistical Weighting

The weights w_{y_i} and $w_{x_{ji}}$ are defined by Equations 3.3.5 and 3.3.6:

$$w_{y_i} \equiv \frac{1}{\sigma_{y_i}^2},$$

(3.3.5)

$$w_{x_{ji}} \equiv \frac{1}{\sigma_{x_{ji}}^2}.$$

(3.3.6)

This type of weighting is often called "statistical weighting."

The reason for using statistical weighting can best be explained by considering Equation 3.3.7:

$$S \equiv \sum_{i=1}^{n} [w_{y_i} R_{y_i}^2 + \sum_{j=1}^{m} w_{x_{ji}} R_{x_{ji}}^2].$$

(3.3.7)

The method of least squares is based upon an adjustment of the parameters a_k such that the sum S is minimized. From Equation 3.3.7 it is seen that S is the sum of n components:

$$S = \sum_{i=1}^{n} s_i,$$

(3.5.1)

where

$$s_i \equiv w_{y_i} R_{y_i}^2 + \sum_{j=1}^{m} w_{x_{ji}} R_{x_{ji}}^2,$$

(3.5.2)

or more simply,

$$s_i \equiv s_{y_i} + \sum_{j=1}^{m} s_{x_{ji}}.$$

(3.5.3)

If the function f (i.e., Equation 3.3.2) adequately represents the data, the

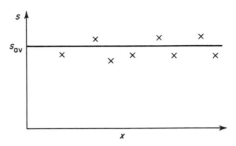

Figure 3.4. Typical distribution of the values of s_i. Note that the values are randomly distributed about an average value.

minimization procedure inherent in the method of least squares will adjust the parameters a_k such that the values of s_i will be randomly distributed about an average value. If there is only one independent variable (i.e., $m = 1$) this random distribution can be illustrated graphically as shown in Figure 3.4. It should be emphasized that a random distribution of the values of s_i will be observed regardless of the choice of weighting function.

For most experiments it can be assumed that the values of R_{y_i} and $R_{x_{j_i}}$ are uncorrelated. For this reason one would expect that the values of s_{y_i} would be randomly distributed about an average value as well as the values of $s_{x_{j_i}}$. For a case in which there is only one independent parameter (i.e., $m = 1$), typical results are shown in Figure 3.5.

The residuals R_{y_i} and $R_{x_{j_i}}$ are *not* randomly distributed about average values. One would expect that the larger values of the residuals would correspond to data points with larger uncertainties or standard deviations. Comparing Equations 3.5.2 and 3.5.3, we obtain:

$$s_{y_i} = w_{y_i} R_{y_i}^2, \tag{3.5.4}$$

and

$$s_{x_{j_i}} = w_{x_{j_i}} R_{x_{j_i}}^2. \tag{3.5.5}$$

Substituting Equations 3.3.5 and 3.3.6 into Equations 3.5.4 and 3.5.5 gives:

$$s_{y_i} = \frac{R_{y_i}^2}{\sigma_{y_i}^2}, \tag{3.5.6}$$

and

$$s_{x_{j_i}} = \frac{R_{x_{j_i}}^2}{\sigma_{x_{j_i}}^2}. \tag{3.5.7}$$

From Equations 3.5.6 and 3.5.7 it is seen that if the values of s_{y_i} and $s_{x_{j_i}}$

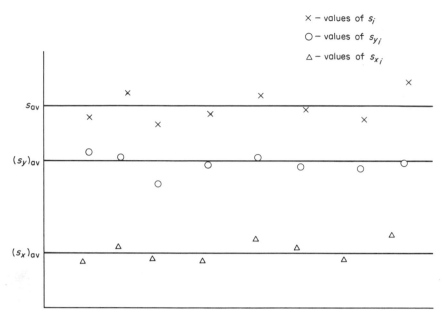

Figure 3.5. Typical distribution of the values of s_i, s_{y_i}, and s_{x_i}. Note that the values of s_{y_i} and s_{x_i}, as well as s_i, are randomly distributed about average values.

are randomly distributed about average values, then the ratios R_{y_i}/σ_{y_i} and $R_{x_{ji}}/\sigma_{x_{ji}}$ are randomly distributed about average values. This situation is logical and is the result of the adoption of statistical weighting (i.e., Equations 3.3.5 and 3.3.6).

In many treatments of the least squares problem, Equations 3.3.5 and 3.3.6 are replaced by:

$$w_{y_i} \equiv \frac{\sigma^2}{\sigma_{y_i}{}^2}, \tag{3.5.8}$$

and

$$w_{x_{ji}} \equiv \frac{\sigma^2}{\sigma_{x_{ji}}{}^2}, \tag{3.5.9}$$

where σ^2 is an arbitrary constant. The values of a_k and s_{a_k} determined by the least squares analysis are not affected by the choice of σ^2.

The constant σ^2 is included in all classical treatments of the method of least squares because this constant can be used to facilitate the computational procedure. For example, if the values of σ_{y_i} are approximately equal to 10^{-4}, then selection of $\sigma^2 = 10^{-8}$ will result in values of w_{y_i} which are approximately equal to 1. If a digital computer is used to perform the analysis, such manipulations are rarely necessary. It is the opinion of the author that digital computers should be used to perform all but the very simplest least squares analyses. The selection of an arbitrary value of σ^2 is therefore unnecessary, and Equations 3.3.5 and 3.3.6 will be used rather than Equations 3.5.8 and 3.5.9. Use of Equations 3.3.5 and 3.3.6 avoids unnecessary complication in the development of prediction analysis.

3.6 Solution for the General Case

The solution for the general least squares problem includes the set of values of a_k which minimizes S. The appropriate definition of S for the general case is Equation 3.3.7:

$$S \equiv \sum_{i=1}^{n} [w_{y_i} R_{y_i}^2 + \{w_{x_{ji}} R_{x_{ji}}^2\}]. \tag{3.3.7}$$

The Gauss brackets $\{\ \}$ imply a summation over j.

The "conditional function" F^i is defined as:

$$F^i \equiv y_i - f(x_{1i}, x_{2i}, \cdots, x_{mi}; a_1, a_2, \cdots, a_p). \tag{3.6.1}$$

From Equation 3.3.2:

$$F^i = 0, \qquad i = 1, 2, \cdots, n. \tag{3.6.2}$$

Derivatives of the function F are denoted by subscripts:

$$F_{x_j}{}^i \equiv \frac{\partial F^i}{\partial x_{ji}}; \quad F_y{}^i \equiv \frac{\partial F^i}{\partial y_i} = 1; \quad F_{a_k}{}^i \equiv \frac{\partial F^i}{\partial a_k}. \tag{3.6.3}$$

The purpose of the analysis is to determine the values a_k. The first step required for a solution is to estimate these values of a_k. The following notation will be used:

$$a_{k0} \equiv \text{the initial guess of the value of } a_k.$$

The choice of the initial values a_{k0} ($k = 1, 2, \cdots, p$) can be critical if the problem is nonlinear. For nonlinear cases, convergence to a solution is dependent upon the function f, the data, and the chosen values of a_{k0}. Whether or not a particular analysis will converge to a solution is difficult to predict. Experience at the Los Alamos Scientific Laboratory has shown that some cases have converged when the initial guesses a_{k0} were as much as a factor of 10^5 different from the values a_k. Conversely, they failed to obtain convergence for cases in which the initial guesses a_{k0} were within 20% of the values of a_k.[7] It can be said, however, that the better the initial

[7] R. H. Moore and R. K. Zeigler, "The Solution of the General Least Squares Problem with Special Reference to High-Speed Computers," *LA-2367*, March 1960.

guesses, the greater the probability of convergence. An example of the sensitivity of convergence to the choice of initial guesses is included in Section 3.7.

The numerical values of F^i and the derivatives of F^i can be estimated from the data and the initial guesses. The estimated values of F^i are denoted as $F_0{}^i$:

$$F_0{}^i \equiv \text{estimated value of } F^i,$$

$$F_0{}^i = Y_i - f(X_{1i}, X_{2i}, \cdots, X_{mi}; a_{10}, a_{20}, \cdots, a_{p0}). \qquad (3.6.4)$$

The values of $F_0{}^i$ are *not* equal to zero. These values of $F_0{}^i$ can be related to the derivatives of F^i, the residuals, and the differences between the values of a_{k0} and a_k. To do this, Equation 3.6.1 is expanded in a Taylor's series. Using Equation 3.6.2 and neglecting higher-order terms of the Taylor's series, we obtain:

$$F_0{}^i = F_y{}^i R_{y_i} + \sum_{j=1}^{m} F_{x_j}{}^i R_{x_{ji}} + \sum_{k=1}^{p} F_{a_k}{}^i A_k, \qquad i = 1, 2, \cdots, n \qquad (3.6.5)$$

where

$$A_k \equiv a_{k0} - a_k. \qquad (3.6.6)$$

If S is at its minimum value, small variations in the residuals will not affect the sum S. Denoting a small variation as δ gives:

$$\tfrac{1}{2}\delta S = \sum_i \left[w_{y_i} R_{y_i} \delta R_{y_i} + \{ w_{x_{ji}} R_{x_{ji}} \delta R_{x_{ji}} \} \right] = 0. \qquad (3.6.7)$$

The values of $F_0{}^i$ are not dependent upon the residuals R_{y_i} and $R_{x_{ji}}$ (see Equation 3.6.4). If the residuals are varied in a manner such that the conditional functions F^i are satisfied, the differential of Equation 3.6.5 is:

$$F_y{}^i \delta R_{y_i} + \sum_{j=1}^{m} F_{x_j}{}^i \delta R_{x_{ji}} + \sum_{k=1}^{p} F_{a_k}{}^i \delta A_k = 0, \qquad i = 1, 2, \cdots, n. \qquad (3.6.8)$$

The variations in A_k (i.e., δA_k) are required because changes in the residuals can be caused only by changes in the values of a_k (and therefore according to Equation 3.6.6, the values of A_k).

Using the method of Lagrange multipliers, the n Equations 3.6.8 are each multiplied by a different multiplier λ_i:

$$\lambda_i \left(F_y{}^i \delta R_{y_i} + \sum_{j=1}^{m} F_{x_j}{}^i \delta R_{x_{ji}} + \sum_{k=1}^{p} F_{a_k}{}^i \delta A_k \right) = 0, \qquad i = 1, 2, \cdots, n. \qquad (3.6.9)$$

There are n values of λ_i (i.e., $\lambda_1, \lambda_2, \cdots, \lambda_n$). Subtracting all n Equations

3.6.9 from 3.6.7, we obtain:

$$\sum_{i=1}^{n} (w_{y_i} R_{y_i} - \lambda_i F_y{}^i) \delta R_{y_i} + \{ \sum_{i=1}^{n} (w_{x_{ji}} R_{x_{ji}} - \lambda_i F_{x_j}{}^i) \delta R_{x_{ji}} \}$$

$$- \sum_{k=1}^{p} (\sum_{i=1}^{n} \lambda_i F_{a_k}{}^i) \delta A_k = 0. \quad (3.6.10)$$

To satisfy Equation 3.6.10, the coefficients of all the variations must be zero. Thus,

$$R_{y_i} = \frac{\lambda_i}{w_{y_i}} F_y{}^i, \qquad i = 1, 2, \cdots, n, \qquad (3.6.11)$$

$$R_{x_{ji}} = \frac{\lambda_i}{w_{x_{ji}}} F_{x_j}{}^i, \qquad i = 1, 2, \cdots, n; j = 1, 2, \cdots, m, \quad (3.6.12)$$

$$\sum_{i=1}^{n} \lambda_i F_{a_k}{}^i = 0, \qquad k = 1, 2, \cdots, p. \qquad (3.6.13)$$

Substituting Equations 3.6.11 and 3.6.12 into Equation 3.6.5 gives:

$$\frac{\lambda_i}{w_{y_i}} (F_y{}^i)^2 + \sum_{j=1}^{m} \frac{\lambda_i}{w_{x_{ji}}} (F_{x_j}{}^i)^2 + \sum_{k=1}^{p} F_{a_k}{}^i A_k = F_0{}^i, \quad i = 1, 2, \cdots, n. \quad (3.6.14)$$

Equations 3.6.13 and 3.6.14 constitute a set of $n + p$ equations. These equations can be simplified by the following substitution:

$$L_i \equiv \frac{(F_y{}^i)^2}{w_{y_i}} + \sum_{j=1}^{m} \frac{(F_{x_j}{}^i)^2}{w_{x_{ji}}}, \qquad i = 1, 2, \cdots, n. \qquad (3.6.15)$$

It can be seen from Equation 3.6.1 that the derivative $F_y{}^i$ is equal to one. In addition, Equations 3.3.5 and 3.3.6 can be substituted into Equation 3.6.15 to further simplify the expression for L_i:

$$L_i = \sigma_{y_i}{}^2 + \sum_{j=1}^{m} (F_{x_j}{}^i \cdot \sigma_{x_{ji}})^2. \qquad (3.6.16)$$

If we substitute Equation 3.6.15 into Equation 3.6.14:

$$L_i \lambda_i = F_0{}^i - \sum_{k=1}^{p} F_{a_k}{}^i A_k, \qquad i = 1, 2, \cdots, n. \qquad (3.6.17)$$

Solving for λ_i, we obtain:

$$\lambda_i = L_i^{-1}(F_0{}^i - \sum_{k=1}^{p} F_{a_k}{}^i A_k), \qquad i = 1, 2, \cdots, n. \qquad (3.6.18)$$

If Equation 3.6.18 is substituted into Equation 3.6.13:

$$A_1 \sum_i L_i^{-1} F_{a_1}{}^i F_{a_1}{}^i + A_2 \sum_i L_i^{-1} F_{a_1}{}^i F_{a_2}{}^i + \cdots$$

$$+ A_p \sum_i L_i^{-1} F_{a_1}{}^i F_{a_p}{}^i = \sum_i L_i^{-1} F_{a_1}{}^i F_0{}^i$$

$$\vdots$$

$$A_1 \sum_i L_i^{-1} F_{a_p}{}^i F_{a_1}{}^i + \cdots$$

$$+ A_p \sum_i L_i^{-1} F_{a_p}{}^i F_{a_p}{}^i = \sum_i L_i^{-1} F_{a_p}{}^i F_0{}^i. \tag{3.6.19}$$

Equations 3.6.19 constitute the p equations required to solve for the p unknown values of A_k.

The notation of Equations 3.6.19 may be simplified considerably. If we drop the subscript i, and denote a summation over i by square brackets [ ]:

$$A_1 \left[\frac{F_{a_1} F_{a_1}}{L} \right] + A_2 \left[\frac{F_{a_1} F_{a_2}}{L} \right] + \cdots + A_p \left[\frac{F_{a_1} F_{a_p}}{L} \right] = \left[\frac{F_{a_1} F_0}{L} \right]$$

$$\vdots \qquad\qquad\qquad\qquad \vdots \tag{3.6.20}$$

$$A_1 \left[\frac{F_{a_p} F_{a_1}}{L} \right] + \cdots \qquad + A_p \left[\frac{F_{a_p} F_{a_p}}{L} \right] = \left[\frac{F_{a_p} F_0}{L} \right].$$

An even further simplification may be made if the following notation is adopted:

$$C_{kl} \equiv \left[\frac{F_{a_k} F_{a_l}}{L} \right], \qquad \begin{matrix} k = 1, 2, \cdots, p, \\[1mm] l = 1, 2, \cdots, p. \end{matrix} \tag{3.6.21}$$

$$V_k \equiv \left[\frac{F_{a_k} F_0}{L} \right], \qquad k = 1, 2, \cdots, p. \tag{3.6.22}$$

Substituting Equations 3.6.21 and 3.6.22 into Equation 3.6.20 gives:

$$A_1 C_{11} + A_2 C_{12} + \cdots + A_p C_{1p} = V_1$$

$$\vdots \qquad\qquad\qquad\qquad \vdots \tag{3.6.23}$$

$$A_1 C_{p1} + \cdots \qquad + A_p C_{pp} = V_p$$

It should be noticed that the elements C_{kl} constitute a symmetrical matrix or array. That is,

$$C_{kl} = C_{lk}. \tag{3.6.24}$$

Matrix notation yields the ultimate simplicity in expressing the set of equations. Defining the vectors A and V, we obtain:

$$A \equiv (A_1, A_2, \cdots, A_p), \tag{3.6.25}$$

$$V \equiv (V_1, V_2, \cdots, V_p), \tag{3.6.26}$$

and the matrix C:

$$C \equiv \begin{vmatrix} C_{11} & C_{12} & \cdots & C_{1p} \\ C_{p1} & & & C_{pp} \end{vmatrix}. \tag{3.6.27}$$

Equation 3.6.23 is simply:

$$CA = V. \tag{3.6.28}$$

The matrix C will be referred to as the *coefficient matrix*. Equation 3.6.28 is a matrix equation and the elements of the unknown vector A are determined by premultiplying both sides of the equation by the *inverse coefficient matrix* C^{-1}:

$$C^{-1}CA = A = C^{-1}V. \tag{3.6.29}$$

Equation 3.6.29 implies that the individual elements A_k of the vector A are calculated as follows:

$$A_k = \sum_{j=1}^{p} C_{kj}^{-1} V_j. \tag{3.6.30}$$

The unique property of the inverse coefficient matrix C^{-1} is that when it is multiplied by the coefficient matrix C, the product is the *unity matrix* I:

$$C^{-1}C = I, \tag{3.6.31}$$

where the unity matrix I is a p by p matrix with all elements equal to zero, except the diagonal elements which are equal to 1:

$$I \equiv \begin{pmatrix} 1 & 0 & 0 & \cdots & 0 \\ 0 & 1 & 0 & \cdots & 0 \\ 0 & 0 & 1 & & 0 \\ 0 & \cdots & & \cdots & 1 \end{pmatrix}. \tag{3.6.32}$$

In other words:

$$\sum_{k=1}^{p} C_{jk}^{-1} C_{kl} = \delta_{jl}, \tag{3.6.33}$$

where $\delta_{jl} = 1$ when $j = l$ and equals 0 when $j \neq l$.

Many computer facilities have subroutine available for "inverting" a matrix. If, however, a matrix inversion routine is not available, the reader should realize that methods particularly suitable for computers have been developed. One such method is discussed in Section 4.8.

The procedure for solution of the values a_k can be summarized:

1. Collect all the data (i.e., the values of Y_i, σ_{y_i}, X_{ji}, and $\sigma_{x_{ji}}$ for $i = 1$, $2, \cdots, n; j = 1, 2, \cdots, m$).
2. Make initial guesses a_{k0} for the unknown parameters a_k.
3. Using the data and the initial guesses, compute the numerical values of the derivatives of F with respect to the dependent and independent variables (i.e., the values of $F_y{}^i$ and $F_{x_j}{}^i$).
4. Using Equation 3.6.15, compute the values of L_i.
5. Using the data and the initial guesses, compute the values of the derivatives of F with respect to the unknown parameters (i.e., the values of $F_{a_k}{}^i$) and the values of $F_0{}^i$.
6. Compute the values of the elements of the matrix C and the vector V using Equations 3.6.21 and 3.6.22.
7. Invert the "coefficient matrix" C. A matrix inversion procedure is discussed in Section 4.8.
8. Using Equation 3.6.30, compute the elements A_k of the vector A.
9. Using Equation 3.6.6, the values of A_k and the initial guesses a_{k0}, compute the values of a_k. (The modified form of Equation 3.6.6 as expressed by Equations 3.8.1 and 3.8.2 may be used if a G factor is required to facilitate convergence to a solution.)

At this point it would appear that the job has been completed, but unfortunately, this is true only for the linear cases (see Section 3.4, Case II). For the nonlinear cases, Equation 3.6.5 was written only after it had been assumed that higher-order terms in the Taylor's series expansion of F were negligible. Inasmuch as these terms often are not negligible, further iterations are required.

10. Use the values of a_k computed in Step 9 as the initial guesses for the next iteration.
11. Using the new values of a_{k0}, repeat the calculation from Step 3.
12. Continue the process until all values of a_k converge to fixed values. Convergence is normally tested by computing the fractional change in the values of a_k from iteration to iteration, and then comparing the fractional changes to an arbitrary value ϵ. If ϵ is greater than the fractional changes for all values of a_k, the procedure is terminated. That is, if

$$\left| \frac{\Delta a_k}{a_k} \right| = \left| \frac{A_k}{a_k} \right| < \epsilon, \qquad k = 1, 2, \cdots, p, \qquad (3.6.34)$$

then the computation is assumed to have converged to the least squares values of a_k. The choice of ϵ is dependent upon the particular problem and the time required per iteration. Obviously, the smaller the value of ϵ, the greater the number of iterations required and the larger the computer time per problem. The analyst must balance all factors and decide upon a value of ϵ that is reasonable for his particular requirements.

3.7 A Specific Application of the Method of Least Squares

In this section the method of least squares is applied to a specific problem.[8] The problem is an example of the most general type of analysis (i.e., uncertainty in the independent as well as the dependent variable, more than one independent variable, and a nonlinear function f).

For this problem, the values of the dependent variable η_i theoretically should be related to the values of the independent variables ξ_{1i} and ξ_{2i} according to Equation 3.7.1:

$$\eta_i = \frac{(1 + \alpha_1^2 \xi_{1i})(1 + \alpha_2^2 \xi_{1i})}{(1 + \alpha_1^2 \xi_{2i})(1 + \alpha_2^2 \xi_{2i})}, \qquad i = 1, 2, \cdots, n. \qquad (3.7.1)$$

An experiment is performed and the measured values of the dependent variable (Y_i), the independent variables (X_{1i} and X_{2i}), and the fractional uncertainties (σ_{yi}/Y_i, σ_{x1i}/X_{1i} and σ_{x2i}/X_{2i}) are tabulated in Table 3.1.

The parameters a_1 and a_2 are the least squares estimates of the true

TABLE 3.1. Experimental data

i	Y_i	σ_{yi}/Y_i	X_{1i}	σ_{x1i}/X_{1i}	X_{2i}	σ_{x2i}/X_{2i}
1	0.7500	0.0144	0.01372	0.0056	0.02579	0.0057
2	0.5667	0.0147	0.01372	0.0056	0.04591	0.0065
3	0.4000	0.0155	0.01372	0.0056	0.07409	0.0070
4	0.8750	0.0142	0.02400	0.0086	0.03200	0.0068
5	0.7000	0.0146	0.02400	0.0086	0.04534	0.0057
6	0.5750	0.0150	0.02400	0.0086	0.06400	0.0054
7	0.3800	0.0153	0.02400	0.0086	0.08801	0.0055
8	0.5750	0.0150	0.02598	0.0093	0.06660	0.0122
9	0.2967	0.0262	0.02598	0.0093	0.13430	0.0134
10	0.1550	0.0187	0.02598	0.0093	0.22908	0.0140
11	0.0900	0.0210	0.02598	0.0093	0.35094	0.0143

[8] Prediction analysis of a similar problem is discussed in Chapter 9.

values of the parameters α_1 and α_2. Initial guesses for the values of a_1 and a_2 (i.e., a_{10} and a_{20}) must be made.

The equation relating the calculated values of y_i, x_{1i}, x_{2i}, a_1, and a_2 is chosen to be of the same form as Equation 3.7.1:

$$y_i = \frac{(1 + a_1^2 x_{1i})(1 + a_2^2 x_{1i})}{(1 + a_1^2 x_{2i})(1 + a_2^2 x_{2i})}.$$ (3.7.2)

The conditional function F^i is determined by substituting Equation 3.7.2 into Equation 3.6.1:

$$F^i = y_i - \frac{(1 + a_1^2 x_{1i})(1 + a_2^2 x_{1i})}{(1 + a_1^2 x_{2i})(1 + a_2^2 x_{2i})}.$$ (3.7.3)

The derivatives of F^i can be determined by differentiating Equation 3.7.3. Dropping the subscript i to simplify the notation, we obtain:

$$F_y = 1,$$ (3.7.4)

$$F_{x_1} = -\frac{a_1^2 + a_2^2 + 2a_1^2 a_2^2 x_1^2}{(1 + a_1^2 x_2)(1 + a_2^2 x_2)}$$ (3.7.5)

$$F_{x_2} = \frac{a_1^2 + a_2^2 + 2a_1^2 a_2^2 x_2^2}{(1 + a_1^2 x_2)(1 + a_2^2 x_2)} \cdot \frac{(1 + a_1^2 x_1)(1 + a_2^2 x_1)}{(1 + a_1^2 x_2)(1 + a_2^2 x_2)}$$ (3.7.6)

$$F_{a_1} = \frac{2a_1(x_2 - x_1)}{(1 + a_1^2 x_2)^2} \cdot \frac{(1 + a_2^2 x_1)}{(1 + a_2^2 x_2)}$$ (3.7.7)

$$F_{a_2} = \frac{2a_2(x_2 - x_1)}{(1 + a_2^2 x_2)^2} \cdot \frac{(1 + a_1^2 x_1)}{(1 + a_1^2 x_2)}$$ (3.7.8)

These derivatives can be estimated at every point by using the observed values of X_{1i} and X_{2i} (as tabulated in Table 3.1) and the initial guesses a_{10} and a_{20}.

The next step in the procedure is to compute the values of L_i using Equation 3.6.16:

$$L_i = \sigma_{y_i}^2 + (F_{x_1}{}^i \sigma_{x_{1i}})^2 + (F_{x_2}{}^i \sigma_{x_{2i}})^2.$$ (3.7.9)

The values of $F_0{}^i$ are then determined:

$$F_0{}^i = Y_i - \frac{(1 + a_{10}^2 X_{1i})(1 + a_{20}^2 X_{1i})}{(1 + a_{10}^2 X_{2i})(1 + a_{20}^2 X_{2i})}.$$ (3.7.10)

The next step in the procedure is to compute the elements of the matrix

C and the vector V from Equations 3.6.21 and 3.6.22:

$$C_{11} = \sum_{i=1}^{n} \frac{F_{a_1}{}^i F_{a_1}{}^i}{L_i}, \tag{3.7.11}$$

$$C_{12} = C_{21} = \sum_{i=1}^{n} \frac{F_{a_1}{}^i F_{a_2}{}^i}{L_i}, \tag{3.7.12}$$

$$C_{22} = \sum_{i=1}^{n} \frac{F_{a_2}{}^i F_{a_2}{}^i}{L_i}, \tag{3.7.13}$$

$$V_1 = \sum_{i=1}^{n} \frac{F_{a_1}{}^i F_0{}^i}{L_i}, \tag{3.7.14}$$

$$V_2 = \sum_{i=1}^{n} \frac{F_{a_2}{}^i F_0{}^i}{L_i}. \tag{3.7.15}$$

Inversion of a 2 by 2 matrix (i.e., 2 rows and 2 columns) is a simple algebraic manipulation. Furthermore, since the matrix is symmetric:

$$C_{11}{}^{-1} = \frac{C_{22}}{C_{11}C_{22} - C_{12}{}^2}, \tag{3.7.16}$$

$$C_{22}{}^{-1} = \frac{C_{11}}{C_{11}C_{22} - C_{12}{}^2}, \tag{3.7.17}$$

$$C_{12}{}^{-1} = C_{21}{}^{-1} = \frac{-C_{12}}{C_{11}C_{22} - C_{12}{}^2}. \tag{3.7.18}$$

The values of A_1 and A_2 are determined using Equation 3.6.30:

$$A_1 = C_{11}{}^{-1}V_1 + C_{12}{}^{-1}V_2, \tag{3.7.19}$$

$$A_2 = C_{21}{}^{-1}V_1 + C_{22}{}^{-1}V_2. \tag{3.7.20}$$

Using Equation 3.6.6, the values of a_1 and a_2 are estimated:

$$a_1 = a_{10} - A_1, \tag{3.7.21}$$

$$a_2 = a_{20} - A_2. \tag{3.7.22}$$

The values of a_1 and a_2 as determined by Equations 3.7.21 and 3.7.22 are then used as initial guesses for the next iteration until convergence is achieved (or until the solution diverges, indicating failure:).

Using the data in Table 3.1 and initial guesses $a_{10} = 5.0$ and $a_{20} = 2.0$, the values of a_1 and a_2 are tabulated from iteration to iteration in Table 3.2. From Table 3.2 it can be seen that after the third iteration, the changes

TABLE 3.2. Values of of a_1 and a_2 from iteration to iteration

Iteration	a_1	a_2
Initial Guesses	5.0	2.0
1	5.3641	1.5616
2	5.2881	1.6186
3	5.2859	1.6202
4	5.2859	1.6202

in the values were less than 0.0001. The least squares values of a_1 and a. as determined from the data in Table 3.1 are therefore 5.2859 and 1.6202 respectively.

3.8 Convergence

A basic question associated with all nonlinear least squares analyses is whether or not all the values of a_k will converge to a solution. The example considered in Section 3.7 converged after three iterations. Not all problems will converge, however, and it is difficult to predict "a priori" whether or not a particular analysis will converge.

Convergence is dependent upon many factors. If, for example, the value of F_0 or one of the partial derivative F_{a_k} becomes "unbounded" for one of the data points, convergence would most probably not be achieved. If one of the derivatives were zero for all n data points, one row and one column of the C matrix would be zero and no solution would be obtained. In addition, if two or more of the partial derivatives F_{a_k} were highly correlated, the probability that the matrix C would be "singular" (i.e., the determinant of C would be zero) is increased. If C is a "singular matrix," no solution will be obtained.

The choice of the initial guesses has an important effect upon convergence. To illustrate the influence of the initial guesses, a "computer experiment" was performed. Using the same function as considered in Section 3.7 (i.e., Equation 3.7.1), a fictitious set of data was generated. The set of data included 8 data points, the errors in X_{1i} and X_{2i} were assumed to be zero, and the values of X_{1i} were the same for all points. Every point fell exactly on the curve represented by $a_1 = 17.32$ and $a_2 = 5.00$. In other words, if convergence was to be achieved, the final values of a_1 and a_2 would be 17.32 and 5.00, and the value of S would be zero. The data are tabulated in Table 3.3.

Every data point is on the curve:

$$Y_i = \frac{(1 + 300X_{1i})(1 + 25X_{1i})}{(1 + 300X_{2i})(1 + 25X_{2i})} .$$

Note that $17.32^2 = 300$ and $5^2 = 25$.

TABLE 3.3. Fictitious data

i	Y_i	σ_{y_i}/Y_i	X_{1i}	$\sigma_{x_{1i}}/X_{1i}$	X_{2i}	$\sigma_{x_{2i}}/X_{2i}$
1	0.9106	0.0160	0.001305	0.00	0.00171	0.00
2	0.7901	0.0196	0.001305	0.00	0.00239	0.00
3	0.6630	0.0249	0.001305	0.00	0.00333	0.00
4	0.5453	0.0320	0.001305	0.00	0.00455	0.00
5	0.4439	0.0410	0.001305	0.00	0.00604	0.00
6	0.3600	0.0518	0.001305	0.00	0.00780	0.00
7	0.2921	0.0644	0.001305	0.00	0.00983	0.00
8	0.2376	0.0787	0.001305	0.00	0.01212	0.00

For any iterative solution, convergence can also be effected by altering the change in parameters from iteration to iteration. The solution proceeds as outlined in Sections 3.6 and 3.7, but the final step is varied. For the example considered in Section 3.7, Equations 3.7.21 and 3.7.22 are replaced by:

$$a_1 = a_{10} - A_1/G, \qquad (3.8.1)$$

$$a_2 = a_{20} - A_2/G. \qquad (3.8.2)$$

Equations 3.7.21 and 3.7.22 are thus special cases of Equations 3.8.1 and 3.8.2 (i.e., $G = 1.0$). The factor G is set by the analyst and will influence the rate of convergence as well as the basic question of whether or not the solution will converge.

For the set of data tabulated in Table 3.3 convergence was tested in the computer experiment for several combinations of initial guesses and values of G. The results are tabulated in Table 3.4.

TABLE 3.4. Testing convergence for various combinations of a_{10}/a_1, a_{20}/a_2, and G

Case	a_{10}/a_1	a_{20}/a_2	G	Results
1	0.75	0.60	1	Converges
2	0.75	0.60	3	Converges
3	0.75	0.60	5	Converges
4	0.25	0.25	1	Diverges
5	0.50	2.00	1	Diverges
6	2.00	0.50	1	Diverges
7	2.00	2.00	1	Diverges
8	0.50	0.50	1	Diverges
9	0.55	0.55	1	Converges
10	0.60	0.60	1	Converges
11	2.00	0.50	3	Converges
12	0.50	1.00	1	Converges
13	1.00	0.20	1	Converges
14	0.50	0.50	3	Converges

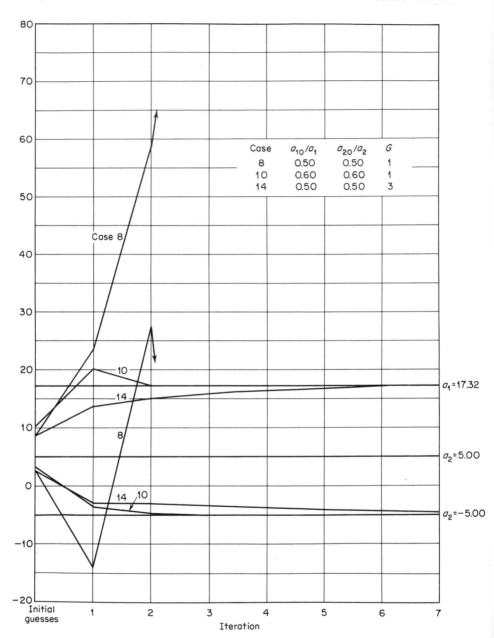

Figure 3.6. Influence of the initial guesses and G upon convergence.

An analysis of Table 3.4 shows that the further the initial guesses a_{k0} are from the least squares values a_k, the greater is the probability of divergence. For example, a comparison of Cases 8, 9, and 10 shows that, as the values of a_{10}/a_1 and a_{20}/a_2 increase from 0.5 to 0.55 and 0.60, the solution passes from a region of divergence to a region of convergence. This point is illustrated in Figure 3.6. The values of a_1 and a_2 are plotted from iteration to iteration for Cases 8, 10, and 14. A comparison of Cases 8 and 10 shows that initial guesses closer to the true value can lead to convergence.

A comparison of Cases 8 and 14 in Figure 3.6 shows that increasing the value of G can also lead to convergence. A further example illustrating the influence of G can be seen by comparing Cases 6 and 11.

Case 13 illustrates an interesting point. One initial guess was close to its least squares value; the second initial guess was far from its least squares value, and the solution converged.

The influence of G upon the rate of convergence is seen in Figure 3.7. The values of a_1 and a_2 are plotted from iteration to iteration for Cases 1, 2, and 3. For $G = 1.0$, the values of a_1 and a_2 were very close to the least squares values after only three iterations. For G values of 3.0 and 5.0, even after six iterations, the values of a_1 and a_2 were still significantly different from the true least squares values. The choice of a value for G must therefore be a compromise between the points illustrated in Figure 3.6 and 3.7. In Figure 3.6 the choice of $G = 3.0$ as compared to $G = 1.0$ for the same initial guesses was shown to result in convergence. Conversely, as shown in Figure 3.7, for a different choice of initial guesses, convergence was attained with $G = 1.0$ at a faster rate than for larger values of G.

It should again be emphasized that the probability of convergence differs from problem to problem. It is often difficult to predict whether or not convergence will be an issue. If the initial analyses show that convergence is difficult to obtain, the corrective steps that should first be undertaken are to improve the method of choosing the initial guesses and then to increase the value of G. If convergence is still difficult to obtain, the number of parameters that are varied in the analysis can be reduced. For example, in the preceding analysis the value of a_1 could be fixed, and the value of a_2 which minimizes S for the given value of a_1 could then be determined using a one-parameter rather than a two-parameter analysis. The procedure could be repeated for enough values of a_1 to locate the combination of a_1 and a_2 which minimizes S. The least square values of a_1 and a_2 determined in this manner can then be used to estimate the standard deviations σ_{a_1} and σ_{a_2} according to the procedure discussed in Section 3.10.

Another approach to the problem of achieving convergence is to limit the changes in the parameters to specified maximum values. This is accomplished by simply testing the magnitude of A_k before altering a_{k0} for the

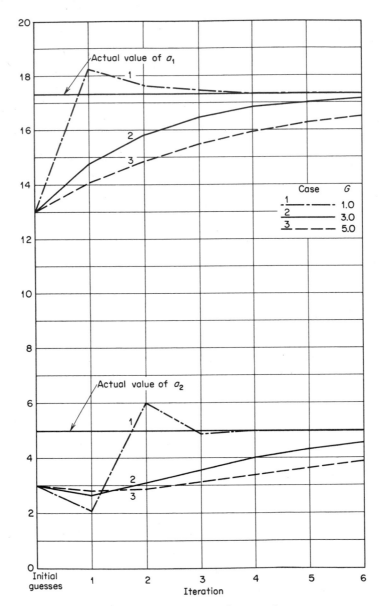

Figure 3.7. Influence of G upon the rate of convergence.

next iteration. If A_k is larger than a specified maximum value, the value of a_{k0} is not changed. Often this procedure will lead to convergence for a problem which would otherwise have diverged.

3.9 Solution for the Linear Case

The linear least squares problem was defined in Case II of Section 3.4. The least squares solution discussed in Section 3.6 applies to linear as well as nonlinear problems. If, however, a problem is linear, the method of solution can be considerably simplified. Linear least squares theory is applicable to many problems and in this section a simplified method of solution for these problems is discussed.

From Equations 3.4.2 and 3.6.1, the conditional function F^i can be determined for the linear case:

$$F^i = y_i - \sum_{k=1}^{p} a_k \phi_k(\psi_i),$$ (3.9.1)

where ψ denotes the independent variables (see Equation 3.4.3). The derivatives of F^i are determined by differentiating Equation 3.9.1. The derivative $F_y{}^i$ is simply equal to 1. No equations are needed for F_{x_i} because the values of $\sigma_{x_{ji}}$ are assumed to be negligible. The derivatives $F_{a_k}{}^i$ are:

$$F_{a_k}{}^i = -\phi_k(\psi_i), \qquad k = 1, 2, \cdots, p.$$ (3.9.2)

Since the $\sigma_{x_{ji}}$'s are zero for the linear cases, the expression for L_i is simply:

$$L_i = w_{y_i}{}^{-1} = \sigma_{y_i}{}^2, \qquad i = 1, 2, \cdots, n.$$ (3.9.3)

For the linear case, the solution is independent of the initial guesses, so the simplest procedure is to set the initial guesses a_{k0} equal to zero. The values of $F_0{}^i$ are thus:

$$F_0{}^i = Y_i, \qquad i = 1, 2, \cdots, n.$$ (3.9.4)

The elements of the coefficient matrix C and the vector V are:

$$C_{kl} = \sum_{i=1}^{n} \frac{\phi_k(\psi_i)\phi_l(\psi_i)}{\sigma_{y_i}{}^2},$$ (3.9.5)

$$V_k = \sum_{i=1}^{n} \frac{\phi_k(\psi_i)Y_i}{\sigma_{y_i}{}^2}.$$ (3.9.6)

Note that according to Equation 3.9.2, the sign of V_k should be negative. The values of A_k are determined using Equation 3.6.30. Since the initial guesses a_{k0} were chosen to be zero, from Equation 3.6.6, the least squares values a_k would equal the values $-A_k$. Since the sign has been changed for all values of V_k, the sign also changes for A_k and therefore:

$$a_k = A_k. \tag{3.9.7}$$

Iterations are *not* required. The values of a_k so determined constitute the least squares solution.

It should be noted that no difficulties will be encountered if linear problems are treated in a general computer program of the type discussed in Section 4.7. It is not even necessary to "warn" the computer that the problem is linear. If, however, a special computer program is to be written for a linear problem, the fact that no iterations are required permits a simpler program.

3.10 Uncertainties of the Least Squares Parameters

The uncertainties (or standard deviations) of the least squares parameters (i.e., the values of σ_{a_k}) are of utmost significance to the analyst. Success or failure of an experiment usually hinges upon these values. For example, consider a particular experiment in which the objective is to determine a_1 to an accuracy of 1%. The analyst can judge upon the success or failure of the experiment by considering the resulting value of σ_{a_1}/a_1. If this value is equal to or less than 0.01, the experiment can most probably be considered as a success. For values greater than 0.01, the analyst is faced with the choice of accepting results inferior to his original criteria, or extending or repeating the experiment.

The importance of the uncertainty associated with experimental results is now recognized by almost all scientists and engineers. Most results are quoted with their uncertainty (or standard deviation). In most cases, failure to mention the uncertainty of a particular result greatly detracts from the usefulness of the work. For this reason, it is essential to determine not only the least squares values of a_k for a given experiment, but also to estimate the uncertainties σ_{a_k}. In this section the formalism for estimating the values of σ_{a_k} is developed.

From Figure 3.3 it is seen that the relationship between the true and observed values of the variables can be expressed as:

$$Y_i = \eta_i + E_{y_i}, \tag{3.10.1}$$

$$X_{ji} = \xi_{ji} + E_{x_{ji}}. \tag{3.10.2}$$

The "errors" E_{y_i} and $E_{x_{ji}}$ will cause the least squares values of the parameters to be different from the true values α_k. The uncertainties σ_{a_k} are defined as the root-mean-squares of these differences:

$$\sigma_{a_k} \equiv \left[(a_k - \alpha_k)^2{}_m \right]^{1/2}, \qquad k = 1, 2, \cdots, p. \qquad (3.10.3)$$

The subscript m refers to the "mean value" of the square of the difference $(a_k - \alpha_k)$ and is the average value that would be obtained if the experiment were repeated many times.

The problem is to develop a method for estimating the values of σ_{a_k}. Equation 3.6.29 is a matrix equation for determining the A vector. Matrix multiplication implies that the individual elements of the A vector are calculated as follows:

$$A_l = \sum_{k=1}^{p} C_{lk}{}^{-1} V_k. \qquad (3.10.4)$$

Substituting Equation 3.6.22 (i.e., the defining equation for V_k) into Equation 3.10.4, we obtain:

$$A_l = \sum_{k=1}^{p} C_{lk}{}^{-1} \sum_{i=1}^{n} L_i{}^{-1} F_{a_k}{}^i F_0{}^i. \qquad (3.10.5)$$

When convergence is achieved, the elements A_k are zero, and the values of $F_0{}^i$ are determined using the least squares values of the parameters (i.e., the a_k's):

$$0 = \sum_{k=1}^{p} C_{lk}{}^{-1} \sum_{i=1}^{n} L_i{}^{-1} F_{a_k}{}^i F_0{}^i, \qquad (3.10.6)$$

where

$$F_0{}^i = Y_i - f(X_{1i}, \cdots, X_{mi}; a_1, \cdots, a_p). \qquad (3.10.7)$$

If the "errors" E_{y_i} and $E_{x_{j\,i}}$ were zero, the values of $F_0{}^i$ would be zero, and the final iteration would yield the true values α_k:

$$0 = \eta_i - f(\xi_{1i}, \cdots, \xi_{mi}; \alpha_1, \cdots, \alpha_p). \qquad (3.10.8)$$

Subtracting Equation 3.10.8 from Equation 3.10.7, substituting Equations 3.10.1 and 3.10.2 into the resulting equation, and assuming that higher terms in a Taylor's series expansion of $f(X_{1i}, \cdots, X_{mi}, a_1, \cdots, a_p) - f(\xi_{1i}, \cdots, \xi_{mi}; \alpha_1, \cdots, \alpha_p)$ are negligible, we obtain:

$$F_0{}^i = E_{y_i} - \sum_{j=1}^{m} \frac{\partial f}{\partial x_j} \cdot (E_{x_{j\,i}}) - \sum_{k=1}^{p} \frac{\partial f}{\partial a_k} \cdot (a_k - \alpha_k). \qquad (3.10.9)$$

From Equation 3.6.1 it can be seen that the derivatives $\partial f/\partial x_j$ and $\partial f/\partial a_k$ are equal to the derivatives of F with a change in sign (i.e., $-F_{x_j}$ and $-F_{a_k}$). Making these substitutions, then substituting Equation 3.10.9

into Equation 3.10.6, bringing the a_k terms to the left side of the equation, and finally squaring both sides, we obtain:

$$Q = [\sum_{k=1}^{p} C_{lk}^{-1} \sum_{i=1}^{n} L_i^{-1} F_{a_k}{}^i \sum_{j=1}^{p} F_{a_j}{}^i (a_j - \alpha_j)]^2$$

$$= [\sum_{k=1}^{p} C_{lk}^{-1} \sum_{i=1}^{n} L_i^{-1} F_{a_k}{}^i (E_{y_i} + \sum_{j=1}^{m} F_{x_j}{}^i E_{x_{ji}})]^2. \quad (3.10.10)$$

The symbol Q is introduced to facilitate the following analysis. Using Equation 3.6.21, that is:

$$C_{kj} = \sum_{i=1}^{n} L_i^{-1} F_{a_k}{}^i F_{a_j}{}^i, \quad (3.6.21)$$

the left side of Equation 3.10.10 can be simplified:

$$Q = [\sum_{k=1}^{p} C_{lk}^{-1} \sum_{j=1}^{p} (a_j - \alpha_k) C_{kj}]^2 = [\sum_{j=1}^{p} (a_j - \alpha_j) \sum_{k=1}^{p} C_{lk}^{-1} C_{kj}]^2 \quad (3.10.11)$$

From Equation 3.6.33, it is seen that Equation 3.10.11 reduces still further:

$$Q = [\sum_{j=1}^{p} (a_j - \alpha_j) \delta_{jl}]^2 = (a_l - \alpha_l)^2, \quad (3.10.12)$$

where $\delta_{jl} = 1$ when $j = l$ and $\delta_{jl} = 0$ when $j \neq l$.

Turning our attention to the right side of Equation 3.10.10, the following substitutions are useful:

$$H_l{}^i \equiv \sum_{k=1}^{p} C_{lk}^{-1} F_{a_k}{}^i, \quad (3.10.13)$$

$$E_i = E_{y_i} + \sum_{j=1}^{m} F_{x_j}{}^i E_{x_{ji}}. \quad (3.10.14)$$

Substituting Equations 3.10.13 and 3.10.14 into the right side of Equation 3.10.10 gives:

$$Q = (\sum_{i=1}^{n} L_i^{-1} E_i H_l{}^i)^2, \quad (3.10.15)$$

$$Q = (L_1^{-1} E_1 H_l{}^1)^2 + (L_2^{-1} E_2 H_l{}^2)^2 + \cdots + (L_n^{-1} E_n H_l{}^n)^2$$

$$+ 2 \frac{E_1}{L_1} \frac{E_2}{L_1} H_l{}^1 H_l{}^2 + \text{other cross-product terms.} \quad (3.10.16)$$

The values of Q will be distributed about a "mean value" Q_m if the experiment is repeated many times. The cross-product terms in Equation 3.10.16 will take on positive and negative values, but for most experiments the "mean value" of $E_i E_j$ will be zero. (In statistical language the values of E_i and E_j are "uncorrelated.") After the cross-product terms of Equation 3.10.16 are dropped, an expression for Q_m can be developed from Equations 3.10.12 and 3.10.16:

$$Q_m = (a_l - \alpha_l)^2{}_m = (L^{-1}E^2)_m \sum_{i=1}^{n} L_i{}^{-1}H_l{}^i H_l{}^i, \qquad (3.10.17)$$

where $(L^{-1}E^2)_m$ is the "mean value" of $L_i{}^{-1}E_i{}^2$ over all values of i. This result is still not in a convenient form and can be further simplified. Using Equation 3.10.13 gives:

$$\sum_{i=1}^{n} L_i{}^{-1}H_l{}^i H_l{}^i = \sum_{i=1}^{n} L_i{}^{-1} \sum_{k=1}^{p} C_{lk}{}^{-1}F_{a_k}{}^i \sum_{j=1}^{p} C_{lj}{}^{-1}F_{a_j}{}^i. \qquad (3.10.18)$$

Rearranging Equation 3.10.18, we obtain:

$$\sum_{i} L_i{}^{-1}H_l{}^i H_l{}^i = \sum_{k} C_{lk}{}^{-1} \sum_{j} C_{lj}{}^{-1} \sum_{i} L_i{}^{-1}F_{a_j}{}^i F_{a_k}{}^i. \qquad (3.10.19)$$

If we again use Equation 3.6.21:

$$\sum_{i} L_i{}^{-1}H_l{}^i H_l{}^i = \sum_{k} C_{lk}{}^{-1} \sum_{j} C_{lj}{}^{-1}C_{jk}. \qquad (3.10.20)$$

Using Equation 3.6.33, the final simplification can be made:

$$\sum_{i} L_i{}^{-1}H_l{}^i H_l{}^i = \sum_{k} C_{lk}{}^{-1}\delta_{lk} = C_{ll}{}^{-1}. \qquad (3.10.21)$$

Substituting Equation 3.10.21 into Equation 3.10.17 gives:

$$(a_l - \alpha_l)_m{}^2 = (L^{-1}E^2)_m C_{ll}{}^{-1}. \qquad (3.10.22)$$

Before proceeding with the analysis, it is worthwhile to reflect upon the derivation of Equation 3.10.22. The derivation is based upon the assumption that Equation 3.10.9 is reasonable. Unless the errors are very large, this equation will be reasonable for most experiments. Equation 3.10.9 was then substituted into Equation 3.10.6, and the remainder of the derivation was primarily algebra. An additional assumption that the errors are uncorrelated was also required in the derivation. This assumption is valid for most practical problems.[9]

The next step in the analysis is to consider the term $(L^{-1}E^2)_m$. Using Equations 3.10.9, and 3.10.14 and the substitutions F_{x_i} for $-\partial f/\partial x_i$ and

---

[9] Systematic errors are usually correlated. The subject of systematic errors is considered in Section 4.9, and a specific problem which includes systematic errors is discussed in Chapter 9.

F_{a_k} for $-\partial f/\partial a_k$, we then obtain:

$$F_0{}^i - \sum_{k=1}^{p} F_{a_k}{}^i (a_k - \alpha_k) = E_{y_i} + \sum_{j=1}^{m} F_{x_j}{}^i E_{x_{ji}} \equiv E_i. \quad (3.10.23)$$

Squaring both sides of Equation 3.10.23 gives:

$$(F_0{}^i)^2 - 2 \sum_k F_0{}^i F_{a_k}{}^i (a_k - \alpha_k) + \sum_k (F_{a_k}{}^i)^2 (a_k - \alpha_k)^2$$

$$+ 2F_{a_1}{}^i F_{a_2}{}^i (a_1 - \alpha_1)(a_2 - \alpha_2) + \text{other cross-product terms} = E_i{}^2.$$

$$(3.10.24)$$

If the experiment is repeated many times, the values of $a_k - \alpha_k$ will fluctuate about zero, and the average value will be zero. The average values of the cross-product terms are *not*, however, equal to zero. For many experiments the parameters are "correlated." The average values of the products are called the "covariances" and equations for the covariances can be derived in a manner similar to the derivation of Equation 3.10.22. The result of such a derivation is:

$$\rho_{lk}\sigma_{a_l}\sigma_{a_k} \equiv [(a_l - \alpha_l)(a_k - \alpha_k)]_m = (L^{-1}E^2)_m C_{lk}{}^{-1}. \quad (3.10.25)$$

The symbol ρ_{lk} in statistical terminology is called the *correlation coefficient* of the parameters a_l and a_k. The product $\rho_{lk}\sigma_{a_k}$ is called the *covariance* of the parameters a_l and a_k. The covariances are discussed at greater length in Section 3.12.

The "mean value" of $L_i^{-1}E_i{}^2$ can be determined from Equation 3.10.24:

$$[L_i^{-1}(F_0{}^i)^2]_m + \left[L_i^{-1} \sum_{j=1}^{p} F_{a_j}{}^i (a_j - \alpha_j) \sum_{k=1}^{p} F_{a_k}{}^i (a_k - \alpha_k)\right]_m = (L_i^{-1}E_i{}^2)_m.$$

$$(3.10.26)$$

We are interested in the "mean value" of $L^{-1}E^2$ over the entire range of i. This "overall mean value" can be related to the "mean values" $(L_i^{-1}E_i{}^2)_m$:

$$n \cdot (L^{-1}E^2)_m = \sum_{i=1}^{n} (L_i^{-1}E_i{}^2)_m. \quad (3.10.27)$$

Summing Equation 3.10.26 over all values of i and using Equation 3.10.27, we obtain

$$n \cdot (L^{-1}E^2)_m = \sum_{i=1}^{n} [L_i^{-1}(F_0{}^i)^2]_m$$

$$+ \sum_{i=1}^{n} \left[L_i^{-1} \sum_j F_{a_j}{}^i (a_j - \alpha_j) \sum_k F_{a_k}{}^i (a_k - \alpha_k)\right]_m. \quad (3.10.28)$$

An "unbiased estimate" of the first term on the right side of Equation 3.10.28 can be made from the data. This "unbiased estimate" is simply the value obtained from the experiment under consideration:

$$\sum_{i=1}^{n} [L_i^{-1}(F_0{}^i)^2]_m \approx \sum_{i=1}^{n} L_i^{-1}(F_0{}^i)^2. \qquad (3.10.29)$$

It is next shown that $\sum L_i^{-1}(F_0{}^i)^2$ is equal to the "weighted sum of the squares of the residuals" S:

$$S = \sum_{i=1}^{n} w_{y_i} R_{y_i}{}^2 + \{ \sum_{i=1}^{n} w_{x_{j i}} R_{x_{j i}}{}^2 \}, \qquad (3.10.30)$$

where the Gauss brackets $\{ \ \}$ indicate a summation over j. Substituting Equations 3.6.11 and 3.6.12 for R_{y_i} and $R_{x_{j i}}$, substituting Equation 3.6.15 for L_i, and substituting Equation 3.6.18 for λ_i, we obtain:

$$S = \sum_{i=1}^{n} \frac{\lambda_i^2}{w_{y_i}} (F_y{}^i)^2 + \left\{ \sum_{i=1}^{n} \frac{\lambda_i^2}{w_{x_{j i}}} (F_x{}^i)^2 \right\}, \qquad (3.10.31)$$

$$= \sum_i \lambda_i^2 L_i, \qquad (3.10.32)$$

$$= \sum_i L_i^{-1}(F_0{}^i - \sum_k F_{a_k}{}^i A_k)^2. \qquad (3.10.33)$$

After convergence to a solution, the values of A_k are zero, so Equation 3.10.33 reduces to:

$$S = \sum_{i=1}^{n} \frac{1}{L_i} (F_0{}^i)^2. \qquad (3.10.34)$$

We next consider the second term on the right side of Equation 3.10.28. Substituting Equation 3.10.34 into Equation 3.10.29 and then into Equation 3.10.28, we obtain:

$$n(L^{-1}E^2)_m - S \approx [\sum_i L_i^{-1} \sum_j F_{a_i}{}^i(a_j - \alpha_j) \sum_k F_{a_k}{}^i(a_k - \alpha_k)]_m. \qquad (3.10.35)$$

Equation 3.10.35 is not an exact equality because S is only an "unbiased estimate" of

$$\sum_i [L_i^{-1}(F_0{}^i)^2]_m.$$

Using Equations 3.10.22 and 3.10.25 gives:

$$n(L^{-1}E^2)_m - S \approx (L^{-1}E^2)_m \sum_i L_i^{-1} \sum_j F_{a_i}{}^i \sum_k C_{jk}{}^{-1} F_{a_k}{}^i. \qquad (3.10.36)$$

Using Equation 3.6.21 and rearranging the right side of Equation 3.10.36, we obtain

$$n(L^{-1}E^2)_m - S \approx (L^{-1}E^2)_m \sum_j \sum_k C_{jk}{}^{-1}C_{kj},$$

and then using Equation 3.6.33 gives:

$$n(L^{-1}E^2)_m - S \approx (L^{-1}E^2)_m \sum_{j=1}^{p} 1 = p(L^{-1}E^2)_m. \qquad (3.10.37)$$

Solving for $(L^{-1}E^2)_m$ from Equation 3.10.37 gives:

$$(L^{-1}E^2)_m \approx \frac{S}{n-p}. \qquad (3.10.38)$$

The sum S is the weighted sum of the squares of the residuals and $n-p$ is the number of data points minus the number of unknown parameters. In statistical terminology, $n-p$ is called the *number of degrees of freedom*, and is denoted as k in Section 2.5.

Although it will not be proven, an interesting facet of Equation 3.10.38 should be mentioned. If the experiment is repeated many times, the values of S would be distributed according to a chi-squared distribution with $n-p$ degrees of freedom. The "mean value" of this distribution is equal to the "number of degrees of freedom"[10] and therefore the "mean value" of $S/(n-p)$ is equal to one. This fact will be used in the development of Prediction Analysis in Chapter 4.

The final equation for estimating the uncertainties (or standard deviations) is obtained by substituting Equation 3.10.38 into Equation 3.10.22 and then using Equation 3.10.3:

$$\sigma_{a_k} \approx \left(\frac{S}{n-p}\right)^{1/2} \cdot (C_{kk}{}^{-1})^{1/2}, \qquad k = 1, 2, \cdots, p. \qquad (3.10.39)$$

The square of the standard deviation is called the *variance* and is obtained by simply squaring Equation 3.10.39:

$$\text{Variance} \equiv \sigma_{a_k}{}^2 \approx \frac{S}{n-p} C_{kk}{}^{-1}. \qquad (3.10.40)$$

The *covariance* σ_{lk} can also be expressed in terms of $S/(n-p)$:

$$\text{Covariance} \equiv \sigma_{lk} = \rho_{lk}\sigma_{a_l}\sigma_{a_k} \approx \frac{S}{n-p} C_{lk}{}^{-1}, \qquad l \neq k. \qquad (3.10.41)$$

The symbol ρ_{lk} denotes the *correlation coefficient*. The utility of the covariances is discussed in Section 3.12.

---

[10] A. Hald, "Statistical Theory with Engineering Applications," John Wiley & Sons, Inc., New York, 1952, p. 256.

TABLE 3.5. Values[a] of $t_{\beta/2, \, n-p}$ for the *t-distribution*.

Degrees of Freedom, $n - p$	$\beta/2 \equiv$ One half of fraction outside the "confidence interval."							
	0.005	0.010	0.025	0.050	0.10	0.15	0.20	0.25
1	63.657	31.821	12.706	6.314	3.078	1.963	1.376	1.000
2	9.925	6.965	4.303	2.920	1.886	1.386	1.061	0.816
3	5.841	4.541	3.182	2.353	1.638	1.250	0.978	0.765
4	4.604	3.747	2.776	2.132	1.533	1.190	0.941	0.741
5	4.032	3.365	2.571	2.015	1.476	1.156	0.920	0.727
6	3.707	3.143	2.447	1.943	1.440	1.134	0.906	0.718
7	3.499	2.998	2.365	1.895	1.415	1.119	0.896	0.711
8	3.355	2.896	2.306	1.860	1.397	1.108	0.889	0.706
9	3.250	2.821	2.262	1.833	1.383	1.100	0.883	0.703
10	3.169	2.764	2.228	1.812	1.372	1.093	0.879	0.700
15	2.947	2.602	2.131	1.753	1.341	1.074	0.866	0.691
20	2.845	2.528	2.086	1.725	1.325	1.064	0.860	0.687
25	2.787	2.485	2.060	1.708	1.316	1.058	0.856	0.684
30	2.750	2.457	2.042	1.697	1.310	1.055	0.854	0.683
∞	2.576	2.326	1.960	1.645	1.282	1.036	0.842	0.674

[a] It should be noted that $u_{(\beta/2)} = t_{(\beta/2, \, \infty)}$. The $\beta/2 = 0.005$ column refers to a "percent confidence" of 99%, $\beta/2 = 0.010$ to 98%, etc.

A *confidence interval* can be set on the values of α_k. If the uncertainties σ_{a_k} were known exactly, the "confidence interval" for α_k would be:

$$a_k - \sigma_{a_k} \cdot u_{\beta/2} < \alpha_k < a_k + \sigma_{a_k} \cdot u_{\beta/2}, \qquad (3.10.42)$$

where $u_{\beta/2}$ is the $100\beta/2$ percentage point of the "normal distribution" and $100\,(1 - \beta)$ is the *percent confidence* desired. In other words, if we wanted 95% confidence that α_k would be within the quoted range, from Table 3.5, $u_{0.025}$ equals 1.960, and the "confidence interval" is $a_k -$ $1.960\sigma_{a_k} < \alpha_k < a_k + 1.960\sigma_{a_k}$. There is a 95% probability that α_k is within this range.

The exact values of σ_{a_k} are not known. We only know the *unbiased estimates* s_{a_k} (sometimes called the *estimated standard errors* of the parameters):

$$s_{a_k} \equiv \text{unbiased estimate of } \sigma_{a_k},$$

and

$$s_{a_k}{}^2 = \frac{S}{n - p} C_{kk}^{-1}. \qquad (3.10.43)$$

Similarly, the "unbiased estimate" of the covariances can be determined:

$$s_{lk} = r_{lk}s_{a_l}s_{a_k} = \frac{S}{n-p} C_{lk}^{-1}, \qquad (3.10.44)$$

where r_{lk} is the "unbiased estimate" of the correlation coefficient ρ_{lk}.

If s_{a_k} is used in place of σ_{a_k}, the "confidence interval" is:

$$a_k - s_{a_k} \cdot t_{(\beta/2,n-p)} < \alpha_k < a_k + s_{a_k} \cdot t_{(\beta/2,n-p)}, \qquad (3.10.45)$$

where $t_{(\beta/2, n-p)}$ is the $100\beta/2$ percentage point of the *t-distribution* of $n - p$ degrees of freedom. The t-distribution approaches the normal distribution as $n - p$ approaches infinity (see Table 3.5). The values of $t_{(0/025,n-p)}$ are 2.571, 2.042, and 1.960 for values of $n - p$ of 5, 30, and infinity. It can be seen that the t-distribution approaches the normal distribution rapidly.

It should be understood why the values of $t_{(\beta/2,n-p)}$ decrease as the number of degrees of freedom $(n - p)$ increases. Since s_{a_k} is an estimate of σ_{a_k}, there is a "degree of uncertainty" associated with this estimate. The larger values of $t_{(\beta/2,n-p)}$ reflect this increased uncertainty. Of course, as $n - p$ increases, the values of s_{a_k} become better and better estimates of the values of σ_{a_k} and, therefore, $t_{(\beta/2,n-p)}$ decreases as $n - p$ increases.

3.11 Relative Uncertainties

The least squares problem as stated in Section 3.3 is based upon the assumption that the uncertainties σ_{y_i} and $\sigma_{x_{ii}}$ can be estimated. For some experiments it is possible to estimate the values of σ_{y_i} relative to an unknown reference value, but no estimate of the absolute values of σ_{y_i} can be made. For example, if all values of σ_{y_i} are the same but the magnitude of these uncertainties is unknown, then:

$$\sigma_{y_i} = \sigma_{\text{ref}}. \qquad (3.11.1)$$

Another possibility is that the values of σ_{y_i} are proportional to y_i but the absolute values are unknown. The equation for σ_{y_i} might then be:

$$\sigma_{y_i} = \sigma_{\text{ref}} \cdot y_i/y_{\text{ref}} \qquad (3.11.2)$$

If the uncertainties of the independent variables (i.e., the $\sigma_{x_{ii}}$'s) can be neglected, the method of least squares can be applied not only to the problem of determining the least squares values a_k, but also to the problem of estimating σ_{ref}.

The method of solution is to first arbitrarily select a value for σ_{ref} and then, using the appropriate expression for σ_{y_i} (in the above examples,

Equation 3.11.1 or 3.11.2), compute a fictitious set of values for σ_{y_i}. Since the $\sigma_{x_{y_i}}$'s are assumed to be negligible, Equation 3.6.16 reduces to:

$$L_i = \sigma_{y_i}^2. \tag{3.11.3}$$

Proceeding in the manner described in Section 3.6, values of a_k and s_{a_k} are computed. Although at first glance it seems remarkable, the values of a_k and s_{a_k} will be the same, regardless of the choice of σ_{ref}.

The explanation for this independence of the results to the choice of σ_{ref} is straightforward. Since all values of σ_{y_i} are proportional to σ_{ref}, then L_i is proportional to σ_{ref}^2 and, from Equations 3.6.21 and 3.6.22, all elements of the matrix C and the vector V are inversely proportion to σ_{ref}^2. The elements of the inverse matrix C^{-1} must therefore be proportion to σ_{ref}^2. Since the $\sigma_{x_{ii}}$'s are negligible, the sum S is computed using Equation 3.4.1. Substituting Equation 3.3.5 into Equation 3.4.1 gives:

$$S = \sum_{i=1}^{n} w_{y_i} R_{y_i}^2 = \sum_{i=1}^{n} \frac{R_{y_i}^2}{\sigma_{y_i}^2}. \tag{3.11.4}$$

The sum S is therefore inversely proportional to σ_{ref}^2. From Equation 3.6.29, the vector A is seen to equal the product $C^{-1}V$, and, from the preceding argument, this product must be independent of σ_{ref}. The elements of A are independent of σ_{ref} and thus the values of a_k independent of σ_{ref}. Similarly, from Equation 3.10.43, it is seen that s_{a_k} is proportional to the product SC_{kk}^{-1} and, from the preceding argument, this product is independent of σ_{ref}.

To estimate the correct value of σ_{ref} (if this quantity is indeed of interest), use is made of the fact that the mean value of S is $(n - p)$ if the "correct weights" are used. (See the discussion following Equation 3.10.38.) If the assumed value of σ_{ref} (denoted as $(\sigma_{\text{ref}})_{\text{as.}}$) is far from the correct value of σ_{ref} (denoted as $(\sigma_{\text{ref}})_{\text{cor.}}$), then the value of S computed using Equation 3.11.4 will also be far from $n - p$. To estimate the correct value of σ_{ref}, use is made of the conclusion that S is inversely proportional to σ_{ref}^2:

$$(\sigma_{\text{ref}})_{\text{cor.}}^2 = \frac{S}{n - p} \cdot (\sigma_{\text{ref}})_{\text{as.}}^2. \tag{3.11.5}$$

For example, if a value of 0.1 is assumed for σ_{ref}, if $n - p = 15$, and if the computed value of S turns out to be 60.0, then from Equation 3.11.5 it is seen that the best estimate of the correct value of σ_{ref} is 0.2. In other words, the values of σ_{y_i} used throughout the analysis were probably underestimated by a factor of two.

3.12 Confidence Intervals

Determination of the values of a_k and the unbiased estimates of σ_{a_k} (i.e., the values of s_{a_k}) is not necessarily the final step in the least squares analysis of experimental data. For some experiments it is of interest to determine *confidence intervals* for the function f. As will be seen, "confidence intervals" are useful for checking the plausibility of the *observed values* of the dependent variable (i.e., the Y_i's). They are also useful for estimating the uncertainty of interpolated and extrapolated values of f.

Neglecting higher order terms, the *variance* of the function f for a particular combination of values x_j is:

$$\sigma_f^2 = \left(\frac{\partial f}{\partial a_1}\right)^2 \sigma_{a_1}^2 + \cdots + \left(\frac{\partial f}{\partial a_p}\right)^2 \sigma_{a_p}^2$$

$$+ 2\left(\frac{\partial f}{\partial a_1}\right)\left(\frac{\partial f}{\partial a_2}\right)\rho_{12}\sigma_{a_1}\sigma_{a_2} + \text{other cross-product terms},\quad (3.12.1)$$

$$\sigma_f^2 = \sum_{j=1}^{p}\sum_{k=1}^{p}\left(\frac{\partial f}{\partial a_j}\right)\left(\frac{\partial f}{\partial a_k}\right)\rho_{jk}\sigma_{a_j}\sigma_{a_k},\quad (3.12.2)$$

where $\rho_{jk} = 1$ when $j = k$. From Equations 3.6.1, 3.6.3, and 3.12.2:

$$\sigma_f^2 = \sum_{j=1}^{p}\sum_{k=1}^{p}F_{a_j}F_{a_k}\rho_{jk}\sigma_{a_j}\sigma_{a_k}.\quad (3.12.3)$$

Since the *true values* of the uncertainties σ_{a_k} and the *correlation coefficients* ρ_{jk} are unknown, the unbiased estimates s_{a_k} and r_{jk} must be used. The result will be an unbiased estimate of σ_f^2 and will be denoted as s_f^2:

$$s_f^2 = \sum_{j=1}^{p}\sum_{k=1}^{p}F_{a_j}F_{a_k}r_{jk}s_{a_j}s_{a_k},\quad (3.12.4)$$

where $r_{jk} = 1$ when $j = k$.

Substituting Equations 3.10.43 and 3.10.44 into Equation 3.12.4 gives:

$$s_f^2 = \frac{S}{n-p}\sum_{j=1}^{p}\sum_{k=1}^{p}F_{a_j}F_{a_k}C_{jk}^{-1}.\quad (3.12.5)$$

To determine a confidence interval, the *percent confidence* is chosen, and from Table 3.5, the appropriate value of $t_{(\beta/2, n-p)}$ is determined. For a given combination of values of x_{ji}, the value of f is determined using Equation 3.3.2 and the value of s_f is determined using Equation 3.12.5. The confidence interval for the particular point is:

$$f(x_1, \cdots, x_m; a_1, \cdots, a_p) - s_f \cdot t_{(\beta/2, n-p)} < f(x_1, \cdots, x_m; \alpha_1, \cdots, \alpha_p)$$

$$< f(x_1, \cdots, x_m; a_1, \cdots, a_p) + s_f \cdot t_{(\beta/2, n-p)}.\quad (3.12.6)$$

The values of x_j in Equation 3.12.6 should be considered as any arbitrary set. They are neither *observed*, *calculated*, nor *true values*. The analyst merely assumes a set of values of the independent variables and for this set calculates an "unbiased estimate" of the variance σ_f^2. If the experiment includes one independent variable, graphical presentation of the results is simplified. "Confidence bands" are drawn on the graph by connecting the upper and lower limits of the ranges determined using Equation 3.12.6. Typical results are shown in Figure 3.8.

As an example, consider a case in which a straight line is fit to a set of data and yields values of $a_1 = 10.7$ and $a_2 = 2.81$, where

$$y = f(x; a_1, a_2) = a_1 + a_2 x. \tag{3.12.7}$$

Furthermore, assume that the "inverse coefficient matrix" is:

$$C^{-1} = \begin{vmatrix} 1.310 & -0.218 \\ -0.218 & 0.582 \end{vmatrix}, \tag{3.12.8}$$

and $S/(n - p) = 1.1$. If we use Equation 3.10.43, $s_{a_1}^2 = 1.1 \times 1.31 = 1.44$, and $s_{a_2}^2 = 1.1 \times 0.58 = 0.64$. The values of s_{a_1} and s_{a_2} are therefore 1.20 and 0.80, respectively. The value of r_{12} is computed using Equation 3.10.44 and is $-1.1 \times 0.218/1.20 \times 0.80 = -0.25$.

It is required to determine the "best estimate" of y corresponding to $x = 2$. Using Equation 3.12.7 and the "least squares values" of a_1 and a_2 gives:

$$y = f(2; 10.7, 2.81) = 10.7 + 2.81 \times 2.0 = 16.32. \tag{3.12.9}$$

The unbiased estimate of σ_y is denoted as s_f and is determined using

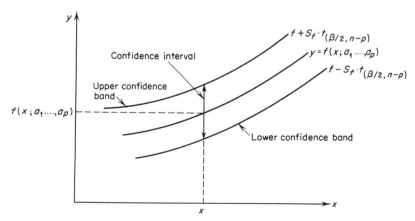

Figure 3.8. A *least squares* curve and the corresponding *confidence band*.

Equation 3.12.5:

$$s_f^2 = \frac{S}{n-p} (F_{a_1}^2 C_{11}^{-1} + F_{a_2}^2 C_{22}^{-1} + 2F_{a_1}F_{a_2}C_{12}^{-1}), \quad (3.12.10)$$

$$s_f^2 = \frac{S}{n-p} (C_{11}^{-1} + x^2 C_{22}^{-1} + 2xC_{12}^{-1}), \quad (3.12.11)$$

because $F_{a_1} = -1$ and $F_{a_2} = -x$. Substituting the correct values into Equation 3.12.11 gives:

$$s_f^2 = 1.1(1.310 + 4.0 \times 0.582 - 4.0 \times 0.218) = 3.04. \quad (3.12.12)$$

The unbiased estimate of $y \pm \sigma_y$ for $x = 2$ is therefore 16.32 ± 1.75. To determine the confidence interval for 95% confidence, the number of degrees of freedom must be known. If the number of data points is 10, then $n - p = 10 - 2 = 8$ and, from Table 3.6, $t_{(0.025,8)} = 2.306$. The "confidence interval" is therefore $16.32 - 2.306 \times 1.75 < y < 16.32 + 2.306 \times 1.75$, or $12.30 < y < 20.34$. In other words, the probability is 95% that the true value of y corresponding to a value of x equal to 2.0 falls within the given range.

It is interesting to note that the *confidence intervals* determined by Equation 3.12.6 can be reduced to arbitrarily small values if systematic errors do not affect the experiment. By properly planning the experiment the s_{a_k}'s can be made arbitrarily small and therefore from Equation 3.12.4 it is seen that s_f can be made arbitrarily small. There is, of course, a practical limitation upon the design of every experiment. For this reason practical minima exists for the values of s_{a_k} and therefore for s_f and the confidence interval.

The value of s_f as determined by Equation 3.12.4 or 3.12.5 is of considerable importance for a certain type of experiment. Often the calculated curve rather than the values of the unknown parameters is the more interesting result of the experiment. For example, the purpose of the experiment might be to determine a curve which can be used for *interpolation* or *extrapolation* for any combination of the independent variables. The value of s_f determined using Equation 3.12.4 or 3.12.5 is the *unbiased estimate of the uncertainty of the interpolated and extrapolated values.*

The concept of a confidence interval can also be applied to the problem of testing the plausibility of the *observed values* of the dependent variable (i.e., the Y_i's). In Section 3.3 the notation σ_{y_i} was used to denote the uncertainty or standard deviation of Y_i. If it can be assumed that the values of Y_i are normally distributed about the true values η_i, and if the values of $\sigma_{x_{ji}}$ can be neglected, then the corresponding values of f can be tested by the following confidence interval:

$$Y_i - \sigma_{y_i} \cdot u_{\beta/2} < f(X_{1i}, \cdots, X_{mi}; a_1, \cdots, a_p) < Y_i + \sigma_{y_i} \cdot u_{\beta/2}. \quad (3.12.13)$$

TABLE 3.6. χ^2 as a function of $n - p$ and P.[a]

Degrees of Freedom, $n - p$	Values of P										
	0.99	0.98	0.95	0.90	0.80	0.50	0.20	0.10	0.05	0.02	0.01
1	0.00016	0.00063	0.0039	0.0158	0.0642	0.455	1.642	2.706	3.841	5.412	6.635
2	0.0201	0.0404	0.103	0.211	0.446	1.386	3.219	4.605	5.991	7.824	9.210
3	0.115	0.185	0.352	0.584	1.005	2.366	4.642	6.251	7.815	9.837	11.341
4	0.297	0.429	0.711	1.064	1.649	3.357	5.989	7.779	9.488	11.668	13.277
5	0.554	0.752	1.145	1.610	2.343	4.351	7.289	9.236	11.070	13.388	15.086
6	0.872	1.134	1.635	2.204	3.070	5.348	8.558	10.645	12.592	15.033	16.812
7	1.239	1.564	2.167	2.833	3.822	6.346	9.803	12.017	14.067	16.622	18.475
8	1.646	2.032	2.733	3.490	4.594	7.344	11.030	13.362	15.507	18.168	20.090
9	2.088	2.532	3.325	4.168	5.380	8.343	12.242	14.684	16.919	19.679	21.666
10	2.558	3.059	3.940	4.865	6.179	9.342	13.442	15.987	18.307	21.161	23.209
15	5.229	5.985	7.261	8.547	10.307	14.339	19.311	22.307	24.996	28.259	30.578
20	8.260	9.237	10.851	12.443	14.578	19.337	25.038	28.412	31.410	35.020	37.566
25	11.524	12.697	14.611	16.473	18.940	24.337	30.675	34.382	37.652	41.566	44.314
30	14.953	16.306	18.493	20.599	23.364	29.336	36.250	40.256	43.773	47.962	50.892

[a] For $n - p$ greater than 30, $(2\chi^2)^{1/2} - [2(n - p) - 1]^{1/2}$ may be used as a "normal deviate" with "unit variance." For example, for 50 degrees of freedom, the value of χ^2 for $P = 0.99$ is determined by setting the "normal deviate" equal to $- 2.326$ (see Table 3.5, $n - p = \infty$, $\beta/2 = 0.01$). That is, $(2\chi^2)^{1/2} - (99)^{1/2} = -2.326$. Similarly, the value of χ^2 for $P = 0.01$ is determined as $(2\chi^2)^{1/2} = (99)^{1/2} + 2.326$.

Note that the values of χ^2 for $P = 0.50$ are slightly less than $n - p$. $P = 0.50$ corresponds to the "medians" of the distributions. The values of χ^2 for the "means" of the distributions are exactly equal to $n - p$.

The factor $u_{\beta/2}$ is equivalent to $t_{(\beta/2,\infty)}$ and is determined from Table 3.5 for the desired *percent confidence*.

This confidence interval is very useful. By determining a confidence interval" for each data point, the analyst can check for spurious data. If for example, one or several data points fall outside the confidence interval corresponding to 95% confidence, these data points might be considered as "suspects." The analyst should examine these points to see if errors were made in the recording or initial data processing operations. If the errors in the independent variables are not negligible, he should also check to see if changes in the values of x_{ji} commensurate with the uncertainties $\sigma_{x_{ji}}$ could account for the location of the suspect points outside the confidence interval. By bringing "possible spurious data" to the attention of the analyst, the confidence interval plays an important role in the analysis. If, for example, a real mistake is located, the mistake must be corrected and the analysis repeated. If the mistake cannot be corrected without repeating the measurement, the analyst must decide whether he should reject the data point and then proceed with a new analysis in which the spurious point is excluded, or repeat the measurement before re-analyzing the data.

The choice of the percent confidence is dependent upon the particular experiment. If the percent confidence is too low, the analyst will have to check many data points. If the percent confidence is too high, spurious data might go undetected. The analyst must make his decision on the basis of the importance of the experiment, the application of the results, the number of data points, and other factors.

It should be emphasized that the *testing of data* to see if points fall within the confidence intervals can be accomplished as part of the computer analysis of the experimental data. For each set of *observed values* of the independent variables (i.e., $X_{1i}, X_{2i}, \cdots, X_{mi}; i = 1, 2, \cdots, n$) the corresponding value of f can be determined and then tested according to Equation 3.12.13. Special notation can be included in the *computer program output* for all points which fall outside their appropriate confidence interval. In fact, the test can be repeated for *several values of percent confidence*. If a large number of points are outside their confidence intervals, a possible explanation is that the values of σ_{y_i} have been underestimated. This possibility is considered in Section 3.13.

3.13 Goodness of the Fit

Upon completion of the analysis it is worthwhile to check whether or not a "good fit" has been obtained. Several techniques are available for facilitating this decision.

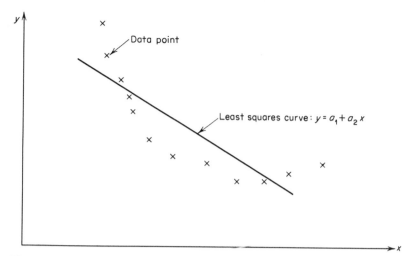

Figure 3.9. Fitting a straight line to data which exhibit quadratic behavior.

As explained in Section 3.12, the confidence interval technique permits a test of the plausibility of each data point. If a particular data point is outside the "confidence interval," an attempt should be made to understand why. Statistically the point can be outside the confidence interval, but the possibility of an error must also be considered. The most obvious sources of error are mistakes in the measurements, miscopied data, or errors in the preliminary data processing. These obvious sources of error should first be checked before considering the more serious sources of error.

If the analysis under consideration is the initial analysis using a particular computer program or subroutine, the possibility of a program error should be checked. For example, the functional form of the partial derivatives used throughout the analysis should be re-examined.

The most serious source of error is an incorrect choice for the function f. The best method for testing the validity of the function f is to see if the residuals are *randomly distributed*. If the residuals are not "randomly distributed" the function f is most probably not suitable for the data in question. For example, consider a set of data in which y should be a quadratic function of x. If a straight line is fit to the data, the results might appear as in Figure 3.9. Note that the first three data points have positive residuals R_{y_i}, the next seven points have negative residuals, and the final two points have positive residuals. The residuals do not appear to be randomly distributed and the analyst would be justified in reappraising his original choice of a straight line for the function f.

If values of experimental uncertainties (i.e., the values of σ_{y_i} and $\sigma_{x_{ii}}$) can be estimated the resulting value of S computed using Equation

3.3.7 offers the analyst an additional clue concerning the "goodness of the fit." The sum S should be distributed according to the "chi-squared distribution" (see Section 2.5). A rather surprising property of this distribution is that it is "only dependent upon the number of degrees of freedom." Defining P as the fractional probability that S will exceed a specified value of χ^2, the values of χ^2 as a function of P and $n - p$ are tabulated in Table 3.6. Comparing the value of S with the appropriate values of χ^2 from Table 3.6 will indicate whether or not a good fit has been obtained.

Consider an example in which the "number of degrees of freedom" is 15, and it is desired to determine the value of χ^2 which is only exceeded in 5% of all cases. From Table 3.6, for $n - p = 15$ and $P = 0.05$, $\chi^2 = 24.996$. In other words, one would expect S to exceed 24.996 in 5% of all cases in which $n - p = 15$ and "the χ^2 distribution is applicable." The value of $S/(n - p)$ would exceed $24.996/15 = 1.667$ in 5% of these cases.

The "χ^2 distribution" offers a test for the value of S resulting from a least squares analysis. The usual procedure is to set a "confidence interval" based upon a specific value of "percent confidence." If the resulting value of S is outside the "confidence interval," the possibility of an error should be considered. If $P = 0.05$ is chosen as the upper limit for the "confidence interval," then from the preceding example, a value of S greater than 24.996 or $S/(n - p)$ greater than 1.667 would indicate that there is cause for concern.

An important source of error which would cause S to fall outside the "confidence interval" is an under- or overestimate of the experimental uncertainties (i.e., the values of σ_{y_i} and $\sigma_{x_{ii}}$). For example, if the uncertainties were underestimated, the "weights" would be overestimated (note Equations 3.3.5 and 3.3.6) and the resulting value of S would be too large. Other sources of error are one or several poor data points or a poor choice for the function f.

An important check on the "goodness of fit" is a comparison of the results with theoretical or other experimental results. From theory, the analyst might know that a specific parameter should be within a given range. If the parameter as determined by the analysis is outside the range, the analyst should attempt to understand the discrepancy. Discrepancies with other experimental results should also be examined.

As already mentioned, a source of error might be a mistake in the computer program. The "debugging procedure" for all new computer programs should include a test to see that the values of a_k determined by the analysis are in fact the least squares values. To perform this test, one may determine the value of S for several combinations of a_k other than the least squares combination. If the least squares combination does, in fact, yield the minimum value of S, the analyst may be confident that the least squares solution is being obtained. An even simpler procedure is to compute the least squares values of a_k for a problem which has been solved elsewhere.

In summary, three powerful statistical tools are available for testing "goodness of the fit." These are:

1. The *confidence interval* at each data point.
2. The *randomness* of the residuals.
3. The χ^2 *test* for the value of S.

Each of these tests should be considered as "indicators." That is, failure to pass any of the tests should indicate that something might be wrong. Checking the data and the other factors affecting the analysis should then be initiated. In addition to the statistical methods for testing "goodness of fit," comparison of the results with theoretical or other experimental results should be made whenever possible.

3.14 Summary

An understanding of the method of least squares is important as background material for understanding prediction analysis. The method of least squares has been known since the nineteenth century, and much has been written on the subject. In most textbooks and references in the areas of statistics and numerical analysis the method of least squares is mentioned, but is rarely treated in detail for the more general cases.

In this chapter an attempt has been made to present the method of least squares as applied to curve fitting in a general form. It is recognized that most of the readers of this book will not be statisticians or mathematicians, so the treatment includes discussions of many of the points which often confound the experimentalist who is interested only in using the method as a tool to extract information from his experimental data.

At the same time an attempt has been made to present the method in a rigorous manner. Sections 3.6 and 3.10, in particular, are heavily flavored mathematically. This mathematical detail should not, however, frighten the casual user of the method of least squares because very little mathematics is required for solution of a real problem.

The complete procedure for performing a least squares analysis can be summarized:

1. Select the function f (i.e., Equation 3.3.1 or 3.3.2) which will be used to relate the dependent and independent variables.
2. List the function F (Equation 3.6.1) and all its derivatives (Equation 3.6.3).

3. Tabulate all the data. This includes Y_i, σ_{y_i}, X_{ji}, $\sigma_{x_{ji}}$, $i = 1, 2, \cdots, n$ (all data points) and $j = 1, 2, \cdots, m$ (all independent variables). Relationships between Y_i and σ_{y_i} (the standard deviation or uncertainty of Y_i) and between X_{ji} and $\sigma_{x_{ji}}$ are usually used to estimate the standard deviation (or uncertainty) of the data. For example, if it is estimated that the Y_i's are measured to 1% accuracy, then $\sigma_{y_i} = 0.01 Y_i$ is the appropriate relationship.

4. Make initial guesses a_{k0} for the least squares parameters a_k.

5. Using the expressions for the derivatives of F with respect to the dependent and independent variables (i.e., $F_y{}^i$ and $F_{x_j}{}^i$ of Equation 3.6.3), the data, and the initial guesses, compute the numerical values of $F_y{}^i$ and $F_{x_j}{}^i$.

6. Using Equation 3.6.16, compute the values of L_i.

7. Using the expressions for the derivatives of F with respect to the unknown parameters (i.e., $F_{a_k}{}^i$ of Equation 3.6.3), the data, and the initial guesses, compute the numerical values of $F_{a_k}{}^i$.

8. Using Equations 3.6.21 and 3.6.22, compute all the elements C_{kl} and V_k of the matrix C and the vector V.

9. Invert the C matrix. A matrix inversion procedure is discussed in Section 4.8.

10. Using Equation 3.6.30, compute the elements A_k of the vector A.

11. Using Equation 3.6.6 (or a modified version such as 3.8.1 and 3.8.2), the values of A_k and the initial guesses a_{k0}, compute the values of a_k.

12. Using Equation 3.6.34, test the values of A_k/a_k for convergence.

13. If convergence is not obtained, substitute the values of a_k for the initial guesses a_{k0} and repeat the procedure from Step 5. If the particular problem does not show signs of converging after several iterations, the techniques discussed in Section 3.8 might be necessary.

14. When convergence has been obtained, calculate the value of S using Equation 3.10.34.

15. Using Equations 3.10.43 and 3.10.44, calculate the "unbiased estimates" of the variances and covariances of the parameters.

16. If confidence intervals are required for the least square parameters, Equation 3.10.45 and Table 3.5 should be used.

17. If confidence intervals are required for the dependent variable y, the equations developed in Section 3.12 are applicable.

18. If it is required to test the "goodness of fit" of the analysis, the techniques discussed in Section 3.13 are applicable.

The subject of least squares computer programs is considered in Section 4.7. A suggested program block diagram is included (Figure 4.4) and programming details are discussed.

Problems

1. An experiment is performed in which the variables y and x are to be related by a straight line:

$$y = a_1 + a_2x.$$

The experimental data is included in the tabulation.

Experimental Data

Data Point	y	σ_y	x	σ_x
1	4.03	0.04	3.00	0.00
2	4.95	0.05	4.00	0.00
3	6.04	0.06	5.00	0.00
4	6.95	0.07	6.00	0.00
5	8.08	0.08	7.00	0.00

Compute the least squares values of a_1 and a_2, and the unbiased estimates of σ_{a_1}, σ_{a_2}, and σ_{12}.

2. The least squares curve determined in Problem 1 is to be used to compute the values of y corresponding to $x = 8.00$. What is the value of y and the unbiased estimate of the standard deviation of this value? (Example 2 of Section 5.2 is similar to this problem.)

3. Fit the data in Problem 1 using the following equation:

$$y = a_1 + a_2x + a_3x^2.$$

Compute the least squares values of a_1, a_2, a_3, and the unbiased estimates of σ_{a_1}, σ_{a_2}, and σ_{a_2}, and σ_{a_3}.

Computer Projects

1. An experiment is performed in which the variables y and x are to be related by the following equation:

$$y = a_1 \exp(-a_2x) + a_3.$$

The experimental data are included in the following table:

Data Point	y	x	Data Point	y	x
1	76275	0.25	11	6983	2.75
2	56763	0.50	12	5432	3.00
3	44203	0.75	13	4567	3.25
4	33827	1.00	14	3655	3.50
5	26662	1.25	15	3160	3.75
6	20846	1.50	16	2601	4.00
7	16628	1.75	17	2328	4.25
8	13085	2.00	18	1955	4.50
9	10599	2.25	19	1842	4.75
10	8411	2.50	20	1581	5.00

The uncertainties for the values of y_i are estimated using "counting statistics" [i.e., $\sigma_{y_i} = (y_i)^{1/2}$]. The uncertainties for all the values of x_i are assumed to be 0.01. Compute the least squares values of a_1, a_2, and a_3, and the unbiased estimates of σ_{a1}, σ_{a2}, and σ_{a3}. Does the mathematical model adequately represent the experimental data?

2. Study the problem of convergence using the data of the previous problem and the following procedure:

a. Assume an initial guess of a_3 equal to the least squares value computed in the previous problem.

b. For a given combination of initial guesses a_{10} and a_{20} and for a given value of G, determine whether or not the least squares analysis converges to a solution, and, if so, how many iterations are required to achieve convergence. Use the criteria $\epsilon = 0.001$ (see Equation 3.6.34).

c. By repeating Step (b) several times, determine how far the guesses a_{10} and a_{20} can be from the least squares values before the analysis diverges.

d. Repeat Step (c) for several values of G.

Chapter 4

Prediction Analysis

4.1 Introduction

Prediction analysis is a general method for predicting the accuracy of results that should be obtained from a proposed experiment. The method is applicable to any experiment amenable to analysis by the method of least squares. This includes a significant fraction of all experiments being performed in all areas of science and technology. Prediction analysis of a proposed experiment gives the experimenter a picture of how the accuracy of the results should be dependent upon the experimental variables. Using this information, the experimenter has a basis for answering the following important questions:

1. Can the proposed experiment yield results with sufficient accuracy to justify the experiment?
2. What choice or choices of the experimental variables (e.g., number of data points, accuracy of the individual data points, etc.) will satisfy the accuracy requirements of the experiment?

In this chapter, the method of prediction analysis is developed and discussed. The following chapters will be devoted to specific applications of the method.

4.2 Experimental Uncertainty

Experiments are never exact. An element of experimental uncertainty exists for even the most carefully planned and executed experiments. Experimental results are rarely useful unless some estimate of this uncertainty is made.

Uncertainty in the results of experiments is generally caused by uncertainty in the measurements of the individual data points. Theoretically, if the individual uncertainties could be reduced to zero, the results could also be determined exactly. In practice, however, measuring instruments have a limiting accuracy. One can improve upon the limiting accuracy of the instrument by repeating the measurement several times and then using the average value as the data point. Nevertheless, it is impossible to reduce the uncertainty to zero.

In addition, some variables have an inherent statistical distribution. Such variables are called "stochastic" or "random variables" and are discussed in Section 2.2. For this type of variable, even if the measuring instrument is perfectly accurate, the measurement would be expected to vary if repeated.

In Section 3.10, the formalism for estimating the uncertainty of experimental results was developed. The true uncertainty of the least squares value a_k is denoted as σ_{a_k}. The σ_{a_k}'s are measures of the differences between the a_k's and the true values α_k (see Equation 3.10.3). The σ_{a_k}'s cannot be determined, but "unbiased estimates" of them can be made. The unbiased estimate of σ_{a_k} is denoted as s_{a_k}. The values of s_{a_k} are determined as a part of the least squares analysis of the experimental data.

It is extremely useful in the planning stage of an experiment to be able to predict the values of σ_{a_k}. Formalism for predicting these values is developed in this chapter. The predicted values of the uncertainties σ_{a_k} are denoted as $\hat{\sigma}_{a_k}$. These values should not be confused with the unbiased estimates s_{a_k}. The predicted values are determined *before* the experiment has been performed. The unbiased estimates are based upon an analysis of the experimental data and must therefore be determined *after* the experiment has been performed.

In Section 3.3 the problems associated with "systematic errors" were discussed. The unbiased estimate s_{a_k} is a realistic measure of the root-mean-square difference between a_k and α_k only if systematic errors do not affect the measurements. In Section 3.13 several tests for "goodness of fit" were discussed. These tests can often indicate the presence of systematic errors. The method of prediction analysis must be modified if the presence of systematic errors is to be considered in the analysis. Treatment of systematic errors is discussed in Section 4.9.

4.3 Predicting the Values of σ_{a_k}

Predicted values of the various quantities will be distinguished from the true or experimentally determined values, by the use of a caret (i.e., ˆ) above the existing symbol for the true or experimentally determined quantity. For example, the predicted value of σ_{a_k} will be denoted as $\hat{\sigma}_{a_k}$.

The predicted value of σ_{a_k} (i.e., $\hat{\sigma}_{a_k}$) is defined as the mean value of the unbiased estimate s_{a_k} that would be expected if the experiment were repeated many times. From Equation 3.10.43:

$$\hat{\sigma}_{a_k}^2 \equiv \hat{s}_{a_k}^2 = \frac{\hat{S}}{n-p} \hat{C}_{kk}^{-1}. \tag{4.3.1}$$

The quantity $\hat{S}$ is the mean value of S that would be obtained if the experiment were repeated many times. In Section 3.13 it was stated that S should be distributed according to the "chi-squared distribution." The mean value of this distribution is equal to the number of degrees of freedom (see Equation 2.5.15). For a case in which there are n data points and the analysis is for p unknown parameters, the number of degrees of freedom is equal to $n - p$. Thus:

$$\hat{S} = n - p, \tag{4.3.2}$$

and therefore Equation 4.3.1 reduces to:

$$\hat{\sigma}_{a_k}^2 = \hat{C}_{kk}^{-1}. \tag{4.3.3}$$

Similarly, from Equation 3.10.44,

$$\hat{\sigma}_{lk} = \hat{\rho}_{lk}\hat{\sigma}_{a_l}\hat{\sigma}_{a_k} = \hat{C}_{lk}^{-1}. \tag{4.3.4}$$

Equations 4.3.3 and 4.3.4 are the basic equations of prediction analysis. To predict the values of σ_{a_k} and σ_{lk} it is seen from these equations that the elements of the inverse coefficient matrix C^{-1} must be predicted. The procedure for making this prediction is discussed in Section 4.4.

Often it is of interest to predict the "confidence intervals" for the proposed experiment. The predicted confidence intervals for the parameters a_k are:

$$a_k - \hat{\sigma}_{a_k} \cdot t_{(\beta/2,\, n-p)} < \alpha_k < a_k + \hat{\sigma}_{a_k} \cdot t_{(\beta/2,\, n-p)}. \tag{4.3.5}$$

The values of $t_{(\beta/2,\, n-p)}$ are obtained by using Table 3.5. The predicted confidence intervals for values of f are:

$$f(x_1, \cdots, x_m; a_1, \cdots, a_p) - \hat{\sigma}_f \cdot t_{(\beta/2,\, n-p)} < f(x_1, \cdots, x_m; \alpha_1, \cdots, \alpha_p)$$
$$< f(x_1, \cdots, x_m; a_1, \cdots, a_p) + \hat{\sigma}_f \cdot t_{(\beta/2,\, n-p)}. \tag{4.3.6}$$

where

$$\hat{\sigma}_f^2 = \sum_{l=1}^{p} \sum_{k=1}^{p} \hat{F}_{a_l}\hat{F}_{a_k}\hat{C}_{lk}^{-1}. \tag{4.3.7}$$

As illustrated by Example 4 of Section 4.5, the confidence interval provides a more realistic comparison of experiments than the uncertainty when the number of degrees of freedom (i.e., $n - p$) is small. For $n - p$

greater than 15, the value of $t_{(\beta/2,n-p)}$ changes by not more than 10%, and therefore comparisons on the basis of confidence interval or uncertainty will result in essentially the same conclusions. It can be shown that if only the locations of the axes of the coordinate system are changed, the values of $\hat{C}_{lk}^{-1}$ will change, but $\hat{\sigma}_f$ for a given point will remain the same (see Example 2, Section 5.2).

4.4 The Predicted Elements of the Matrix C^{-1}

To determine the predicted values of σ_{a_k} and σ_{lk}, the predicted values of the elements of the inverse coefficient matrix C^{-1} must be evaluated (see Equations 4.3.3 and 4.3.4). To evaluate these elements, the predicted values of the elements of the coefficient matrix C are first determined and then the matrix C is inverted.

The calculation of the elements of the matrix C is based upon Equation 3.6.21:

$$C_{jk} = \sum_{i=1}^{n} L_i^{-1} F_{a_j}{}^i F_{a_k}{}^i. \qquad (3.6.21)$$

Using the notation introduced in Section 4.3 for indicating "predicted values," we obtain:

$$\hat{C}_{jk} = \sum_{i=1}^{n} \hat{L}_i^{-1} \hat{F}_{a_j}{}^i \hat{F}_{a_k}{}^i. \qquad (4.4.1)$$

Predicting the value of the elements C_{jk} thus requires prediction of the values of L_i, $F_{a_j}{}^i$, and $F_{a_k}{}^i$ at n points, and then summation of the products according to Equation 4.4.1. The total number of elements in the $\hat{C}$ matrix is p^2 because $j = 1, 2, \cdots, p$ and $k = 1, 2, \cdots, p$. Inasmuch as the $\hat{C}$ matrix is symmetrical, the total number of elements that must be evaluated equals $p + (p - 1) + \cdots + 1 = (p^2 + p)/2$. The exact number of operations required to invert the matrix $\hat{C}$ is dependent upon the method chosen for inversion, but in general it is proportional to p^3.

The first step required to evaluate the elements of the matrix $\hat{C}$ is to assume values for all the parameters and independent variables of the proposed experiment. That is, values are assumed for $a_1, a_2, \cdots, a_p$, and X_{ji}, $i = 1, 2, \cdots, n$ (all points), and $j = 1, 2, \cdots, m$ (all independent variables). Values of Y_i are then assumed to be equal to the calculated values y_i as determined by using Equation 3.3.2. Values of $\sigma_{x_{ji}}$ and σ_{y_i} are determined using assumed relationship with the values of X_{ji} and Y_i (for example, Equations 4.5.2, 4.5.3, and 4.5.4).

It is worthwhile to consider the question: "If the purpose of the experiment is to determine values of a_k, how can values of these parameters be assumed?" For most experiments the parameters to be determined can be estimated from theoretical considerations or from the results of similar or preliminary experiments.[1] In fact, it is the rare experiment in which absolutely no estimate of the results can be made. The rough estimates of the values of a_k are often good enough so that the values of $\hat{\sigma}_{a_k}$ determined by the prediction analysis are indicative of the values of s_{a_k} that will result from the experiment. If, however, the values of $\hat{\sigma}_{a_k}$ are strongly dependent upon the values of a_k, the prediction analysis can be repeated for as many sets of a_k's as are required to cover the possible range of values. The results can then be presented as a "function of the values of a_k" as well as the experimental variables. In Part II of the book, it is shown that a useful method for presenting the results of a prediction analysis is in terms of dimensionless groups. These groups often contain the parameters a_k, as well as the values of $\hat{\sigma}_{a_k}$.

Once values of the variables, parameters, and uncertainties have been assumed, two approaches exist for determining the elements of the matrix $\hat{C}^{-1}$. The terminology "direct approach" and "analytical approach" are used to distinguish these two possibilities.

1. The Direct Approach

The *direct approach* implies direct computation of the elements of the matrix $\hat{C}$. If the computation is to be repeated for a large number of combinations of experimental variables and parameters, a digital computer should be used to facilitate the computation.

The steps required for applying the direct approach to the evaluation of the elements of the matrix $\hat{C}^{-1}$ are as follows:

  (a) Using Equation 3.6.16, compute all n values of $\hat{L}_i$.
  (b) Compute all n values of the derivative $\hat{F}_{a_1}{}^i$ (see Equation 3.6.3).
  (c) Repeat Step (b) for $k = 2, \cdots, p$.
  (d) Using Equation 4.4.1, evaluate all elements $\hat{C}_{jk}$.
  (e) Invert the matrix $\hat{C}$.

If a digital computer is to be used to facilitate the analysis, at first glance the amount of storage space in the computer memory required just for the values of $\hat{L}_i$ and $\hat{F}_{a_k}{}^i$ seems large. If, for example, $n = 100$ and $p = 6$, it appears that 100 numbers would have to be stored for the $\hat{L}_i$'s

---

[1] According to G. E. P. Box and H. L. Lucas, "If we are to suppose that effective design is possible at all, we must also assume therefore that something is known about the values of the parameters in advance. In practical problems it will almost invariably be the case that some such information is available." "Design of Experiments in Non-linear Situations," *Biometrika*, Vol. 46, June 1959.

and 600 numbers would have to be stored for the $\hat{F}_{a_k}{}^i$'s. If the required storage is excessive for a particular computer, the order of the steps can be changed to avoid the need for storage of most of the numbers:

(a) Using Equation 3.6.16, compute the value of $\hat{L}_1$.

(b) Compute all p values of $\hat{F}_{a_k}{}^1$ ($k = 1, 2, \cdots, p$). See Equation 3.6.3.

(c) Evaluate the p^2 products $(1/\hat{L}_1)\hat{F}_{a_j}{}^1 F_{a_k}{}^1$ and store them in the memory locations reserved for $\hat{C}_{jk}$. (Since $\hat{C}$ is a symmetric matrix, the computation can be speeded by filling only the elements on or above the diagonal. This point is considered in Section 4.7.)

(d) Repeat Steps (a) through (c) for $i = 2, \cdots, n$. Step (c) is performed by adding the new values of the products to the sum already existing in the memory locations reserved for $\hat{C}_{jk}$. (Upon completion of Step (d), the elements of $\hat{C}$ below the diagonal are set equal to the elements above the diagonal because the matrix is symmetrical.)

(e) Invert the matrix $\hat{C}$.

Using this procedure, the storage requirements for $\hat{L}_i$ and $\hat{F}_{a_k}{}^i$ are reduced for the preceding example from a total of 700 numbers to one number for $\hat{L}$ and six numbers for $\hat{F}_{a_k}$. In other words, storage space for only 7 of the 700 numbers is required. The numbers in these seven locations will, however, change one hundred times throughout the course of the analysis.

The amount of computer time required for calculation of all the values $\hat{L}_i$, $\hat{F}_{a_k}{}^i$, and $\hat{C}_{jk}$ is proportional to the number of points n. Most prediction analyses require very little computer time, but often it is necessary to repeat the calculation for many combinations of the variables and parameters. Computer time therefore can be a problem. To save time, a modified version of the *direct approach* may be employed. This approach is to use only some of the points. If q points (where q is less than n) are used, Equation 4.4.2 is used in place of Equation 4.4.1:

$$\hat{C}_{jk} \approx \frac{n}{q} \cdot \sum_{\nu=1}^{q} \hat{L}_{\nu}{}^{-1} \hat{F}_{a_k}{}^{\nu} \hat{F}_{a_j}{}^{\nu}. \tag{4.4.2}$$

When $q = n$, Equation 4.4.2 reduces to Equation 4.4.1.

A criterion must be chosen for selecting the q points. If, for example, there is only one independent variable (i.e., $m = 1$) and the values of x_i are evenly spaced:

$$x_{i+1} - x_i = \Delta x, \qquad i = 1, 2, \cdots, n - 1, \tag{4.4.3}$$

then a criterion for selecting the q values of x might be:

$$x_{\nu} = x_1 + \frac{\nu - 1}{q - 1}(x_n - x_1), \qquad \nu = 1, \cdots, q. \tag{4.4.4}$$

If $q = 3$, the three values of x are x_1, $x_1 + \frac{1}{2}(x_n - x_1)$, and x_n. These three values of x are then used to compute the values of y_ν, $\hat{L}_\nu$, and $\hat{F}_{a_k}{}^\nu$. The numerical value of q required to ensure that the numerical result using Equation 4.4.2 is approximately equal to the result using Equation 4.4.1, is dependent upon the function f (i.e., Equation 3.3.1 or 3.3.2). For very simple functions, such as a straight line (i.e., $y = a_1 + a_2x$), a value of $q = 2$ or 3 will yield values of $\hat{\sigma}_{a_1}$ and $\hat{\sigma}_{a_2}$ which are reasonably close to the values that would be determined using Equation 4.4.1. For more complicated functions, such as a sine series (i.e., $y = \sum_{k=1}^{p} a_k \sin k\pi x$), a larger value of q is required.

2. The Analytical Approach

The *analytical approach* implies development of analytical expressions for the elements of the matrix $\hat{C}$. The expressions will usually not yield results which are exactly equal to the results as determined by the direct approach. Nevertheless, the accuracy can often be excellent and, furthermore, having the results in the form of analytical expressions is extremely useful. For example, planning an "optimum experiment" (e.g., an experiment in which the $\hat{\sigma}_{a_k}$'s are minimized for a specified set of conditions) is often facilitated if analytical expressions for the $\hat{\sigma}_{a_k}$'s are available (see Section 4.10).

The steps required for applying the analytical approach to the evaluation of the elements of the matrix $\hat{C}^{-1}$ are as follows:

(a) Select a mathematical model for estimating the elements of the matrix $\hat{C}$:

$$\hat{C}_{jk} = n \cdot (\hat{L}^{-1}\hat{F}_{a_j}\hat{F}_{a_k})_{\text{average}}. \tag{4.4.5}$$

If the number of independent variables is only one (i.e., $m = 1$), and if the values of x_i are evenly spaced, then Equation 4.4.3 is valid and a simple integral expression for the $\hat{C}_{jk}$'s can be used.[2] As n approaches infinity,

$$\hat{C}_{jk} \to n \cdot \int_{x_a}^{x_b} \hat{L}^{-1}\hat{F}_{a_k}\hat{F}_{a_j}\, dx \bigg/ (x_b - x_a), \tag{4.4.6}$$

---

[2] It should be emphasized that any approximation which yields an estimate of the C_{jk}'s is valid. The need for accuracy for the predicted values $\hat{C}_{jk}$ should, however, be realized. The matrix inversion procedure often requires evaluation of small differences of large numbers. Thus errors in the values $\hat{C}_{jk}$ are magnified when these values are used to determine the elements $\hat{C}_{jk}^{-1}$.

If the x_i's are not evenly spaced, it is more difficult to obtain an analytical solution. For such cases one must usually resort to the direct approach.

where

$$x_a = x_1 - \frac{\Delta x}{2}, \tag{4.4.7}$$

$$x_b = x_n + \frac{\Delta x}{2}. \tag{4.4.8}$$

Alternatively, the integration can be performed over the dependent variable y. As n approaches infinity,

$$\hat{C}_{jk} \to n \cdot \int_{y_a}^{y_b} \hat{L}^{-1}\hat{F}_{a_k}\hat{F}_{a_j} \frac{dx}{dy} \cdot dy \Big/ (x_b - x_a), \tag{4.4.9}$$

where y_a and y_b are the values of the dependent variable corresponding to x_a and x_b. That is,

$$y_a = f(x_a; a_1, a_2, \cdots, a_p), \tag{4.4.10}$$

and

$$y_b = f(x_b; a_1, a_2, \cdots, a_p). \tag{4.4.11}$$

The value of $\hat{C}_{jk}$ determined using Equation 4.4.6 or 4.4.9 approaches the true value (determined using Equation 4.4.1) as n approaches infinity or as $\Delta x/(x_b - x_a)$ approaches zero. For many problems these equations yield a good approximation of 4.4.1 even for small values of n.

(b) Invert the matrix $\hat{C}$ to determine analytical expressions for the elements of $\hat{C}^{-1}$.

If the number of independent variables is one (i.e., $m = 1$) and the number of unknown parameters is not more than two (i.e., $p = 1$ or 2), then the analytical approach will often yield manageable expressions for the $\hat{\sigma}_{a_k}$'s. If p is equal to three, the equations for the $\hat{\sigma}_{a_k}$'s are more complicated and often the solutions for the integrals required in Equation 4.9.6 cannot be expressed analytically. For such cases it is usually advisable to proceed to a solution using the direct approach. For p greater than three, the analytical approach is rarely possible, and, even if possible, the expressions obtained for the $\hat{\sigma}_{a_k}$'s are extremely cumbersome. For cases in which m is greater than one, the complexity of the analysis increases and this tends to favor the direct approach.

4.5 An Example of Prediction Analysis—The Straight-Line Experiment

The *straight-line experiment* refers to any experiment in which the dependent variable is related to a single independent variable by the following equation:

$$y = a_1 + a_2 x. \tag{4.5.1}$$

In Chapter 5, this function is referred to as the *first-order polynomial*. The analysis of this class of experiments does not depend upon the physical nature of the variables x and y or the parameters a_1 and a_2.

In Section 1.3 the uncertainties or standard deviations were related to the variables by Equations 1.3.2 and 1.3.3:

$$\sigma_{x_i} = g(x_i), \tag{1.3.2}$$

$$\sigma_{y_i} = h(y_i). \tag{1.3.3}$$

The various combinations of choices for the functions g and h represents the sub-cases for a class of experiments. The most common choices are *constant uncertainty, constant fractional uncertainty, and counting or Poisson statistics*. The functions g and h for these three cases are:

Constant uncertainty:
$$\sigma_{x_i} = K_{cx},$$
$$\sigma_{y_i} = K_{cy}. \tag{4.5.2}$$

Constant fractional uncertainty:
$$\sigma_{x_i} = K_{fx} \cdot |x_i|,$$
$$\sigma_{y_i} = K_{fy} \cdot |y_i|. \tag{4.5.3}$$

Counting statistics[3]:
$$\sigma_{x_i} = K_{sx} \cdot (x_i)^{1/2},$$
$$\sigma_{y_i} = K_{sy} \cdot (y_i)^{1/2}. \tag{4.5.4}$$

The parameters K_{cx}, K_{cy}, K_{fx}, K_{fy}, K_{sx}, and K_{sy} are parameters which must be specified. For a particular case, one must specify the equations for both σ_{x_i} and σ_{y_i}. It should be emphasized that the functions g and h need not be limited to the forms shown in Equations 4.5.2, 4.5.3, or 4.5.4. In Chapter 9, for example, a problem is considered in which the function h is more complicated

Three cases of the "straight-line experiment" are analyzed in this section. Because this class of experiments is important in almost all areas of science and technology, analyses of still other cases are included in Chapter 5. The three cases which are considered in this section are:

Case I. Constant uncertainty for both the x and y variables.

Case II. Constant fractional uncertainty for the y variable and negligible uncertainty for the x variable (i.e., $\sigma_{x_i} = 0$).

Case III. Counting statistics for the y variable and negligible uncertainty for the x variable.

---

[3] In most experiments in which the number of events per unit time is counted, Equation 4.5.4 is applicable. Experimentalists often refer to this type of experiment as a "counting experiment," and Equation 4.5.4 is referred to as "counting statistics." The statistician refers to Equation 4.5.4 as "Poisson statistics" (see Equation 2.5.9) because the Poisson distribution is applicable to such problems.

Case I. *Constant uncertainty for both the x and y variables.*

For this case, the values of σ_{x_i} and σ_{y_i} are:

$$\sigma_{x_i} = K_{cx},$$
$$\sigma_{y_i} = K_{cy}. \tag{4.5.5}$$

The purpose of the proposed experiment is to fit Equation 4.5.1 to the experimental data to determine the least squares values of a_1 and a_2. The purpose of the prediction analysis is to predict the uncertainties σ_{a_1} and σ_{a_2} before the experiment has been performed.

In Section 4.4, both "direct" and "analytical" approaches to the problem were discussed. For this problem, an analytical solution is possible, and therefore the first step in the analysis is to develop analytical expressions for the matrix elements $\hat{C}_{jk}$.

For one independent variable, Equation 3.6.16 reduces to a simpler form:

$$L_i = \sigma_{y_i}^2 + (F_x^i \cdot \sigma_{x_i})^2. \tag{4.5.6}$$

The term F_x^i is the derivative of F with respect to x evaluated at x_i. The function F is defined by Equation 3.6.1. For this example it is:

$$F \equiv y - a_1 - a_2 x,$$

and therefore,

$$F_x^i = -a_2. \tag{4.5.7}$$

Substituting Equations 4.5.5 and 4.5.7 into Equation 4.5.6 gives[4]:

$$\hat{L}_i = K_{cy}^2 + (a_2 \cdot K_{cx})^2 \equiv L. \tag{4.5.8}$$

Since all values of $\hat{L}_i$ are the same, the subscript i may be dropped.

The values for $\hat{F}_{a_1}^i$ and $\hat{F}_{a_2}^i$ are determined by taking the appropriate derivatives of F:

$$\hat{F}_{a_1}^i = -1, \tag{4.5.9}$$
$$\hat{F}_{a_2}^i = -x_i. \tag{4.5.10}$$

The elements of the matrix $\hat{C}$ are determined by substituting Equations 4.5.8, 4.5.9, and 4.5.10 into Equation 4.4.1:

$$\hat{C}_{11} = \sum_{i=1}^{n} \frac{1}{L} = \frac{n}{L}, \tag{4.5.11}$$

$$\hat{C}_{21} = \hat{C}_{12} = \sum_{i=1}^{n} \frac{x}{L} = \frac{n}{L} x_{av}, \tag{4.5.12}$$

$$\hat{C}_{22} = \sum_{i=1}^{n} \frac{x^2}{L} = \frac{n}{L} (x^2)_{av}, \tag{4.5.13}$$

where x_{av} is the average value of x and $(x^2)_{av}$ is the average value of x^2.

[4] The caret notation (i.e., $\hat{L}_i$) is used to denote predicted values of L_i.

The "predicted values" of $\sigma_{a_1}^2$ and $\sigma_{a_2}^2$ (i.e., $\hat{\sigma}_{a_1}^2$ and $\hat{\sigma}_{a_2}^2$) are equal to the diagonal elements of the inverse coefficient matrix $\hat{C}^{-1}$ (see Equation 4.3.3). Equations for the diagonal elements of $\hat{C}^{-1}$ are derived using the expressions for elements of the inverse of a 2 × 2 matrix (i.e., Equations 3.7.16 and 3.7.17). Since the matrix is symmetric:

$$\hat{\sigma}_{a_1}^2 = \hat{C}_{11}^{-1} = \frac{\hat{C}_{22}}{\hat{C}_{22} \cdot \hat{C}_{11} - (\hat{C}_{12})^2},$$
(4.5.14)

$$\hat{\sigma}_{a_2}^2 = \hat{C}_{22}^{-1} = \frac{\hat{C}_{11}}{\hat{C}_{22} \cdot \hat{C}_{11} - (\hat{C}_{12})^2},$$
(4.5.15)

Substituting Equations 4.5.11, 4.5.12, and 4.5.13 into Equations 4.5.14 and 4.5.15 gives:

$$\hat{\sigma}_{a_1}^2 = \frac{L}{n} \cdot \frac{(x^2)_{av}}{(x^2)_{av} - (x_{av})^2},$$
(4.5.16)

$$\hat{\sigma}_{a_2}^2 = \frac{L}{n} \cdot \frac{1}{(x^2)_{av} - (x_{av})^2} = \frac{\hat{\sigma}_{a_1}^2}{(x^2)_{av}}.$$
(4.5.17)

No simplifying assumptions have been made in the derivations of Equations 4.5.16 and 4.5.17. To develop more useful analytical expressions, equations for x_{av} and $(x^2)_{av}$ must be derived. If it is assumed that the values of x_i are evenly spaced (i.e., Equation 4.4.3 is valid), then x_{av} and $(x^2)_{av}$ can be related to simple integrals:

$$x_{av} = \int_{x_a}^{x_b} x \, dx \bigg/ (x_b - x_a),$$
(4.5.18)

$$(x^2)_{av} \to \int_{x_a}^{x_b} x^2 \, dx \bigg/ (x_b - x_a), \qquad \text{as } n \to \infty,$$
(4.5.19)

where the limits of integration x_a and x_b are defined by Equations 4.4.7 and 4.4.8. From these equations and Equation 4.4.3, it can be seen that:

$$x_b - x_a = n \cdot \Delta x.$$
(4.5.20)

It should be noted that the average value of x as determined by Equation 4.5.18 is exactly equal to the value that would be obtained by direct averaging. The value of $(x^2)_{av}$ is not the same as would be obtained by direct averaging. Equation 4.5.19 will, however, yield a good approximation of $(x^2)_{av}$, especially if n is large. In fact, the value as determined by Equation 5.4.19 approaches the true value as n approaches infinity. Integrating Equations 4.5.18 and 4.5.19 gives:

$$x_{av} = \frac{1}{2} \frac{x_b^2 - x_a^2}{x_b - x_a} = \tfrac{1}{2}(x_b + x_a),$$
(4.5.21)

and as n approaches infinity,

$$(x^2)_{av} \to \frac{1}{3} \frac{x_b{}^3 - x_a{}^3}{x_b - x_a} = \tfrac{1}{3}(x_b{}^2 + x_a x_b + x_a{}^2). \tag{4.5.22}$$

If Equations 4.5.21 and 4.5.22 are substituted into Equations 4.5.16 and 4.5.17, as n approaches infinity,

$$\hat{\sigma}_{a_1}{}^2 \to \frac{4L}{n} \frac{x_b{}^2 + x_a x_b + x_a{}^2}{(x_b - x_a)^2}, \tag{4.5.23}$$

and

$$\hat{\sigma}_{a_2}{}^2 \to \frac{12L}{n} \frac{1}{(x_b - x_a)^2}. \tag{4.5.24}$$

Equation 4.5.24 can be put into a useful alternative form by substituting Equation 4.5.20 into 4.5.24:

$$\hat{\sigma}_{a_2}{}^2 \to \frac{12L}{n^3 \Delta x^2}. \tag{4.5.25}$$

To express Equation 4.5.23 in a more convenient form, the equality of Equation 4.5.22 is used. As $n \to \infty$,

$$\hat{\sigma}_{a_1}{}^2 \to \frac{4L}{n} \frac{x_b{}^3 - x_a{}^3}{(x_b - x_a)^3},$$

$$\hat{\sigma}_{a_1}{}^2 \to \frac{4L}{n} \frac{x_b{}^3 - x_a{}^3}{x_b{}^3 - x_a{}^3 + 3x_b x_a(x_a - x_b)}, \tag{4.5.26}$$

$$\hat{\sigma}_{a_1}{}^2 \to \frac{4L}{n} \left[1 - \frac{3x_b x_a(x_b - x_a)}{x_b{}^3 - x_a{}^3} \right]^{-1}. \tag{4.5.27}$$

Defining r_x as follows and using Equations 4.5.20 and 4.5.21, we obtain:

$$r_x \equiv \frac{x_b + x_a}{2(x_b - x_a)} = \frac{x_{av}}{n \cdot \Delta x}, \tag{4.5.28}$$

and then substituting Equation 4.5.28 into Equation 4.5.27 gives:

$$\hat{\sigma}_{a_1}{}^2 \to \frac{4L}{n} \left[1 - \frac{3r_x{}^2 - \tfrac{3}{4}}{3r_x{}^2 + \tfrac{1}{4}} \right]^{-1} = \frac{4L}{n} (3r_x{}^2 + \tfrac{1}{4}). \tag{4.5.29}$$

From Equations 4.5.29 and 4.5.8 it can be seen that $\hat{\sigma}_{a_1}$ is a function of K_{cy}, K_{cx}, a_2, n, and r_x. From Equations 4.5.25 and 4.5.8 it can be seen that $\hat{\sigma}_{a_2}$ is a function of K_{cy}, K_{cx}, a_2, n, and Δx. It should be noted that for this

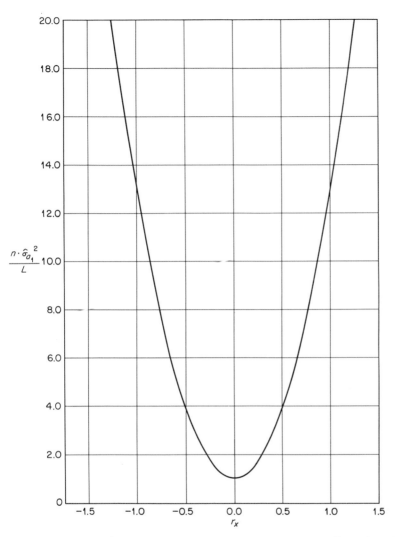

Figure 4.1. $n\hat{\sigma}_{a_1}^2/L$ as a function of r_x. The ratio $r_x \equiv x_{av}/(n \cdot \Delta x)$ and x_{av} is the average value of the range for x. The curve is based upon Equation 4.5.29 and is exact only for n equal to infinity. It is, however, a good approximation even for small values of $n - 2$.

case (i.e., constant uncertainty for both the x and y variables), neither $\hat{\sigma}_{a_1}$ nor $\hat{\sigma}_{a_2}$ is a function of a_1.

Using Equation 4.5.29, the dimensionless ratio $n \cdot \hat{\sigma}_{a_1}^2/L$ is plotted as a function of r_x in Figure 4.1. From Figure 4.1 it is seen that the minimum value of $n \cdot \hat{\sigma}_{a_1}^2/L$ corresponds to an r_x equal to zero (i.e., $x_a = -x_b$). This is a reasonable result and simply states that if the values of a_1, a_2,

K_{cx}, K_{cy}, n, and Δx are fixed, the y intercept can be measured to the best accuracy if the average value of x (i.e., x_{av}) is zero.

To illustrate the usefulness of the results, several examples are considered.

Example 1—Assume that the values of y can be measured to an uncertainty of about 1.0, and the possible range of values of a_2 is 5.0 to 10. Also, assume that values of x can be measured to about 0.1 and must fall within the range 4.0 to 20. If the purpose of the experiment is to measure a_1 to ± 0.5, and a_2 to at least 1% accuracy, how many data points are required if the points are to be evenly spaced within the given range?

The most conservative estimate of the value of n required to measure a_1 to a given accuracy is made by taking the largest possible value of r_x. As Δx approaches zero, r_x approaches $\frac{1}{2}(x_1 + x_n)/(x_n - x_1)$ which, for this example, equals 0.75. The value of $n \cdot \hat{\sigma}_{a_1}^2/L$ for $r_x = 0.75$ is determined from Equation 4.5.29 and is equal to 7.75. The largest possible value of L is determined by substituting the largest possible value of a_2 into Equation 4.5.8. The resulting maximum value of L is $1.0^2 + 1.0^2 = 2$. The computed value of n required to determine a_1 to ± 0.5 is therefore $2.0 \times 7.75/0.5^2 = 62$.

To determine a_2 to 1% accuracy, if a_2 is equal to its minimum value (i.e., 5.0), then σ_{a_2} would have to be at most 0.05. Using Equation 4.5.24, and assuming that $x_n - x_1$ is approximately equal to $x_b - x_a$, then the minimum value of n is $12 \times (1.0^2 + 0.5^2)/(0.05 \times 16)^2 = 23.4$ (or, rounding to the next higher integer, is equal to 24).

It is therefore seen that the accuracy requirement for a_1 is more difficult to satisfy than the accuracy requirement for a_2. For this reason the experiment should be planned to include at least 62 data points. If the effort required to measure 62 data points is considered excessive, the experimenter must then reconsider the "ground rules." For example, improved techniques for measuring the values of y_i and x_i more accurately would reduce the required number of data points. Similarly, an extension of the range of values for x (particularly on the lower side) would reduce the required number of data points.

Example 2—Assume that the experiment considered in Example 1 is limited to only ten data points. Upon reconsideration of the "ground rules," it is decided that the simplest procedure is to extend the range of the values for x. Using this procedure, can the experiment be performed in a manner which satisfies the accuracy requirement for a_1?

From Figure 4.1, it is seen that the smallest value of $\hat{\sigma}_{a_1}$ is obtained if r_x is equal to zero. If the range is altered so that r_x is equal to zero, then $n \cdot \hat{\sigma}_{a_1}^2/L$ would be equal to 1.0, and the minimum value of n would be $2.0 \times 1.0/0.5^2 = 8$. Adjustment of the range can therefore reduce the required number of data points to less than ten. If, however, only positive values of x are possible, then the minimum value of $n \cdot \hat{\sigma}_{a_1}^2/L$ corresponds to $r_x = 0.5$ and is equal to 4. The minimum value of n would be 32, which does not satisfy the requirement of a maximum of ten data points.

Example 3—Assume that the experiment considered in Example 1 is limited to only ten data points and the range for the values of x_i has been extended down to 3.0 and up to 30. How accurately would the values of y_i have to be measured to satisfy the accuracy requirements for a_1?

From Equation 4.5.29, the value of $n \cdot \hat{\sigma}_{a_1}{}^2/L$ for $r_x = 0.61$ is 5.49. For $\hat{\sigma}_{a_1} = 0.5$ and $n = 10$, L must therefore be equal to $10 \times 0.5^2/5.49 = 0.456$. Even if K_{cy} was reduced to zero, the value of L would be equal to 1.0 (if a_2 was equal to its maximum possible value). Reducing K_{cy} will not, therefore, reduce the required number of points to ten.

At this point, several courses of action may be followed. These include:

1. Reduce the uncertainty of the measurements of the values of x to a value less than 0.1.
2. Relax the requirement that a_1 must be measured to a maximum uncertainty of 0.5.
3. Increase the acceptable number of data points to a value greater than 10.
4. Cancel the experiment. If each of the first three courses of action is impossible or unacceptable, then the experiment is "doomed to failure" and the experimenter can avoid a waste of time, money, and effort by canceling the experiment. Alternatively, an entirely different method of measuring a_1 might be used which would prove to be acceptable.

Example 4—This example illustrates the applicability of Equation 4.3.5. For comparisons of experiments in which n is small, it is more realistic to compare the alternatives on the basis of "confidence interval" rather than "standard deviation" or "uncertainty." This point is illustrated as follows.

Consider Figure 4.2. Assume that there is a permanent weight of 3 kg attached to the spring and that up to ten 1-kg weights can be added to the permanent weight. The objective of the experiment is to determine the value of the spring constant a_2. Assume that the uncertainty for each value of y is 0.1 cm (i.e., $\sigma_{y_i} = K_{cy} = 0.1$ cm) and the uncertainty for each value of x is negligible. A condition of the method of least squares is that the number of data points must be greater than the number of unknown parameters. The minimum number of data points must therefore be three. In other words, y must be measured at a minimum of three values of x.

The possible values of x are 3.0, 4.0, $\cdots$, 13.0 kg. The minimum value of Δx is 1.0 kg. The requirement that at least three data points must be determined can be satisfied by values of Δx equal to 1.0, 2.0, 3.0, 4.0, and 5.0 kg. It is assumed that the values of x are evenly spaced, $x_1 = 3.0$ kg

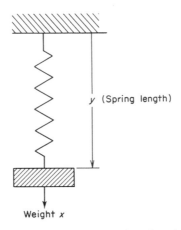

Figure 4.2. An example of a "straight-line" experiment. The spring length y is related to the weight x by Equation 4.5.1 (i.e., $y = a_1 + a_2 x$) where a_1 is the unstretched spring length and a_2 is the spring constant.

and $x_n = 3.0 + (n - 1)\Delta x$ kg. Values of $\hat{\sigma}_{a_2}$ are determined using Equation 4.5.25 and the results for all possible combinations of n and Δx are summarized in Table 4.1.

If the objective of the experiment is to determine a_2 to an uncertainty or standard deviation of less than 0.02 cm/kg, for $\Delta x = 1.0$ kg, n must be at least 7; for $\Delta x = 2.0$ kg, n must be at least 5; for $\Delta x = 3.0$ kg, n must be 4; and for $\Delta x = 4.0$ and 5.0 kg, $n = 3$ is acceptable.

A more meaningful objective is to measure a_2 to within a given confidence interval rather than a given uncertainty. If the values of n are small, the use of a confidence interval, rather than an uncertainty, drastically alters the comparison. The calculated value $\hat{\sigma}_{a_2}$ is the predicted value of s_{a_2}. (The value of s_{a_2} is determined from analysis of the experimental data and is the unbiased estimate of σ_{a_2}.) Even though $\hat{\sigma}_{a_2}$ is the same for two different values of n, the experimentally determined confidence interval for a_2 will decrease with increasing n (see Equation 3.10.45 and Table 3.5).

For example, the values of $\hat{\sigma}_{a_2}$ are approximately the same for the combinations $n = 3$, $\Delta x = 5.0$ kg, and $n = 9$, $\Delta x = 1.0$ kg. The values of s_{a_2} that will result from least squares analysis of experimental data will be approximately the same, but the confidence interval for a_2 will be larger for the $n = 3$ combination. For the two cases, $n - p$ equals 1 and 7, respectively. For 95% confidence, the values of $t_{(\beta/2, n-p)}$ (from Table 3.5) are 12.7 and 2.31, respectively. The confidence interval for the $n = 3$

TABLE 4.1. Values[a] of $\hat{\sigma}_{a_2}$ for all possible combinations of n and Δx

n	Δx (kg)				
	1.0	2.0	3.0	4.0	5.0
3	0.0667	0.0333	0.0223	0.0167	0.0133
4	0.0434	0.0218	0.0145		
5	0.0310	0.0155			
6	0.0236	0.0118			
7	0.0187				
8	0.0154				
9	0.0129				
10	0.0110				
11	0.0095				

[a] The minimum value of $x = 3.0$ kg. The maximum possible value $= 13.0$ kg. For every combination, $x_1 = 3.0$ and $x_n = 3.0 + (n - 1) \cdot \Delta x$ kg. Ten (10) 1-kg weights are available for the experiment. The parameter a_2 (cm/kg) is the spring constant, $\hat{\sigma}_{a_2}$ is the "predicted value" of the standard deviation or uncertainty for a_2. The values in the table were determined using Equation 4.5.25.

TABLE 4.2. Values[a] of $\hat{\sigma}_{a_2} \cdot t_{(0.025,\ n - p)}$ for all possible combinations of n and Δx

n	Δx (kg)				
	1.0	2.0	3.0	4.0	5.0
3	0.847	0.423	0.283	0.212	0.169
4	0.187	0.0881	0.0585		
5	0.0985	0.0493			
6	0.0655	0.0330			
7	0.0480				
8	0.0377				
9	0.0305				
10	0.0254				
11	0.0215				

[a] Values of $\hat{\sigma}_{a_2}$ are taken from Table 4.1 and values of $t_{(0.025,\ n - p)}$ are taken from Table 3.5. The 95% confidence interval is, for example, $a_2 \pm 0.0305$ for $n = 9$ and $\Delta x = 1.0$ kg.

combination will be $12.7/2.31 = 5.5$ times larger than for the $n = 9$ combination: This very large ratio is due to the fact that the $n = 3$ combination exhibits only one degree of freedom. The *ratio of confidence interval widths* will approach the ratio of the uncertainties as the values of n increase. To consider confidence interval rather than uncertainty according to Equation 4.3.5, the comparison should be made on the basis of $\hat{\sigma}_{a_2} \cdot t_{(\beta/2, n-p)}$, rather than just $\hat{\sigma}_{a_2}$.[5] Values of this product are compared in Table 4.2 for all possible combinations of n and Δx.

The comparison of all possible combinations on the basis of a 95% confidence interval half-width is seen in Table 4.2 to favor the higher n combinations. Comparing Tables 4.1 and 4.2, it is seen that consideration of only $\hat{\sigma}_{a_2}$ rather than $\hat{\sigma}_{a_2} \cdot t_{(\beta/2, n-p)}$ will lead to an erroneous conclusion. If, however, all values of n are greater than 15, the percentage difference for two values of $t_{(\beta/2, n-p)}$ is less than 10%, and a comparison on the basis of just $\hat{\sigma}_{a_2}$ is usually acceptable.

Case II. *Constant fractional uncertainty for the y variable and negligible uncertainty for the x variable.*

For this case, the values of σ_{y_i} and σ_{x_i} are:

$$\sigma_{y_i} = K_{fy} \cdot |y_i|,$$

$$\sigma_{x_i} = 0. \tag{4.5.30}$$

The absolute value notation is necessary because there is no physical significance for negative values of σ_{y_i}. This case is a special case of Case II of Section 5.2. In Case II of Section 5.2, σ_{x_i} is equal to K_{cx}.

The appropriate expression for $\hat{L}_i$ is:

$$\hat{L}_i = K_{fy}^2 y_i^2. \tag{4.5.31}$$

Equations 4.5.9, and 4.5.10 are still applicable.

The elements of the matrix $\hat{C}$ are determined by substituting Equations 4.5.31, 4.5.9, and 4.5.10 into Equation 4.4.1:

$$\hat{C}_{11} = \sum_{i=1}^{n} \frac{1}{K_{fy}^2 y_i^2} = \frac{n}{K_{fy}^2} \left(\frac{1}{y^2}\right)_{av}, \tag{4.5.32}$$

$$\hat{C}_{12} = \sum_{i=1}^{n} \frac{x_i}{K_{fy}^2 y_i^2} = \frac{n}{K_{fy}^2} \left(\frac{x}{y^2}\right)_{av}, \tag{4.5.33}$$

$$\hat{C}_{22} = \sum_{i=1}^{n} \frac{x_i^2}{K_{fy}^2 y_i^2} = \frac{n}{K_{fy}^2} \left(\frac{x^2}{y^2}\right)_{av}. \tag{4.5.34}$$

[5] The product $\hat{\sigma}_{ak} t_{(\beta/2, n-p)}$ can be called the *confidence interval half-width* for the parameter a_k.

Proceeding as in Case I, we obtain:

$$\hat{\sigma}_{a_1}^2 = \frac{K_{fy}^2}{n} \cdot \frac{(x^2/y^2)_{av}}{(x^2/y^2)_{av} \cdot (1/y^2)_{av} - (x/y^2)_{av}^2}, \qquad (4.5.35)$$

$$\hat{\sigma}_{a_2}^2 = \frac{K_{fy}^2}{n} \cdot \frac{(1/y^2)_{av}}{(x^2/y^2)_{av} \cdot (1/y^2)_{av} - (x/y^2)_{av}^2}. \qquad (4.5.36)$$

If it is assumed that the values of x are evenly spaced (i.e., Equation 4.4.3 is valid), then integral expressions for $(1/y^2)_{av}$, $(x/y^2)_{av}$ and $(x^2/y^2)_{av}$ can be developed. As n approaches infinity,

$$\left(\frac{1}{y^2}\right)_{av} \rightarrow \int_{x_a}^{x_b} \frac{1}{y^2} \, dx \Big/ (x_b - x_a) = \int_{y_a}^{y_b} \frac{dy}{a_2 y^2} \Big/ (x_b - x_a), \qquad (4.5.37)$$

because the differential dy is equal to $a_2 dx$. The limits of integration y_a and y_b are expressed by Equations 4.4.10 and 4.4.11. If both y_a and y_b have the same signs,[6] then integration of Equation 4.5.37 yields:

$$(1/y^2)_{av} \rightarrow [a_2 \cdot (x_b - x_a)]^{-1} \cdot (y_a^{-1} - y_b^{-1}), \qquad (4.5.38)$$

and therefore,

$$\left(\frac{1}{y^2}\right)_{av} \rightarrow \frac{y_b - y_a}{a_2 \cdot (x_b - x_a) \cdot y_a y_b} = \frac{1}{y_a y_b}, \qquad (4.5.39)$$

because

$$a_2 = \frac{y_b - y_a}{x_b - x_a}. \qquad (4.5.40)$$

Similarly, as $n \rightarrow \infty$,

$$\left(\frac{x}{y^2}\right)_{av} \rightarrow \frac{1}{a_2 \cdot (y_b - y_a)} \cdot \int_{y_a}^{y_b} \frac{y - a_1}{y^2} \, dy, \qquad (4.5.41)$$

$$\left(\frac{x}{y^2}\right)_{av} \rightarrow \frac{1}{a_2 \cdot (y_b - y_a)} \cdot \left(\ln \frac{y_b}{y_a} + \frac{a_1}{y_b} - \frac{a_1}{y_a}\right), \qquad (4.5.42)$$

and

$$\left(\frac{x^2}{y^2}\right)_{av} \rightarrow \frac{1}{a_2^2 \cdot (y_b - y_a)} \cdot \int_{y_a}^{y_b} \frac{(y - a_1)^2}{y^2} \, dy, \qquad (4.5.43)$$

$$\left(\frac{x^2}{y^2}\right)_{av} \rightarrow \frac{1}{a_2^2 (y_b - y_a)} \cdot \left(y_b - y_a - 2a_1 \ln \frac{y_b}{y_a} - \frac{a_1^2}{y_b} + \frac{a_1^2}{y_a}\right), \qquad (4.5.44)$$

[6] This limitation is considered in the discussion for Case II, Section 5.2.

Substituting Equations 4.5.39, 4.5.42, and 4.5.44 into Equations 4.5.32, 4.5.33, and 4.5.34 gives:

$$\hat{C}_{11} \rightarrow \frac{n}{K_{fy}{}^2} \cdot \left(\frac{1}{y_a y_b} \right), \tag{4.5.45}$$

$$\hat{C}_{12} \rightarrow \frac{n}{K_{fy}{}^2} \cdot \left(\frac{1}{a_2(y_b - y_a)} \right) \cdot \left(\ln \frac{y_b}{y_a} + \frac{a_1}{y_a} - \frac{a_1}{y_b} \right), \tag{4.5.46}$$

$$\hat{C}_{22} \rightarrow \frac{n}{K_{fy}{}^2} \cdot \left(\frac{1}{a_2{}^2(y_b - y_a)} \right) \cdot \left(y_b - y_a - 2a_1 \ln \frac{y_b}{y_a} - \frac{a_1{}^2}{y_b{}^2} + \frac{a_1{}^2}{y_a{}^2} \right). \tag{4.5.47}$$

If these equations are substituted into Equations 4.5.14, as $n \rightarrow \infty$,

$$\hat{\sigma}_{a_1}{}^2 = \hat{C}_{11}{}^{-1}$$

$$\rightarrow \frac{K_{fy}{}^2/n}{\dfrac{1}{y_a y_b} - \dfrac{1}{y_b - y_a} \cdot \dfrac{\{\ln (y_b/y_a) - [a_1(y_b - y_a)/y_b y_a]\}^2}{\{(y_b - y_a) - 2a_1 \ln (y_b/y_a) + [a_1{}^2(y_b - y_a)/y_b y_a]\}}}, \tag{4.5.48}$$

or, more simply,

$$\hat{\sigma}_{a_1}{}^2 = \hat{C}_{11}{}^{-1} \rightarrow \frac{K_{fy}{}^2}{n \cdot D}, \tag{4.5.49}$$

where D is the denominator of Equation 4.5.48. Simplifying the notation, we define the following quantities:

$$u \equiv y_a y_b, \tag{4.5.50}$$

$$v \equiv y_b - y_a, \tag{4.5.51}$$

$$w \equiv \ln \frac{y_b}{y_a}. \tag{4.5.52}$$

If we substitute Equations 4.5.50, 4.5.51, and 4.5.52 into the expression for D:

$$D = \frac{1}{u} - \frac{1}{v^2} \cdot \frac{[w - (a_1 v/u)]^2}{[1 - (2a_1 w/v) + (a_1{}^2/u)]}, \tag{4.5.53}$$

$$D = \frac{1}{u} - \frac{2a_1 w}{uv} + \frac{a_1{}^2}{u^2} - \frac{w^2}{v^2} + \frac{2a_1 w}{uv} - \frac{a_1{}^2}{u^2} \bigg/ \left(1 - \frac{2a_1 w}{v} + \frac{a_1{}^2}{u} \right), \tag{4.5.54}$$

$$D = \frac{u^{-1} - (w^2/v^2)}{1 - (2a_1 w/v) + (a_1{}^2/u)}. \tag{4.5.55}$$

If Equations 4.5.55, 4.5.50, 4.5.51 and 4.5.52 are substituted into Equation 4.5.49, as $n \to \infty$.

$$\hat{\sigma}_{a_1}^2 = \hat{C}_{11}^{-1} \to \frac{K_{fy}^2}{n}$$

$$\cdot \left[\left(1 - \frac{2a_1}{y_b - y_a} \ln \frac{y_b}{y_a} + \frac{a_1^2}{y_a y_b} \right) \middle/ \left(\frac{1}{y_a y_b} - \frac{1}{(y_b - y_a)^2} \left(\ln \frac{y_b}{y_a} \right)^2 \right) \right]. \quad (4.5.56)$$

Similarly,

$$\hat{\sigma}_{a_2}^2 = \hat{C}_{22}^{-1} \to \frac{K_{fy}^2}{n} \cdot \frac{a_2^2/y_a y_b}{(y_a y_b)^{-1} - [(y_b - y_a)^2]^{-1}[\ln (y_b/y_a)]^2}. \quad (4.5.57)$$

Equations 4.5.56 and 4.5.57 are good approximations for $\hat{\sigma}_{a_1}^2$ and $\hat{\sigma}_{a_2}^2$, even for small values of n. It should be remembered that Case II is only a special case of the more general Case II of Section 5.2. It should also be remembered that the equations are valid only if the signs of y_a and y_b are the same. In Section 5.2, Case II, the modifications required when the signs of y_a and y_b are different, are developed. The equations are plotted subsequently in Figures 5.4 and 5.6 (for $\omega_f \equiv a_2 K_{cx}/a_1 K_{fy} = 0$).

Example 5—Assume that the values of y can be measured to an uncertainty of about 1%, the possible range of values of a_2 is 5.0 to 10, and the range of values of a_1 is 0.0 to 5. Also, assume that values of x must fall within the range 4.0 to 20. If the purpose of the experiment is to measure a_1 to ± 0.5 and a_2 to at least 1% accuracy, how many data points are required if the points are to be evenly spaced within the given range?

The four limiting combinations of a_1 and a_2 are:

Combination	a_1	a_2
1	0.0	5.0
2	0.0	10.0
3	5.0	5.0
4	5.0	10.0

The experiment should be designed in a manner such that the accuracy requirements will be satisfied for all four limiting combinations of a_1 and a_2.

TABLE 4.3. Solutions for n as determined using Equations 4.5.56 and 4.5.57

Combination	a_1	a_2	y_a	y_b	n (Eq. 4.5.56)	n (Eq. 4.5.57)
1	0.0	5.0	20	100	4.17	5.2
2	0.0	10.0	40	200	16.7	5.2
3	5.0	5.0	25	105	6.0	6.46
4	5.0	10.0	45	205	20.1	5.78

The value of K_{fy} for all combinations is 0.01. If it is assumed that x_a is approximately equal to x_1 (i.e., 4.0) and x_b is approximately equal to x_n (i.e., 20.0), then n can be computed using Equations 4.5.56 and 4.5.57 for all the combinations. For all combinations the value of $\hat{\sigma}_{a_1}$ is 0.5. For combinations 1 and 3, $\hat{\sigma}_{a_2}$ is 0.05 and for 2 and 4, $\hat{\sigma}_{a_2}$ is 0.10. The results are summarized in Table 4.3. From the table it is seen that the accuracy requirement for a_1 for combination 4 is the most difficult requirement to satisfy. If the experiment is designed to include more than 20 data points, the accuracy requirements will be satisfied for all combinations of a_1 and a_2.

Case III. *Counting statistics for the y variable and negligible uncertainty for the x variable.*

For this case, the values of σ_{y_i} and σ_{x_i} are:

$$\sigma_{y_i} = K_{sy} \cdot (y_i)^{1/2},$$

$$\sigma_{x_i} = 0. \tag{4.5.58}$$

The values of y_i can be considered as count rates and must clearly be zero or positive because it is impossible to have a negative number of counts. Case III is a special case of Case III of Section 5.2. In Case III of Section 5.2, σ_{x_i} is equal to K_{cx}.

The appropriate expression for $\hat{L}_i$ is:

$$\hat{L}_i = K_{sy}^2 \cdot y_i. \tag{4.5.59}$$

Equations 4.5.9 and 4.5.10 are still applicable.

The elements of the matrix $\hat{C}$ are determined by substituting Equations 4.5.59, 4.5.9, and 4.5.10 into Equation 4.9.1:

$$\hat{C}_{11} = \sum_{i=1}^{n} \frac{1}{K_{sy}^2 y_i} = \frac{n}{K_{sy}^2} \cdot \left(\frac{1}{y}\right)_{av}, \tag{4.5.60}$$

$$\hat{C}_{12} = \sum_{i=1}^{n} \frac{x_i}{K_{sy}^2 y_i} = \frac{n}{K_{sy}^2} \cdot \left(\frac{x}{y}\right)_{av}, \tag{4.5.61}$$

$$\hat{C}_{22} = \sum_{i=1}^{n} \frac{x_i^2}{K_{sy}^2 \cdot y_i} = \frac{n}{K_{sy}^2} \cdot \left(\frac{x^2}{y}\right)_{av}. \tag{4.5.62}$$

Proceeding as in Case II, we obtain:

$$\hat{\sigma}_{a_1}^2 = \frac{K_{sy}^2}{n} \cdot \frac{(x^2/y)_{av}}{(x^2/y)_{av} \cdot (1/y)_{av} - (x/y)_{av}^2}, \tag{4.5.63}$$

$$\hat{\sigma}_{a_2}^2 = \frac{K_{sy}^2}{n} \cdot \frac{(1/y)_{av}}{(x^2/y)_{av} \cdot (1/y)_{av} - (x/y)_{av}^2}. \tag{4.5.64}$$

If it is assumed that the values of x are evenly spaced (i.e., Equation 4.4.3 is valid), then integral expressions for $(1/y)_{av}$, $(x/y)_{av}$, and $(x^2/y)_{av}$ can be used. As n approaches infinity:

$$\left(\frac{1}{y}\right)_{av} \to \frac{1}{y_b - y_a} \cdot \int_{y_a}^{y_b} \frac{dy}{y}, \tag{4.5.65}$$

$$\left(\frac{1}{y}\right)_{av} \to \frac{1}{y_b - y_a} \cdot \ln \frac{y_b}{y_a}, \tag{4.5.66}$$

$$\left(\frac{x}{y}\right)_{av} \to \frac{1}{a_2 \cdot (y_b - y_a)} \cdot \int_{y_a}^{y_b} \frac{y - a_1}{y} \, dy, \tag{4.5.67}$$

$$\left(\frac{x}{y}\right)_{av} \to \frac{1}{a_2 \cdot (y_b - y_a)} \cdot \left(y_b - y_a - a_1 \ln \frac{y_b}{y_a}\right), \tag{4.5.68}$$

$$\left(\frac{x^2}{y}\right)_{av} \to \frac{1}{a_2^2 \cdot (y_b - y_a)} \cdot \int_{y_a}^{y_b} \frac{(y - a_1)^2}{y} \, dy, \tag{4.5.69}$$

$$\left(\frac{x^2}{y^2}\right)_{av} \to \frac{1}{a_2^2 \cdot (y_b - y_a)} \cdot \left(\tfrac{1}{2}(y_b^2 - y_a^2) - 2a_1(y_b - y_a) + a_1^2 \ln \frac{y_b}{y_a}\right), \tag{4.5.70}$$

$$\left(\frac{x^2}{y^2}\right)_{av} \to \frac{1}{a_2^2} \cdot \left(\tfrac{1}{2}(y_b + y_a) - 2a_1 + \frac{a_1^2}{y_b - y_a} \ln \frac{y_b}{y_a}\right). \tag{4.5.71}$$

Substituting Equations 4.5.66, 4.5.68, and 4.5.71 into Equations 4.5.63 and 4.5.64 gives:

$$\hat{\sigma}_{a_1}^2 \to \frac{K_{sy}^2/n}{\dfrac{\ln (y_b/y_a)}{y_b - y_a} - \dfrac{\{1 - [a_1/(y_b - y_a)] \ln (y_b/y_a)\}^2}{\tfrac{1}{2}(y_b + y_a) - 2a_1 + [a_1^2/(y_b - y_a)] \ln (y_b/y_a)}}. \tag{4.5.72}$$

$$\hat{\sigma}_{a_1}^2 \to \frac{K_{sy}^2}{n} \cdot \frac{y_b + y_a - 4a_1 + [2a_1^2/(y_b - y_a)] \ln (y_b/y_a)}{\dfrac{(y_b + y_a) \ln (y_b/y_a)}{y_b - y_a} - 2}, \tag{4.5.73}$$

$$\hat{\sigma}_{a_2}^2 \to \frac{a_2^2 K_{sy}^2/n}{\left(\tfrac{1}{2}(y_b + y_a) - 2a_1 + \dfrac{a_1^2}{y_b - y_a} \ln \dfrac{y_b}{y_a}\right) - \dfrac{[1 - (a_1/y_b - y_a)] \ln (y_b/y_a)]^2}{(y_b - y_a)^{-1} \ln (y_b/y_a)}}, \tag{4.5.74}$$

$$\hat{\sigma}_{a_2}^2 \to \frac{K_{sy}^2 \cdot a_2^2}{n} \cdot \frac{2 \ln (y_b/y_a)}{(y_b - y_a)} \Bigg/ \left(\frac{(y_b + y_a) \ln (y_b/y_a)}{y_b - y_a} - 2\right). \tag{4.5.75}$$

Equations 4.5.73 and 4.5.75 are good approximations for $\hat{\sigma}_{a_1}{}^2$ and $\hat{\sigma}_{a_2}{}^2$, even for small values of n.

It should be remembered that Case III is only a special case of the more general Case III of Section 5.2. Graphical presentation of Equations 4.5.73 and 4.5.75 are therefore included subsequently in Figures 5.7 and 5.8 (for $\omega_s \equiv a_2{}^2 K_{cx}{}^2/a_1 K_{sy}{}^2 = 0$).

An additional comment should be made concerning the interpretation of the constant K_{sy}. For most counting experiments, the best estimate of the uncertainty or standard deviation of the total number of counts is the square root of this total. If the y_i's are total numbers of counts, then the constant is simply equal to one. Often, however, it is more convenient to express the y_i's as count rates (i.e., counts per unit time). If t represents the time per observation and t_u represents the unit time, then K_{sy} is:

$$K_{sy} = \left(\frac{t_u}{t}\right)^{1/2}. \tag{4.5.76}$$

If, for example, the y_i's are expressed as counts per minute, and each value of y_i is determined by dividing the number of counts recorded in an hour by 60, then the values of σ_{y_i} are:

$$\sigma_{y_i} = (\tfrac{1}{60}y_i)^{1/2}. \tag{4.5.77}$$

In other words, the uncertainties σ_{y_i} are reduced by a factor $(t_u/t)^{1/2}$ as a result of extending the counting periods from t_u to t.

Example 6—Assume that the values of y_i are counts per minute, the possible range of values of a_2 is 5.0 to 10, the range of values of a_1 is 0.0 to 5.0, and the x_i's must fall within the range 4.0 to 20. If the purpose of the experiment is to measure a_1 to ±0.5 and a_2 to at least 1% accuracy, how much time should be devoted to each measurement of y to limit the total number of data points to ten? The points are to be spaced evenly within the given range and the experiment should be designed in a manner such that the accuracy requirements will be satisfied for the four limiting combinations of a_1 and a_2 (as listed in Example 5).

Since the number of data points and the range of x values are fixed, the values of x_a and x_b can be determined exactly. The spacing between points

TABLE 4.4. Solutions for t as determined using Equations 4.5.73 and 4.5.75 to solve for $K_{sy}{}^2$ and then Equation 4.5.76 to solve for t

Combination	a_1	a_2	y_a	y_b	t, min (Eq. 4.5.73)	t, min (Eq. 4.5.75)
1	0.0	5.0	15.55	104.5	83.7	74.6
3	0.0	10.0	31.1	209.0	167.5	37.4
3	5.0	5.0	20.55	109.5	93.0	78.2
4	5.0	10.0	36.1	214.0	188.0	40.0

TABLE 4.5. Summary of Cases I, II, and III[a]

Case	σ_{y_i}	σ_{x_i}	Equation for $\hat{\sigma}_{a_1}{}^2$	Equation for $\hat{\sigma}_{a_2}{}^2$
I	K_{cy}	K_{cx}	4.5.29	4.5.24 and 4.5.25
II	$K_{fy} \cdot \lvert y_i \rvert$	0	4.5.56	4.5.57
III	$K_{sy} \cdot (y_i)^{1/2}$	0	4.5.73	4.5.75

[a] The equations are for experiments in which the values of the independent variable x are evenly spaced (i.e., Equation 4.4.3 is valid).

$\Delta x = (x_n - x_1)/(n - 1) = 16.0/9 = 1.78$. From Equations 4.4.7 and 4.4.8, the values of x_a and x_b are therefore 3.11 and 20.89, respectively.

For all combinations, the value of $\hat{\sigma}_{a_1}$ is 0.5. For combinations 1 and 3, the value of $\hat{\sigma}_{a_2}$ is 0.05, and for 2 and 4, the value of $\hat{\sigma}_{a_2}$ is 0.10. Using Equations 4.5.73 and 4.5.75, the required values of K_{sy} can be determined for each combination. These values can then be converted to counting times using Equation 4.5.76. The results are summarized in Table 4.4.

From Table 4.4 it is seen that the accuracy requirement for a_1 for combination 4 is the most difficult requirement to satisfy. This is the same result that was obtained in Example 4. If at least 188 minutes are devoted to each measurement of the count rates y_i, the accuracy requirements will be satisfied for all combinations of a_1 and a_2.

The three cases considered in Section 4.5 are summarized in Table 4.5. The cases considered in this section are all for the "straight-line" class of experiments. Additional analysis of this class of experiments is included in Chapter 5, all the results being presented in dimensionless form.

4.6 Comparison of Direct and Analytical Solutions

The direct and analytical approaches to prediction analysis were discussed in Section 4.4. The direct approach leads to exact solutions for the $\hat{\sigma}_{a_k}$'s. The solutions resulting from an analytical approach are only exact as the number of data points approaches infinity.

Except for the simplest problems, the direct approach is feasible only if the computations are performed using a digital computer. This limitation is not, however, serious, because most scientists and engineers now have access to digital computers. In fact, the computer codes being used for least squares analysis of experimental data can easily be modified to accommodate prediction analysis. Design of prediction and least squares analysis computer programs is discussed in Section 4.7.

The most serious limitation of the direct approach is that it is often difficult to generalize the results. The results from one combination of experimental variables is rarely sufficient for planning an experiment. To understand how changes in the experimental variables affect the predicted

TABLE 4.6. Average values of $1/y^2$, x/y^2, and x^2/y^2. The values of y are equal to $7.0 + 1.0x$

Data Point	x	y	$1/y^2$	x/y^2	x^2/y^2
1	3.0	10.0	0.010000	0.030000	0.090000
2	4.0	11.0	0.008264	0.033056	0.132224
3	5.0	12.0	0.006945	0.034725	0.173625
4	6.0	13.0	0.005918	0.035508	0.213048
5	7.0	14.0	0.005102	0.035714	0.249998
Sum			0.036229	0.169003	0.858895
Average			0.007246	0.033801	0.171779

uncertainties $\hat{\sigma}_{a_k}$, it is necessary to repeat the computation many times. This process can be time-consuming, even if a digital computer is used.

The analytical approach will lead to approximate solutions. Even so, these equations can be extremely useful. A prediction analysis for any combination of parameters and variables may immediately be obtained using the equations. In addition, the equations may also be useful for *optimizing* the proposed experiment. Sometimes, however, the simplifying assumptions required to achieve an analytical solution render the solu-

TABLE 4.7. Comparison of direct and analytical solutions[a]

	Direct Solution	Analytical Solution
$(1/y^2)_{av}$	0.007246	0.007259
$(x/y^2)_{av}$	0.033801	0.03380
$(x^2/y^2)_{av}$	0.171779	0.17120
$(n/K_{fy}^2)\ C_{11}^{-1}$	1682.0	1708.0
$(n/K_{fy}^2)\ C_{22}^{-1}$	70.9	72.4
$\hat{\sigma}_{a_1}$	0.1834	0.1848
$\hat{\sigma}_{a_2}$	0.0377	0.0380
$\hat{\sigma}_{a_1}/a_1$	0.0262	0.0264
$\hat{\sigma}_{a_2}/a_2$	0.0376	0.0380

[a] The values of the parameters for this example are $a_1 = 7.0$, $a_2 = 1.0$, $n = 5$, $K_{fy} = 0.01$, $x_1 = 3.0$, and $\Delta x = 1.0$.

tion useless. It is often worthwhile to check an analytical solution by comparing it to a direct solution of the same problem. This comparison should establish the limits of applicability of the equations which constitute the analytical solution.

As an example, direct and analytical solutions are compared for a specific problem. Consider a "straight-line" experiment in which Case II of Section 4.5 is applicable. That is, the uncertainties σ_{x_i} are negligible, and the uncertainties σ_{y_i} are proportional to the values of y_i (i.e., $\sigma_{y_i} = K_{fy} \cdot y_i$). The specific values of the parameters and variables are: $a_1 = 7.0$, $a_2 = 1.0$, $K_{fy} = 0.01$, $n = 5$, $x_1 = 3.0$, and $\Delta x = 1.0$. A direct solution for the values of $\hat{\sigma}_{a_1}$ and $\hat{\sigma}_{a_2}$ can be derived using Equations 4.5.35 and 4.5.36. To complete the solution, the values of $(1/y^2)_{av}$, $(x/y^2)_{av}$, and $(x^2/y^2)_{av}$ are determined directly (see Table 4.6).

Solutions for $\hat{\sigma}_{a_1}$ and $\hat{\sigma}_{a_2}$ are tabulated in Table 4.7. These solutions are compared to analytical solutions determined using Equations 4.5.39, 4.5.42, 4.5.44, 4.5.56, and 4.5.57.

For this example, the analytical approach is seen to yield results which are within 1% of the results obtained directly. This close agreement is particularly impressive if one notices that the number of points n is only equal to five. For many problems, however, one does not obtain such excellent agreement. The two solutions will be exactly the same at n equal to infinity, but the rate at which the direct solution approaches its asymptotic solution is dependent upon the choice of parameters. Convergence of the direct solution to the asymptotic (i.e., analytical) solution is considered in greater detail subsequently in Section 5.2 for the "straight-line" or "first-order polynomial" class of experiments.

An additional comparison is of interest. In Table 4.8, values of $\hat{\sigma}_{a_1}$ and $\hat{\sigma}_{a_2}$ as computed using the "modified direct approach" (i.e., Equations 4.4.2 and 4.4.4) are compared to the direct solution. The modified solutions are seen to approach the direct solutions as q approaches n. Even for $q = 2$, the difference is only about 30%. For more complicated functions, however, the discrepancy is usually much greater.

TABLE 4.8. Comparison of direct and modified solutions[a]

q	$\hat{\sigma}_{a_1}$	$\hat{\sigma}_{a_2}$
2	0.1291	0.0272
3	0.1579	0.0327
4	0.1735	0.0357
5	0.1834	0.0377 (direct solution)

[a] The values of the parameters used in the example were $a_1 = 7.0$, $a_2 = 1.0$, $n = 5$, $K_{fy} = 0.01$, $x_1 = 3.0$, and $\Delta x = 1$. Equations 4.4.2 and 4.4.4 were used to evaluate the elements $\hat{C}_{jk}$. The solution for $q = 5$ is the direct solution because $n = 5$.

4.7 Prediction Analysis and Least Squares Analysis Computer Programs

For most experiments, the tasks of performing prediction and least squares analyses can be completed only if a digital computer is available. For both types of analyses, general computer programs can be written which require only slight modifications to be suitable for any experiment. An alternative procedure is to write one program which can be used for both prediction and least squares analysis.

The suggested block diagram[7] of a general prediction analysis computer program is shown in Figure 4.3. The function of each block is as follows:

(1) The first block includes a read instruction for a "set" of problems. The variables and parameters which must be read as input quantities include:

(a)   NP, the number of unknown parameters.

(b)   NQ, the number of known parameters.

(c)   N, the number of data points.

(d)   M, the number of independent variables.

(e)   KCY, a key denoting which parameter is to be cycled in the set. (If more than one parameter is to be cycled, additional keys will be required.)

(f)   KCT, a key denoting the total number of cases in the set.

(g)   KY, a key denoting how the uncertainties σ_{y_i} are to be computed.

(h)   KX(J), J = 1, M, keys denoting how the uncertainties $\sigma_{x_{ji}}$ are to be computed.

(i)   KF, a key denoting the function.

(j)   A(K), K = 1, NP, the unknown parameters.

(k)   B(K), K = 1, NQ, the known parameters.

(l)   CYCLE, the cycling constant.

(m)   CY, the constant required for computing σ_{y_i}.

(n)   CX(J), J = 1, M, the constants required for computing $\sigma_{x_{ji}}$.

(o)   KCOMP, a key denoting whether or not the values of x_{ji} are to be computed or read in. If KCOMP = 0, the values are read in. If KCOMP > 0, then the number of constants needed for the computation of the values of x_{ji} is equal to KCOMP.

(p)   XCOMP(J), J = 1, KCOMP, the constants required for the computation of the values of x_{ji}.

It is worthwhile to arrange the read format in a manner which minimizes the number of data cards required per set. A complete prediction analysis

---

[7] This section is developed with the assumption that those who read it are familar with at least the common terminology of computer programming.

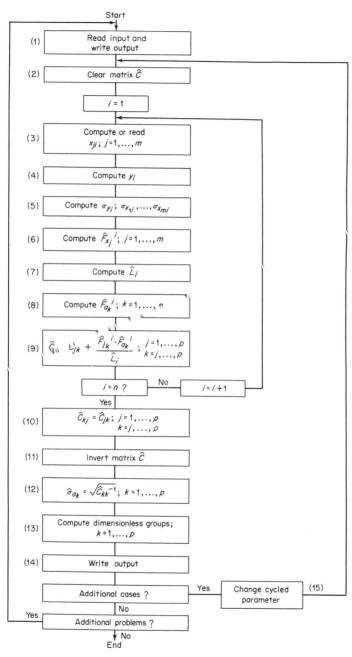

Figure 4.3. Block diagram for a general prediction analysis computer program.

of a particular experiment will often require analysis of a large number of combinations of the experimental variables and parameters. By minimizing the number of cards per set, the job of preparing the data cards for an analysis is simplified.

Another important function of Block (1) is to write on the computer output sheet all the relevant information required to understand what is being computed. It is worthwhile to present the results for all cases in the set in the form of a table. The table heading can be prepared in Block (1).

(2) The matrix $\hat{C}$ is "cleared" in Block (2). Programming this block requires three statements[8]:

$$90 \text{ DO } 95 \text{ J} = 1, \text{ NP}$$

$$\text{DO } 95 \text{ K} = 1, \text{ NP}$$

$$95 \text{ C(J, K)} = 0.$$

Blocks (3) to (9) are within a loop which is cycled from $I = 1$ to N. For cases in which N is large, the time required to complete this loop can be decreased by using Equation 4.4.2 in place of Equation 4.4.1. It should be realized that the results will only be approximate if this procedure is adopted.

(3) Block (3) consists of two paths. If KCOMP$=0$, the program is directed to a READ instruction. If KCOMP>0, then the values of x_{ji} are computed. The necessary equations for computing the values of x_{ji} must be included in the block. If, for example, M is equal to one, and the values of x_i are to be evenly spaced (i.e., Equation 4.4.3 is valid), the recommended procedure for computing the values of x_i is:

$$x_i = x_a + \frac{(i - \frac{1}{2}) \cdot (x_b - x_a)}{n}. \qquad (4.7.1)$$

The experimental variables x_a and $x_b - x_a$ can be read in as XCOMP(1) and XCOMP(2), and are defined by Equations 4.4.7 and 4.4.8. The reason for working in terms of x_a and $x_b - x_a$ rather than x_1 and $x_n - x_1$ can best be explained by referring to Equation 4.4.6. These variables are fundamental if an analytical approach to the analysis is to be attempted. To facilitate comparison of analytical and direct (i.e., computer) solutions, it is recommended that the variables x_a and $x_b - x_a$ be used.

(4) Block (4) should consist of a single statement. If $M = 1$:

$$Y = \text{SFUNC (KF, X)}$$

Within the function routine SFUNC, a library of functions can be stored, each function denoted by a specific value of KF. If, for example, $KF = 1$

---

[8] All programming statements are written in Fortran IV language.

denotes the polynomial function, the function could be stored as follows:

```
      GO TO (1, 2, 3, 4, 5, 6, 7, 8, 9, 10), KF
    1 SFUNC = A(1)
      DO 101 K = 2, NP
  101 SFUNC = SFUNC + A(K) * (X ** (K−1))
      RETURN
        ⋮
      etc.
```

The advantage of computing y_i in a function routine is that additional experiments can be studied by adding to or modifying the routine. The main program does not have to be changed.

(5) Block (5) consists of a few simple statements:

```
      SY = SIG(Y, KY, CY)
      DO 110 J = 1, M
  110 SX(J) = SIG(X(J), KX(J), CX(J))
```

The equations for computing σ_{y_i} (i.e., SY) and σ_{x_j} (i.e., SX(J)) are contained in the function routine SIG.

To perform the computations presented in Chapters 5 through 8, the function routine SIG was programmed as follows:

```
      FUNCTION SIG (XX, KK, CC)
      GO TO (1, 2, 3), KK
    1 SIG = CC
      GO TO 100
    2 SIG = CC * XX
      GO TO 100
    3 SIG = CC * SQRT(XX)
  100 RETURN
      END
```

If, for example, KY or KX(J) = 1, the value of SY or SX(J) is set equal to the constant CY or CX(J). If KY or KX(J) = 2, SY or SX(J) is set equal to a value proportional to Y or X(J). If KY or KX(J) = 3, SY or SX(J) is set equal to a value proportional to the square root of Y r X(J). The constants CY and CX(J) are the constants of proportionality.

The routine could, of course, be extended to include additional options for computing σ_{y_i} and $\sigma_{x_{ji}}$.

(6) Block (6) consists of two statements:

$$\text{DO } 120 \text{ J} = 1, \text{ M}$$

$$120 \text{ DX(J)} = \text{DERX(KF, J)}$$

The function routine DERX is similar to the routine SFUNC. The necessary expressions for the M derivatives $\hat{F}_{x_i}$ are stored in the routine.

(7) Block (7) consists of the following statements:

$$\text{DL} = \text{SY} ** 2$$

$$\text{DO } 130 \text{ J} = 1, \text{ M}$$

$$130 \text{ DL} = \text{DL} + (\text{SX(J)} * \text{DX(J)}) ** 2$$

This computation is based upon Equation 3.6.16 and DL refers to $\hat{L}_i$.

(8) Block (8) consists of two statements:

$$\text{DO } 140 \text{ K} = 1, \text{ NP}$$

$$140 \text{ FA(K)} = \text{DERA(KF, K)}$$

The function routine DERA is similar to the routine SFUNC. The necessary expressions for the NP derivatives $\hat{F}_{a_k}$ are stored in the routine.

(9) Block (9) consists of the following statements:

$$\text{DO } \quad 150 \text{ J} = 1, \text{ NP}$$

$$\text{DO } \quad 150 \text{ K} = \text{J}, \text{ NP}$$

$$150 \text{ C(J, K)} = \text{C(J, K)} + \text{FA(J)} * \text{FA(K)}/\text{DL}$$

Since the matrix $\hat{C}$ is symmetrical, there is no need to fill the entire matrix within the loop $I = 1$, N. The indexing is arranged so that only the elements on and above the diagonal are filled.

(10) Once the loop $I = 1$, N has been completed, the elements below the diagonal of the $\hat{C}$ matrix must be filled. To accomplish this, three statements are required:

$$\text{DO } \quad 160 \, J = 1, \text{ NP}$$

$$\text{DO } \quad 160 \, K = \text{J}, \text{ NP}$$

$$160 \text{ C(K, J)} = \text{C(J, K)}$$

(11) The matrix $\hat{C}$ is inverted in Block (11) by calling the inverse matrix subroutine:

$$\text{CALL INVM (NP)}$$

The reader is referred to the comments on matrix inversion in Section 4.8.

(12) In Block (12) the predicted uncertainties σ_{a_k} are computed:

$$\text{DO} \quad 170 \ \text{K} = 1, \ \text{NP}$$

$$170 \ \text{SA}(\text{K}) = \text{SQRT}(\text{CINV}(\text{K, K}))$$

where $\text{CINV}(\text{K, K})$ denotes the kth diagonal element of the matrix $\hat{C}^{-1}$, and $\text{SA}(\text{K})$ denotes $\hat{\sigma}_{a_k}$.

(13) In Part II of the book, it is shown that the most convenient method for presenting the results are in terms of dimensionless groups. The equations for the dimensionless groups depend upon the type of uncertainty for the variable y (i.e., the key KY) and the index k. The computations of the dimensionless groups is performed in the subroutine DIMEN:

$$\text{CALL DIMEN (KY)}$$

Within the subroutine, the dimensionless groups for all the $\hat{\sigma}_{a_k}$'s are computed.

(14) In Block (14) the results for each case in the set are printed in the table of results on the computer output sheet.

(15) In Block (15) the cycled parameter is changed and the computation is repeated. One possible form of this block is as follows:

$$\text{GO TO } (200, 200, 200, 200, 200, 210, 220, 230, 240, 250), \ \text{KCY}$$

$$200 \ \text{A}(\text{KCY}) = \text{A}(\text{KCY}) + \text{CYCLE}$$

$$\text{GO TO } 90$$

$$210 \ \text{CY} = \text{CY} + \text{CYCLE}$$

$$\text{GO TO } 90$$

$$220 \ \text{CX}(1) = \text{CX}(1) + \text{CYCLE}$$

$$\text{GO TO } 90$$

$$230 \ \text{XCOMP}(1) = \text{XCOMP}(1) + \text{CYCLE}$$

$$\text{GO TO } 90$$

$$240 \ \text{XCOMP}(2) = \text{XCOMP}(2) + \text{CYCLE}$$

$$\text{GO TO } 90$$

$$250 \ \text{NCYC} = \text{CYCLE}$$

$$\text{N} = \text{N} + \text{NCYC}$$

$$\text{GO TO } 90$$

Location 90 denotes the first statement in Block (2). If KCY =1, 2, 3, 4, or 5, the unknown parameter A(1), A(2), A(3), A(4), or A(5) is increased by an amount equal to CYCLE and then the entire calculation is repeated. If KCY =6, CY is increased; if KCY =7, CX(1) is increased; if KCY =8, XCOMP(1) is increased; if KCY =9, XCOMP(2) is increased; and if KCY = 10, N is increased.

A general least squares analysis computer program similar to the prediction analysis program can be written. A suggested block diagram is shown in Figure 4.4. The function of each block is as follows:

(1) Block (1) of the least squares program is similar to Block (1) of the prediction analysis program. The input data, with the exception of the values of the dependent and independent variables, are read in this block and the initial output is printed.

(2) Block (2) is similar to Block (2) of the prediction analysis program:

$$\text{DO} \quad 96 \text{ J} = 1, \text{ NP}$$

$$\text{DO} \quad 95 \text{ K} = 1, \text{ NP}$$

$$95 \text{ C(J, K)} = 0.$$

$$96 \text{ V(J)} = 0.$$

(3) The values of y_i and x_{ji}, $i = 1, \cdots, n, j = 1, \cdots, m$, are read in Block (3).

(4) Block (4) is similar to Block (5) of the prediction analysis program. There is, however, a basic difference. Since the values of y_i, x_{ji}, σ_{y_i}, and $\sigma_{x_{ji}}$ might be used several times (i..e, until the a_k's converge), these quantities should be stored. The program statements are as follows:

$$\text{SY(I)} = \text{SIG(Y(I), KY, CY)}$$

$$\text{DO} \quad 110 \text{ J} = 1, \text{ M}$$

$$110 \text{ SX(I, J)} = \text{SIG(X(I, J), KX(J), CX(J))}$$

The function routine SIG does not have to be changed from the routine used for the prediction analysis program unless alternative methods for computing the uncertainties are required.

(5) Block (5) is the same as Block (6) of the prediction analysis program.

(6) Block (6) is the same as Block (7) of the prediction analysis program.

(7) Block (7) is the same as Block (8) of the prediction analysis program.

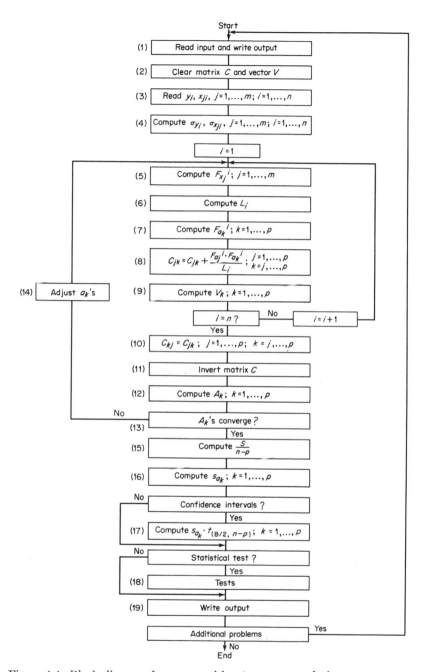

Figure 4.4. Block diagram for a general least squares analysis computer program.

(8) Block (8) is the same as Block (9) of the prediction analysis program.

(9) Block (9) is based upon Equation 3.6.22. If $M=1$:

$$DF = Y(I) - SFUNC(KF, X(I))$$

$$DO \quad 155 \ K=1, \ NP$$

$$155 \ V(K) = V(K) + FA(K) * DF/DL$$

The function routine SFUNC is the same as the routine SFUNC of the prediction analysis program. The symbol DF denotes F_0 which is defined by Equation 3.6.4.

(10) Block (10) is the same for both programs.

(11) Block (11) is the same for both programs.

(12) Block (12) is based upon Equation 3.6.30:

$$DO \quad 165 \ K=1, \ NP$$

$$AA(K) = 0.$$

$$DO \quad 165 \ J=1, \ NP$$

$$165 \ AA(K) = AA(K) + CINV(J, \ K) * V(J)$$

The symbol $AA(K)$ denotes A_k which is defined by Equation 3.6.6.

(13) The test for convergence is based upon Equation 3.6.34. If all the values of A_k (i.e., $AA(K)$) satisfy this equation, then the computation proceeds to Block (15); if not, the a_k's are adjusted in Block (14) and the computation is repeated. It is useful to specify a maximum number of iterations. Once this number is reached a statement, such as "THE ANALYSIS DOES NOT CONVERGE" should be printed and then the program should proceed to the next problem.

(14) The values of a_k (i.e., $A(K)$) are adjusted in Block (14) using the general form of Equations 3.8.1 and 3.8.2:

$$DO \quad 168 \ K=1, \ NP$$

$$168 \ A(K) = A(K) - AA(K)/G$$

The constant G is the *acceleration factor* and is discussed in Section 3.8.

(15) Block (15) is based upon Equation 3.10.34.

(16) Block (16) is based upon Equation 3.10.43.

(17) If confidence intervals are to be computed, values of $t_{(\beta/2, n-p)}$ must either be stored or computed. (See Section 2.5 for a discussion of the t-distribution, and Table 3.5 for values of $t_{(\beta/2, n-p)}$.

(18) If statistical tests of the data and the results are desired, the tests discussed in Section 3.13 could form the basis of computations performed in Block (18).

(19) The results are printed on the computer output sheet in Block (19).

A combined prediction-least squares analysis computer program could also be written, but there is no advantage to combining both types of analyses in one program. There is, however, an advantage in writing the programs in a similar manner. It is easier for the user to master both programs if they bear a resemblance.

4.8 Matrix Inversion

For *symmetric* 2×2 and 3×3 matrices, equations for the elements of the inverse matrix are included in the text. For a 2×2 matrix, Equations 3.7.16 through 3.7.18 are applicable:

$$C_{11}^{-1} = \frac{C_{22}}{C_{11} \cdot C_{22} - C_{12}^2}, \tag{3.7.16}$$

$$C_{22}^{-1} = \frac{C_{11}}{C_{11} \cdot C_{22} - C_{12}^2}, \tag{3.7.17}$$

$$\hat{C}_{21}^{-1} = C_{12}^{-1} = \frac{-C_{21}}{C_{11} \cdot C_{22} - C_{12}^2}. \tag{3.7.18}$$

For a 3×3 matrix, Equations 5.3.10 through 5.3.16 are applicable:

$$C_{11}^{-1} = \frac{C_{22} \cdot C_{33} - (C_{23})^2}{D}, \tag{5.3.10}$$

$$C_{22}^{-1} = \frac{C_{11} \cdot C_{33} - (C_{13})^2}{D}, \tag{5.3.11}$$

$$C_{33}^{-1} = \frac{C_{11} \cdot C_{22} - (C_{12})^2}{D}, \tag{5.3.12}$$

$$\hat{C}_{21}^{-1} = C_{12}^{-1} = \frac{-C_{12} \cdot C_{33} + C_{13} \cdot C_{23}}{D}, \tag{5.3.13}$$

$$\hat{C}_{31}^{-1} = C_{13}^{-1} = \frac{C_{12} \cdot C_{23} - C_{13} \cdot C_{22}}{D}, \tag{5.3.14}$$

$$\hat{C}_{23}^{-1} = C_{23}^{-1} = \frac{-C_{11} \cdot C_{23} + C_{12} \cdot C_{13}}{D}, \tag{5.3.15}$$

where

$$D = C_{11} \cdot [C_{22} \cdot C_{33} - (C_{12})^2] - C_{12} \cdot [C_{12} \cdot C_{33} - C_{13} \cdot C_{23}]$$
$$+ C_{13} \cdot [C_{12} \cdot C_{23} - C_{13} \cdot C_{22}]. \quad (5.3.16)$$

Comparing Equation 3.7.16 through 3.7.18 with 5.3.10 through 5.3.16, it is observed that the complexity of the equations increases rapidly as the order of the matrix increases.

It is extremely useful to have a general method for inverting a matrix that is the same regardless of the order of the matrix. In least squares and prediction analysis computer programs, inversion of the C matrix could then be accomplished in a subroutine based upon the general method. Many computation centers have "matrix inversion" subroutines in their libraries. If, however, one is required to program such an operation, it should be realized that techniques particularly suitable to a computer are available.

A method that is well adapted for use with desk calculators as well as with large digital computers was first suggested by Crout.[9] An excellent description of the method is given by Hildebrand.[10] Although the following discussion will be limited to the Crout method, it should be realized that other methods exist. For example, the Gauss–Jordan method is often used. In fact, the Gauss–Jordan method with the added feature of maximal pivoting,[11] although slower, is more accurate than the Crout method.[12] Accuracy, rather than speed, becomes more important as the size of the matrix increases. (The problem of accuracy arises as a result of round-off errors. The accuracy can be improved if the matrix inversion subroutine is written using double precision arithmetic.) A thorough discussion of matrix theory is given by Hildebrand.[13]

The Crout method is often called the *Crout reduction*. It is not only a method for inverting a matrix but also for solving a set of linear equations. Equation 3.6.23, for example, is a set of linear equations:

$$\left\{ \begin{array}{c} A_1 C_{11} + A_2 C_{12} + \cdots + A_p C_{1p} = V_1 \\ \cdot \qquad\qquad\qquad\qquad \cdot \\ \cdot \qquad\qquad\qquad\qquad \cdot \\ \cdot \qquad\qquad\qquad\qquad \cdot \\ A_p C_{p1} + A_2 C_{p2} + \cdots + A_p C_{pp} = V_p \end{array} \right\} \quad (3.6.23)$$

[9] P. D. Crout, *Trans. AIEE*, 60, 1235-1240 (1941).

[10] F. B. Hildebrand, "An Introduction to Numerical Analysis," McGraw-Hill Book Co., New York, 1956.

[11] E. Stieffel, "An Introduction to Numerical Mathematics," Academic Press, Inc., New York, 1963.

[12] G. Logemann, private communication, 1966.

[13] F. B. Hildebrand, "Methods of Applied Mathematics," Prentice-Hall, Inc., Englewood Cliffs, N. J., 1952.

The A_k's are the unknown parameters of this set of equations. Equation 3.6.28 is the same set of equations re-expressed in matrix notation:

$$CA = V. \qquad (3.6.28)$$

The solution for the "vector" A is determined by premultiplying both sides of Equation 3.6.28 by the "inverse coefficient matrix" C^{-1}:

$$A = C^{-1}V. \qquad (3.6.29)$$

The "Crout reduction" is a series of operations that reduces Equation 3.6.28 to a solution as expressed by Equation 3.6.29. The calculation proceeds in a manner that does not require tabulation or storage of intermediate results. For this reason, it is particularly well suited for use with desk calculators or computers.

The matrix M is defined as the *augmented matrix* of the set of equations:

$$M \equiv \begin{Bmatrix} C_{11} & C_{12} & \cdots & C_{1p} & V_1 \\ C_{21} & C_{22} & \cdots & C_{2p} & V_2 \\ \cdot & & & & \\ \cdot & & & & \\ \cdot & & & & \\ C_{p1} & C_{p2} & \cdots & C_{pp} & V_p \end{Bmatrix}. \qquad (4.8.1)$$

An *auxiliary matrix* M' with the same dimensions as M is determined using a procedure to be explained:

$$M' \equiv \begin{Bmatrix} C_{11}' & C_{12}' & \cdots & C_{1p}' & V_1' \\ C_{21}' & C_{22}' & \cdots & C_{2p}' & V_2' \\ \cdot & & & & \\ \cdot & & & & \\ \cdot & & & & \\ C_{p1}' & C_{p2}' & \cdots & C_{pp}' & V_p' \end{Bmatrix}. \qquad (4.8.2)$$

This matrix is then used to solve for the elements A_k of the vector A. A modification of the procedure is used to determine the elements of the "inverse coefficient matrix" C^{-1}.

To describe the reduction, two definitions are introduced. The "diagonal elements" are those elements whose row and column indices are the same (i.e., $C_{11}, C_{22}, \cdots, C_{pp}$ and $C_{11}', C_{22}' \cdots C_{pp}'$). The "inner product" of a row and column is the sum of the p products of corresponding elements, the elements of a row being ordered from left to right, and the elements of a column from top to bottom.

The order for determining the elements of the matrix M' is:

1. The p elements of the first column.
2. The p remaining elements of the $p + 1$ elements in the first row.
3. The remaining $p - 1$ elements in the second column.
4. The remaining $p - 1$ elements in the second row.
5. The remaining $p - 2$ elements in the third column.
6. The procedure is continued until all the elements of the matrix have been determined.

This procedure is shown schematically in Figure 4.5.

Equations summarizing the rules for determining the elements of M' follow the statements of the rules:

1. Each element on or below the principal diagonal in M' is obtained by subtracting from the corresponding element in M the inner product of its own column and its own row in the square subarray C with all uncomputed elements imagined to be zeros.

2. Each element to the right of the principal diagonal in M' is computed by the same procedure, followed by a division by the diagonal element in its row of M'.

After completion of the matrix M' the elements A_k may be determined. The order for determining these elements is from bottom to top (i.e., A_p first, then A_{p-1}, A_{p-2}, $\cdots$, A_1). The element A_p is equal to V_p' of the matrix M'. Each succeeding element above it, is obtained by subtracting from the corresponding element of the V' column, the inner product of its row in C' and the A column (for which the uncomputed elements are assumed to be zero).

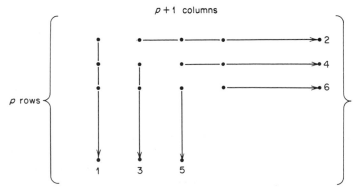

Figure 4.5. Order of calculation for elements of the auxiliary matrix M'.

A summary of the preceding instructions gives:

$$C_{ij}' = C_{ij} - \sum_{k=1}^{j-1} C_{ik}' \cdot C_{kj}', \qquad i \geq j, \qquad (4.8.3)$$

$$C_{ij}' = \frac{1}{C_{ii}'} \left[C_{ij} - \sum_{k=1}^{i-1} C_{ik}' \cdot C_{kj}' \right], \qquad i < j, \qquad (4.8.4)$$

$$V_i' = \frac{1}{C_{ii}'} \left[V_i - \sum_{k=1}^{i-1} C_{ik}' \cdot V_k' \right], \qquad (4.8.5)$$

and

$$A_i = V_i' - \sum_{k=i+1}^{p} C_{ik}' \cdot A_k. \qquad (4.8.6)$$

For prediction and least squares analysis, the coefficient matrix is "symmetrical" (i.e., $C_{ij} = C_{ji}$), and therefore advantage may be taken of an important simplification. For such cases, each element above the principal diagonal is equal to the value of the symmetrically placed element C_{ji}' below the diagonal divided by the diagonal element C_{ii}':

$$C_{ij}' = \frac{C_{ji}'}{C_{ii}'}, \qquad i < j, \quad C \text{ symmetrical.} \qquad (4.8.7)$$

For such cases, Equation 4.8.7 should replace Equation 4.8.4 in the above calculational scheme.

The recommended procedure for inverting the matrix C is a modification of the Crout reduction. If the V vector is replaced with a vector of the form,

$$V = \begin{Bmatrix} 1 \\ 0 \\ \cdot \\ \cdot \\ \cdot \\ 0 \end{Bmatrix}, \qquad (4.8.8)$$

and if the M' matrix and the A vector are determined, then the resulting elements of the A vector will have interesting properties:

$$A_1 C_{11} + A_2 C_{12} + \cdots + A_p C_{1p} = 1,$$
$$A_1 C_{21} + A_2 C_{22} + \cdots + A_p C_{2p} = 0, \qquad (4.8.9)$$
$$\vdots \qquad\qquad \vdots$$
$$A_1 C_{p1} + A_2 C_{p2} + \cdots + A_p C_{pp} = 0.$$

Comparing Equations 4.8.9 and 3.6.33, the elements of the vector A are seen to be equal to the elements of the first row of the matrix C^{-1}:

$$C_{1k}^{-1} = A_k, \qquad k = 1, 2, \cdots, p. \qquad (4.8.10)$$

Similarly, if the V vector is replaced by a vector of the form,

$$V = \left\{ \begin{matrix} 0 \\ 1 \\ 0 \\ \cdot \\ \cdot \\ \cdot \\ 0 \end{matrix} \right\}, \qquad (4.8.11)$$

and if the preceding procedure is repeated, the elements of the vector will be:

$$C_{2k}^{-1} = A_k, \qquad k = 1, 2, \cdots, p. \qquad (4.8.12)$$

Thus, it is seen that the elements of the C^{-1} matrix can be determined by repeating the procedure p times. That is, the vector V is replaced by a unity matrix I (see Equation 3.6.32). The vector A is determined for each column of the I matrix. The jth solution corresponds to the jth row of the matrix C^{-1}:

$$C_{jk}^{-1} = A_k, \qquad k = 1, 2, \cdots, p. \qquad (4.8.13)$$

The Crout reduction also yields the determinant of the "coefficient matrix" C. The determinant is the product of the diagonal elements of the auxiliary matrix M':

$$|C| \equiv \text{determinant of } C = \prod_{k=1}^{p} M_{kk}'. \qquad (4.8.14)$$

The determinant of the C matrix is often a useful quantity. If the value of the determinant is low, the elements of the inverse matrix will usually be high. For least squares analyses of nonlinear functions, a low value for the determinant is often accompanied by convergence problems (see Section 3.8). A low value of the determinant can be caused by a row or column of the C matrix in which all the elements are close to zero. If, for example, one of the derivatives F_{a_k} is nearly zero for all n data points, then from Equation 3.6.21 or 4.4.1 it can be seen that all the elements in the kth row and the kth column of the C matrix will be close to zero. Another cause for a low value of the determinant is a high degree of correlation between two rows or columns of the matrix. This condition will result if two of the derivatives (i.e., F_{a_j} and F_{a_k}) are highly correlated; that is, if the ratio $F_{a_j}{}^i/F_{a_k}{}^i$ is approximately the same for all n data points. It is sometimes worthwhile to compute the value of the determinant of the C matrix in the prediction analysis phase of an experiment. If the fitting function (i.e., Equation 3.3.2) is nonlinear, the value of the determinant is often indicative of whether or not convergence will be a problem during the data analysis phase.

Hildebrand's discussion of the Crout reduction includes methods of checking the calculation for errors. These checks are particularly useful for desk calculations. For computer solutions, however, such checks are unnecessary. What is necessary is a thorough "debugging procedure" for the computer program. To check a matrix inversion routine, the most straightforward method is to invert a simple 3×3 matrix. The resulting matrix and the original matrix should then be tested using Equation 3.6.33. A 3×3 matrix is suggested because it is small enough to permit verification of the results using a slide rule or desk calculator, and yet large enough so that any indexing errors in the program will lead to erroneous results.

As an example of the Crout reduction and matrix inversion, consider the following set of equations:

$$5.0A_1 + 3.0A_2 + 1.0A_3 = 15.0,$$

$$3.0A_1 + 2.0A_2 + 3.0A_3 = 19.0, \qquad (4.8.15)$$

$$1.0A_1 + 3.0A_2 + 1.0A_3 = 11.0.$$

It should be noted that the *coefficient matrix* is symmetrical:

$$C = \begin{Bmatrix} 5.0 & 3.0 & 1.0 \\ 3.0 & 2.0 & 3.0 \\ 1.0 & 3.0 & 1.0 \end{Bmatrix}. \qquad (4.8.16)$$

The *augmented matrix* M is:

$$M = \begin{Bmatrix} 5.0 & 3.0 & 1.0 & 15.0 \\ 3.0 & 2.0 & 3.0 & 19.0 \\ 1.0 & 3.0 & 1.0 & 11.0 \end{Bmatrix}. \qquad (4.8.17)$$

Since the matrix C is symmetrical, Equations 4.8.3, 4.8.5, and 4.8.7 are used to determine the elements of the matrix M':

$$C_{11}' = C_{11} = 5.0, \tag{4.8.18}$$

$$C_{21}' = C_{21} = 3.0, \tag{4.8.19}$$

$$C_{31}' = C_{31} = 1.0, \tag{4.8.20}$$

$$C_{12}' = C_{21}'/C_{11}' = 0.6. \tag{4.8.21}$$

$$C_{13}' = C_{31}'/C_{11}' = 0.2, \tag{4.8.22}$$

$$V_1' = V_1/C_{11}' = 3.0, \tag{4.8.23}$$

$$C_{22}' = C_{22} - C_{21}' \cdot C_{12}' = 0.2, \tag{4.8.24}$$

$$C_{32}' = C_{32} - C_{31}' \cdot C_{12}' = 2.4, \tag{4.8.25}$$

$$C_{23}' = C_{32}'/C_{22}' = 12.0, \tag{4.8.26}$$

$$V_2' = (V_2 - C_{21}' \cdot V_1')/C_{22}' = 50.0, \tag{4.8.27}$$

$$C_{33}' = C_{33} - C_{31}' \cdot C_{13}' - C_{32}' \cdot C_{23}' = -28.0, \tag{4.8.28}$$

$$V_3' = (V_3 - C_{31}' \cdot V_1' - C_{32}' \cdot V_2')/C_{33}' = 4.0. \tag{4.8.29}$$

Substituting Equations 4.8.18 through 4.8.29 into Equation 4.8.2 gives:

$$M' = \begin{Bmatrix} 5.0 & 0.6 & 0.2 & 3.0 \\ 3.0 & 0.2 & 12.0 & 50.0 \\ 1.0 & 2.4 & -28.0 & 4.0 \end{Bmatrix}. \tag{4.8.30}$$

The elements of the vector A are determined using Equation 4.8.6:

$$A_3 = V_3' = 4.0, \tag{4.8.31}$$

$$A_2 = V_2' - C_{23}' \cdot A_3 = 2.0, \tag{4.8.32}$$

$$A_1 = V_1' - C_{12}' \cdot A_2 - C_{13}' \cdot A_3 = 1.0. \tag{4.8.33}$$

Substituting these values into Equation 4.8.15, it is seen that 4.0, 2.0, and 1.0 are indeed the solutions of the set of linear equations.

Substituting the vector $\{1, 0, 0\}$ for the vector V, and then redetermining the M' matrix gives:

$$M' = \begin{Bmatrix} 5.0 & 0.6 & 0.2 & 0.2 \\ 3.0 & 0.2 & 12.0 & -3.0 \\ 1.0 & 2.4 & -28.0 & -0.25 \end{Bmatrix}. \tag{4.8.34}$$

It is worthwhile noting that the only difference between Equations 4.8.30 and 4.8.34 is in the V' column. The next step is to determine the solution for the vector (or first row of the inverse matrix according to Equation 4.8.10):

$$C_{13}^{-1} = A_3 = -\tfrac{1}{4}, \tag{4.8.35}$$

$$C_{12}^{-1} = A_2 = 0, \tag{4.8.36}$$

$$C_{11}^{-1} = A_1 = \tfrac{1}{4}. \tag{4.8.37}$$

These values can be checked using Equation 3.6.33:

$$C_{11}^{-1} \cdot C_{11} + C_{12}^{-1} \cdot C_{21} + C_{13}^{-1} \cdot C_{31} = 1.25 + 0.0 - 0.25 = 1.00, \tag{4.8.38}$$

$$C_{11}^{-1} \cdot C_{12} + C_{12}^{-1} \cdot C_{22} + C_{13}^{-1} \cdot C_{32} = 0.75 + 0.0 - 0.75 = 0.0, \tag{4.8.39}$$

$$C_{11}^{-1} \cdot C_{13} + C_{12}^{-1} \cdot C_{23} + C_{13}^{-1} \cdot C_{33} = 0.25 + 0.0 - 0.25 = 0.0. \tag{4.8.40}$$

Equations 4.8.38, 4.8.39, and 4.8.40 indicate that the values of $\tfrac{1}{4}$, 0 and $-\tfrac{1}{4}$ do indeed constitute the first row of the inverse coefficient matrix. The procedure can be continued to determine the second and third rows of the inverse matrix. The resulting matrix is:

$$C^{-1} = \begin{Bmatrix} \tfrac{1}{4} & 0 & -\tfrac{1}{4} \\ 0 & -\tfrac{1}{7} & \tfrac{12}{28} \\ -\tfrac{1}{4} & \tfrac{12}{28} & -\tfrac{1}{28} \end{Bmatrix}. \tag{4.8.41}$$

It is interesting to note that the inverse matrix C^{-1} possesses the same symmetry as the given matrix C.

The determinant $|C|$ of the given coefficient matrix can be calculated using Equation 4.8.14:

$$|C| = 5.0 \times 0.2 \times (-28.0) = -28. \tag{4.8.42}$$

An examination of Equation 4.8.16, shows that -28.0 is indeed the determinant of the matrix C.

4.9 Treatment of Systematic Errors

The methods of least squares and prediction analysis are based upon the assumption that the errors in the measured values of the dependent and independent variables are statistical in nature. Stated more precisely, if the measurements Y_i and X_{ji} are repeated N times and the average values are computed, it is assumed that the average values approach the "true values" η_i and ξ_{ji} as N approaches infinity. It is also assumed that the average values of the parameters a_k approach the true values α_k if the experiment is repeated many times.

If there is a presence of systematic errors the average values will converge to values other than the true values. Systematic errors can be classified into two types:

1. Errors which cause the measured values of a variable to be consistently different from the true values.
2. Errors which cause the least square values a_k to be consistently different from the true values α_k even though the average values of the measured variables converge to the true values.

Examples of both types of systematic errors can be described. As an example of the first type of error, consider the measurement of the variable temperature for a particular experiment. A measuring instrument is used to measure the temperature for each data point to $\pm 1°C$. One would expect that the measured values of temperature are accurate to $\pm 1°C$. If, however, the measuring instrument is improperly calibrated so that all measurements are $5°C$ too high, a systematic error of the first type exists. As an example of the second type of error, consider an experiment in which the dependent and independent variables are related by the following equation:

$$y = a_1 \exp\left(-\lambda_1 x\right) + a_2. \tag{4.9.1}$$

Prediction analysis of this class of experiments is considered in Section 6.6. The unknown parameters are a_1 and a_2. The decay constant λ_1 is assumed to be known and is treated as an input quantity to the analysis. If the value of λ_1 is in error, even though the average values of y_i and x_i approach the true value η_i and ξ_i, the least squares values of a_1 and a_2 will not approach α_1 and α_2. For this experiment, a systematic error of the second type exists.

The terminology developed in Section 3.10 denotes s_{a_k} as the unbiased estimate of the uncertainty σ_{a_k}. The uncertainty σ_{a_k} is defined by Equation 3.10.3:

$$\sigma_{a_k} \equiv \left[(a_k - \alpha_k)^2{}_m\right]^{1/2}, \qquad k = 1, 2, \cdots, p. \tag{3.10.3}$$

The subscript m refers to the *mean value* of the square of the difference $a_k - \alpha_k$ and is the average value that would be obtained if the experiment were repeated many times. The unbiased estimate s_{a_k} is computed using Equation 3.10.43:

$$s_{a_k}{}^2 = \frac{S}{n-p} C_{kk}{}^{-1}, \tag{3.10.43}$$

where S is the weighted sum of the squares of the residuals, n is the number of data points, p is the number of unknown parameters, and $C_{kk}{}^{-1}$ is the kth diagonal term of the inverse coefficient matrix. A crucial assumption in the derivation of this equation is that the errors E_i and E_j (defined by Equation 3.10.14) are uncorrelated. That is, if E_i is the error for the ith data point and E_j is the error for the jth data point, the average value of the product $E_i E_j$ (i.e., the value obtained by repeating the experiment many times and then averaging the product) is zero. This assumption was used to reduce Equation 3.10.16 to Equation 3.10.17 and is *valid only if there are no systematic errors.*

If systematic errors are present, s_{a_k} is not an accurate estimator of σ_{a_k}. Denoting $\epsilon_{a_k}{}^\nu$ as the uncertainty in a_k caused by the νth source of systematic error, then the unbiased estimate of σ_{a_k} can be computed by the following equation:

$$\text{Unbiased estimate of } \sigma_{a_k}{}^2 \equiv s_{a_k}{}^2 + \sum_{\nu=1}^{q} (\epsilon_{a_k}{}^\nu)^2, \qquad (4.9.2)$$

where q is the total number of sources of systematic error. The form of Equation 4.9.2 is based upon the statistical law of propagation of errors (see Section 2.6).[14]

To determine the unbiased estimate of σ_{a_k}, the procedure is as follows:

1. Compute the value of $s_{a_k}{}^2$ as a part of the least squares analysis of the data.
2. List all possible sources of systematic error and estimate their magnitudes.
3. Compute the partial derivatives $\partial a_k / \partial(\text{source } \nu)$. (Methods for performing these computations are discussed below.)
4. Compute the values of $\epsilon_{a_k}{}^\nu$:

$$\epsilon_{a_k}{}^\nu \approx \frac{\partial a_k}{\partial (\text{source } \nu)} \cdot \text{estimated magnitude of source } \nu. \qquad (4.9.3)$$

5. Compute the unbiased estimate of $\sigma_{a_k}{}^2$ using Equation 4.9.2. The unbiased estimate of σ_{a_k} is simply the square root of the unbiased estimate of $\sigma_{a_k}{}^2$.

The two difficult steps in this procedure are Steps 2 and 3. Step 2 requires a complete understanding of the experiment. There are two basic approaches to performing Step 3—direct and analytical.

1. The *direct approach*—the derivatives can be estimated by simply altering the data or input parameters according to the proposed systematic error and then repeating the least squares analysis. The

---

[14] Equation 4.9.2 assumes that the different types of systematic errors are uncorrelated. This is generally a reasonable assumption.

derivatives can then be related to the changes in the parameters a_k:

$$\frac{\partial a_k}{\partial \,(\text{source } \nu)} \approx \frac{\Delta a_k}{\Delta \,(\text{source } \nu)},\tag{4.9.4}$$

where $\Delta \,(\text{source } \nu)$ refers to a change in the data or input parameters of the source ν type.

2. The *analytical approach*—the derivatives can be evaluated using analytical techniques. This approach, if possible, is the simpler approach. Often, however, analytical expression for the derivatives cannot be obtained and then the direct approach must be applied to the problem.

Treatment of systematic errors can also be included as a part of the prediction analysis of a proposed experiment. Equation 4.9.2 is replaced by Equation 4.9.5:

$$\text{Predicted value of } \sigma_{a_k}{}^2 = \hat{\sigma}_{a_k}{}^2 + \sum_{\nu=1}^{q} (\hat{\epsilon}_{a_k}{}^\nu)^2,\tag{4.9.5}$$

where $\hat{\sigma}_{a_k}$ is the predicted statistical uncertainty and $\hat{\epsilon}_{a_k}{}^\nu$ is the predicted systematic error caused by the νth error source. The procedure for predicting $\sigma_{a_k}{}^2$ is similar to the procedure followed for estimating $\sigma_{a_k}{}^2$:

1. Compute the value of $\hat{\sigma}_{a_k}{}^2$ using one of the approaches discussed in Section 4.4.
2. List all possible sources of systematic error and predict their magnitudes in the proposed experiment.
3. Compute the partial derivatives $\partial a_k / \partial(\text{source } \nu)$ using either the direct or analytical approach.
4. Compute the values of $\hat{\epsilon}_{a_k}{}^\nu$:

$$\hat{\epsilon}_{a_k}{}^\nu = \frac{\partial a_k}{\partial(\text{source } \nu)} \cdot (\text{predicted magnitude of source } \nu).\tag{4.9.6}$$

5. Compute the predicted value of $\sigma_{a_k}{}^2$ using Equation 4.9.5.

As an example of the influence of systematic errors upon the predicted values of σ_{a_k}, the "straight-line experiment" is considered. In Section 4.5, prediction analyses for three cases of this class of experiments are developed. For all these cases, only "statistical errors" are considered. In Section 5.2, the results of Section 4.5 are generalized, but the analyses are also limited to treatments of only statistical errors.

If the possibility of systematic errors exists, the analysis must be amended. For the "straight-line experiment,"[15] the dependent and inde-

---

[15] Referred to as the "first-order polynomial" in Section 5.2.

pendent variables are related by the following equation:

$$y = a_1 + a_2 x. \tag{4.9.7}$$

The predicted statistical errors for the ith data point are denoted as σ_{x_i} and σ_{y_i}. In Sections 4.5 and 5.2, equations are derived for computing $\hat{\sigma}_{a_1}$ and $\hat{\sigma}_{a_2}$. These quantities are the predicted values of the uncertainties σ_{a_1} and σ_{a_2} that arise as a result of the statistical errors σ_{x_i} and σ_{y_i}. Two possible sources of systematic errors are errors in all the values of y and all the values of x. These errors are denoted as ϵ_y and ϵ_x. They are both systematic errors of the first type and can be related to the measured values of y_i and x_i:

$$y_i = \eta_i + \epsilon_y + statistical\ error, \tag{4.9.8}$$

and

$$x_i = \xi_i + \epsilon_x + statistical\ error, \tag{4.9.9}$$

where η_i and ξ_i are the true values of y_i and x_i, respectively. If the measurements are repeated N times and the average values of y_i and x_i are determined, as N approaches infinity,

$$(y_i)_{av} \rightarrow \eta_i + \epsilon_y, \tag{4.9.10}$$

$$(x_i)_{av} \rightarrow \xi_i + \epsilon_x, \tag{4.9.11}$$

where the subscript av refers to the average value of the quantity within the parentheses.

The derivatives $\partial a_k / \partial (\text{source } \nu)$ can be determined analytically for this simple problem. The systematic error in y is referred to as the first source of error, and the systematic error in x is referred to as the second source of error. Four derivatives must be evaluated:

$$\partial a_1 / \partial y, \quad \partial a_2 / \partial y, \quad \partial a_1 / \partial x, \quad \partial a_2 / \partial x.$$

If the statistical errors σ_{x_i} are negligible, the problem is linear and Equations 3.9.5, 3.9.6, 3.6.30, and 3.9.7 can be used. Applying these equations to the "straight-line experiment":

$$C_{kl} = \sum_{i=1}^{n} \frac{\phi_k(x_i)\phi_l(x_i)}{\sigma_{y_i}{}^2}, \tag{4.9.12}$$

$$V_k = \sum_{i=1}^{n} \frac{\phi_k(x_i)y_i}{\sigma_{y_i}{}^2}, \tag{4.9.13}$$

$$a_1 = C_{11}{}^{-1}V_1 + C_{12}{}^{-1}V_2, \tag{4.9.14}$$

$$a_2 = C_{12}{}^{-1}V_1 + C_{22}{}^{-1}V_2. \tag{4.9.15}$$

124

The ϕ_k's for this problem are:

$$\phi_1(x_i) = 1, \tag{4.9.16}$$

$$\phi_2(x_i) = x_i. \tag{4.9.17}$$

Equations 4.9.16 and 4.9.17 were derived by comparing Equations 4.9.7 and 4.9.18:

$$y_i = \sum_{k=1}^{2} a_k \phi_k(x_i). \tag{4.9.18}$$

Substituting Equations 4.9.16 and 4.9.17 into Equations 4.9.12 and 4.9.13 gives

$$C_{11} = \sum_{i=1}^{n} \frac{1}{\sigma_{y_i}^2}, \tag{4.9.19}$$

$$C_{12} = \sum_{i=1}^{n} \frac{x_i}{\sigma_{y_i}^2}, \tag{4.9.20}$$

$$C_{22} = \sum_{i=1}^{n} \frac{x_i^2}{\sigma_{y_i}^2}, \tag{4.9.21}$$

$$V_1 = \sum_{i=1}^{n} \frac{y_i}{\sigma_{y_i}^2}, \tag{4.9.22}$$

$$V_2 = \sum_{i=1}^{n} \frac{x_i y_i}{\sigma_{y_i}^2}. \tag{4.9.23}$$

If a systematic error of magnitude ϵ_y is predicted for all the values y_i, from Equations 4.9.19 through 4.9.23 it is seen that only V_1 and V_2 are affected:

$$V_1' = \sum_{i=1}^{n} \frac{y_i + \epsilon_y}{\sigma_{y_i}^2} = V_1 + \epsilon_y \sum_{i=1}^{n} \frac{1}{\sigma_{y_i}^2}, \tag{4.9.24}$$

$$V_2' = \sum_{i=1}^{n} \frac{x_i(y_i + \epsilon_y)}{\sigma_{y_i}^2} = V_2 + \epsilon_y \sum_{i=1}^{n} \frac{x_i}{\sigma_{y_i}^2}. \tag{4.9.25}$$

The quantities V_1' and V_2' are the values of V_1 and V_2 that would be obtained if the y_i's were increased by an amount equal to ϵ_y. Comparing Equations 4.9.24 and 4.9.25 with Equations 4.9.19 and 4.9.20 yields:

$$V_1' = V_1 + \epsilon_y \cdot C_{11}, \tag{4.9.26}$$

$$V_2' = V_2 + \epsilon_y \cdot C_{12}. \tag{4.9.27}$$

If we substitute Equations 4.9.26 and 4.9.27 into Equations 4.9.14 and 4.9.15:

$$a_1' = C_{11}^{-1} \cdot (V_1 + \epsilon_y \cdot C_{11}) + C_{12}^{-1} \cdot (V_2 + \epsilon_y \cdot C_{12}), \quad (4.9.28)$$

$$a_2' = C_{12}^{-1} \cdot (V_1 + \epsilon_y \cdot C_{11}) + C_{22}^{-1} \cdot (V_2 + \epsilon_y \cdot C_{12}). \quad (4.9.29)$$

Defining Δa_k as $a_k' - a_k$, subtracting Equations 4.9.14 and 4.9.15 from Equations 4.9.28 and 4.9.29, and then using Equation 3.6.33 gives:

$$\Delta a_1 = \epsilon_y \cdot (C_{11}^{-1} \cdot C_{11} + C_{12}^{-1} \cdot C_{12}) = \epsilon_y, \quad (4.9.30)$$

$$\Delta a_2 = \epsilon_y \cdot (C_{12}^{-1} \cdot C_{11} + C_{22}^{-1} \cdot C_{12}) = 0. \quad (4.9.31)$$

Equations 4.9.30 and 4.9.31 state that if all the values of y_i are increased by an amount equal to ϵ_y and the data are re-analyzed, the resulting value of a_1 will be increased by ϵ_y, but the value of a_2 will remain unchanged. This result is logical and could have been stated without resorting to analysis. The purpose of the analysis was to illustrate the method rather than prove this obvious conclusion.

If a systematic error of magnitude ϵ_x is predicted for all the values x_i, from Equations 4.9.19 through 4.9.23 it is seen that C_{12}, C_{22}, and V_2 are affected. Following the same procedure as used for the derivation of Equations 4.9.26 and 4.9.27, we obtain:

$$C_{12}' = C_{12} + \epsilon_x \cdot C_{11}, \quad (4.9.32)$$

$$C_{22}' = C_{22} + 2\epsilon_x \cdot C_{12} + \epsilon_x^2 \cdot C_{11}, \quad (4.9.33)$$

$$V_2' = V_2 + \epsilon_x \cdot V_1. \quad (4.9.34)$$

The terms of the inverse coefficient matrix must be redetermined:

$$(C_{11}^{-1})' = C_{22}'/D', \quad (4.9.35)$$

$$(C_{12}^{-1})' = -C_{12}'/D', \quad (4.9.36)$$

$$(C_{22}^{-1})' = C_{11}/D', \quad (4.9.37)$$

where

$$D' \equiv C_{22}' \cdot C_{11} - (C_{12}')^2. \quad (4.9.38)$$

Substituting Equations 4.9.32 and 4.9.33 into Equation 4.9.38 gives:

$$D' = C_{11} \cdot (C_{22} + 2\epsilon_x \cdot C_{12} + \epsilon_x^2 \cdot C_{11}) - C_{12}^2 - 2\epsilon_x \cdot C_{11}C_{12} - \epsilon_x^2 \cdot C_{11}^2$$

$$= C_{11} \cdot C_{22} - C_{12}^2 \equiv D \quad (4.9.39)$$

Substituting Equations 4.9.32, 4.9.33, and 4.9.39 into Equation 4.9.35

through 4.9.37 gives:

$$(C_{11}^{-1})' = C_{11}^{-1} - 2\epsilon_x \cdot C_{12}^{-1} + \epsilon_x^2 \cdot C_{22}^{-1}, \tag{4.9.40}$$

$$(C_{12}^{-1})' = C_{12}^{-1} - \epsilon_x \cdot C_{22}^{-1}, \tag{4.9.41}$$

$$(C_{22}^{-1})' = C_{22}^{-1}. \tag{4.9.42}$$

Substituting the appropriate equations into Equations 4.9.14 and 4.9.15 and dropping all second-order terms, we obtain:

$$a_1' \approx (C_{11}^{-1} - 2\epsilon_x \cdot C_{12}^{-1}) \cdot V_1 + (C_{12}^{-1} - \epsilon_x \cdot C_{22}^{-1}) \cdot V_2 + C_{12}^{-1} \cdot \epsilon_x V_1, \tag{4.9.43}$$

$$a_2' = (C_{12}^{-1} - \epsilon_x \cdot C_{22}^{-1}) \cdot V_1 + C_{22}^{-1} \cdot (V_2 + \epsilon_x \cdot V_1). \tag{4.9.44}$$

Subtracting Equations 4.9.14 and 4.9.15 from Equations 4.9.43 and 4.9.44 gives:

$$\Delta a_1 \approx -\epsilon_x \cdot (C_{12}^{-1} \cdot V_1 + C_{22}^{-1} \cdot V_2), \tag{4.9.45}$$

$$\Delta a_2 = 0. \tag{4.9.46}$$

Comparing Equations 4.9.45 and 4.9.15 gives

$$\Delta a_1 \approx -\epsilon_x \cdot a_2. \tag{4.9.47}$$

The results are again obvious but illustrate the method.

If we substitute Equations 4.9.30, 4.9.31, 4.9.46, and 4.9.47 into Equation 4.9.5:

$$\text{Predicted value of } \sigma_{a_1}^2 = \hat{\sigma}_{a_1}^2 + \epsilon_y^2 + (\epsilon_x \cdot a_2)^2, \tag{4.9.48}$$

$$\text{Predicted value of } \sigma_{a_2}^2 = \hat{\sigma}_{a_2}^2. \tag{4.9.49}$$

For the simple problem considered above, the analytical approach led to obvious conclusions. For most problems, the analytical approach does not lead to such simple solutions. It should be realized that for many types of systematic errors, the analysis is more complicated. For example, in the preceding problem, if the systematic error in y was a function of the value of y, then the analysis would be more complicated. In addition, the complexity of the analysis increases as the complexity of the mathematical model increases. If analytical solutions for the values of $\hat{\epsilon}_{a_k}^{\nu}$ cannot be obtained, the direct approach will provide the required solutions. Recourse to the direct approach usually requires access to a digital computer.

4.10 Applying Prediction Analysis to the Optimization of an Experiment

It is difficult to visualize the need for optimizing an experiment if one is accustomed to performing simple inexpensive experiments of a routine nature. For such cases, a method of planning based on prediction analysis of the proposed experiment is often of little practical value. The need for a systematic optimization procedure becomes apparent if at least one of the stated objectives of the experiment is difficult to satisfy. For such cases, a method of planning based upon prediction analysis of the proposed experiment can often mean the difference between success and failure.

As an example, consider an experiment to be performed by a group of nuclear physicists which will utilize a very large accelerator. The problem of scheduling experiments for the accelerator is so acute that the group is allotted only one hour of accelerator time in a two-month period. For such a case, it is clear that the physicists should develop a plan whereby the one hour of allotted time will be utilized in an optimum manner.

The formalism for optimizing an experiment utilizes the standard techniques common to all optimization procedures.[16] A mathematical model of the problem is developed. The model must include an equation for the quantity to be minimized or maximized plus equations for all the constraints or objectives. Once the mathematical model has been developed, several methods of solution are possible:

1. For the simplest problems, an analytical solution might be derived. For such problems, the derivatives of the quantity to be minimized or maximized are set equal to zero. A solution for the set of experimental variables is then determined by solving the complete set of equations.

2. For problems not amenable to analytical solution, the quantity to be minimized or maximized may be computed for all combinations of the experimental variables satisfying the constraints or objectives of the problem. The optimum set of the experimental variables is then selected by examining the results of the computations.

3. For those cases in which the total number of combinations of the experimental variables satisfying the constraints or objectives is prohibitively large, modern techniques such as linear or nonlinear programming can be applied to the problem.

It is not within the scope of this book to treat exhaustively the problem of optimizing the design of experiments. However, to appreciate how prediction analysis can be applied to this problem, it is worthwhile to consider several examples.

Example 1—A "straight-line experiment" is to be performed for which the σ_{x_i}'s and the σ_{y_i}'s will be constant for all measurements. Case I of

---

[16] The approach considered by G. E. P. Box and H. L. Lucus is to optimize the design of the experiment by minimizing the determinant of the inverse coefficient matrix C^{-1} ("Design of Experiments in Non-Linear situations," *Biometrika*, Vol. 46, June 1959). Using prediction analysis and performing the necessary computations with the aid of a digital computer, the optimization may be based on more specific objectives.

Section 4.5 is therefore applicable. The purpose of the experiment is to measure a_2 to at least 1% accuracy. The possible range of values of a_2 is 5.0 to 10.0, and therefore σ_{a_2} would have to be at most 0.05. Assume that the values of x must fall within the range 4.0 to 20.0 and are to be evenly spaced. Also, assume that the uncertainties for the measured values of x (i.e., the σ_{x_i}'s) are negligible.

For this example, the purpose of the optimization procedure is to minimize the total cost required to perform the experiment satisfactorily. Assumptions related to the cost of the experiment must therefore be made. Assume that each measurement of y costs \$100, and if y is to be measured to an uncertainty of $\sigma_y = K_{cy} = 1.0$, then the investment in equipment is \$4000. Furthermore, assume that investment cost is inversely proportional to K_{cy}. That is, if K_{cy} is to be reduced to 0.5, then the investment in equipment would have to be increased to \$8000. An equation for the total cost C of the experiment can be written:

$$C = C_1 \cdot n + C_2 \cdot \frac{1.0}{K_{cy}}, \tag{4.10.1}$$

where

$$C_1 = \$100, \quad \text{and} \quad C_2 = \$4000.$$

The appropriate equation for $\hat{\sigma}_{a_2}$ is Equation 4.5.24. Taking the square root of this equation and setting it equal to the maximum permissible value of σ_{a_2} yields:

$$\hat{\sigma}_{a_2} = \left(\frac{12 \cdot L}{n}\right) \cdot \frac{1}{x_b - x_a} = 0.05. \tag{4.10.2}$$

From Equation 4.5.8,

$$L = K_{cy}^2, \tag{4.10.3}$$

because K_{cx} is assumed to be negligible. Substituting Equation 4.10.3 into Equation 4.10.2, and rearranging terms gives:

$$K_{cy} = \frac{\hat{\sigma}_{a_2} \cdot (x_b - x_a) \cdot n^{1/2}}{(12)^{1/2}}. \tag{4.10.4}$$

If Equation 4.10.4 is substituted into Equation 4.10.1:

$$C = C_1 \cdot n + \frac{C_2 \cdot (12)^{1/2}}{\hat{\sigma}_{a_2} \cdot (x_b - x_a) \cdot n^{1/2}}. \tag{4.10.5}$$

From Equations 4.4.3, 4.4.7, and 4.4.8,

$$x_b - x_a = (x_n - x_1) \cdot \left(\frac{n}{n-1}\right). \tag{4.10.6}$$

Equation 4.10.6 could be substituted into Equation 4.10.5, but the problem is simplified if it is assumed that $x_b - x_a \approx x_n - x_1 = 16$. Thus, if it is assumed that $x_b - x_a$ is not a function of n, a simple solution for the optimum value of n may be obtained:

$$\frac{\partial C}{\partial n} = C_1 - \frac{C_2 \cdot (12)^{1/2}}{2\hat{\sigma}_{a_2} \cdot (x_b - x_a)} \cdot n_{\text{opt}}^{-3/2} = 0, \qquad (4.10.7)$$

$$n_{\text{opt}}^{3/2} = \frac{1}{2} \frac{C_2 \cdot (12)^{1/2}}{C_1 \cdot \hat{\sigma}_{a_2} \cdot (x_b - x_a)}. \qquad (4.10.8)$$

If the values for C_1, C_2, $\hat{\sigma}_{a_2}$, and $x_b - x_a$ are substituted into the equation:

$$n_{\text{opt}}^{3/2} = \frac{1}{2} \cdot \frac{4000 \cdot (12)^{1/2}}{100 \cdot 0.05 \cdot 16} = 86.5, \qquad (4.10.9)$$

$$n_{\text{opt}} = 19.5. \qquad (4.10.10)$$

The optimum number of points is between 19 and 20, and from Equation 4.10.4, the optimum value of K_{cy} is 1.02. Purchase of the equipment costing $4000 would therefore be recommended. Using this equipment and 20 data points, the accuracy requirement for a_2 should be satisfied at a total cost of $6000, which is very close to the minimum possible for this experiment.

If the $8000 equipment were purchased, K_{cy} would be reduced to 0.5 and, from Equation 4.10.4, the number of points required for the experiment would be reduced to 5. The total cost of the experiment would therefore be $8500, which is much higher than the cost of the experiment if the cheaper equipment is selected.

For the sake of simplicity, no mention of the reuse of the equipment has been made. If the equipment can be reused in other experiments, then it is invalid to charge the entire purchase price to the experiment. If reuse is to be considered in the analysis, a criterion for charging the experiment for use of the equipment must be selected. To select such a criterion, a knowledge of the principles of economic analysis and accounting are required. It is beyond the scope of this book to discuss economics and accounting. *It is worthwhile, however, to emphasize the importance of financial considerations upon scientific and technological planning. It is practically impossible for the scientist or engineer to divorce himself completely from a consideration of the cost of his experiment.*

Example 2—For this example, the purpose of the optimization procedure is to minimize the total time required to perform the experiment. Assume that the conditions stated for Example 1 are applicable. Additional assumptions must be made concerning the time requirements of the experiment. Assume that the time required to measure y is inversely proportional to

σ_y^2 (i.e., K_{cy}^2), and the set-up time for each point is constant. An equation for the total time T of the experiment can be written:

$$T = T_1 \cdot \frac{n}{K_{cy}^2} + T_2 \cdot n, \qquad (4.10.11)$$

where the first term on the right side of the equation is the total time spent for actual measurements and the second term is the total setup time. Substituting Equation 4.10.4 into Equation 4.10.11 gives:

$$T = T_1 \cdot \frac{12}{\hat{\sigma}_{a_2}^2 \cdot (x_b - x_a)^2} + T_2 \cdot n. \qquad (4.10.12)$$

If it is assumed that $x_b - x_a$ is independent of n, then the first term on the right side of Equation 4.10.12 is independent of n. The total time T therefore approaches a minimum as n approaches zero. In other words, the time required for actual measurements is insensitive to n, but the setup time can be reduced by reducing the number of data points.

It is not necessary to assume that $x_b - x_a$ is independent of n. Substituting Equation 4.10.6 into Equation 4.10.12 gives:

$$T = T_1 \cdot \frac{12 \cdot (n-1)^2}{\hat{\sigma}_{a_2}^2 \cdot (x_n - x_1) \cdot n^2} + T_2 \cdot n. \qquad (4.10.13)$$

Both terms on the right side of Equation 4.10.13 increase with increasing n, and therefore the conclusion that T approaches a minimum as n approaches zero is still valid.

The conclusion for this example is that, to complete the experiment in the shortest possible time, one should measure the values y_i to the best possible accuracy and therefore use the fewest possible data points. One should keep in mind, however, that as n becomes small, the predicted confidence interval on a_2 is no longer proportional to $\hat{\sigma}_{a_2}$ but on the product $\hat{\sigma}_{a_2} \cdot t_{(\beta/2, n-p)}$ (see Equation 4.3.5 and Example 4, Section 4.5).

Examination of Equation 4.10.13 shows that if $T_2 \ll T_1$, then T is not strongly dependent upon n. For such cases it is recommended that n be chosen on the basis of another cirterion. For example, the experiment could be planned on the basis of minimizing cost.

Example 3—For this example, the purpose of the optimization procedure is to minimize the total cost required to perform the experiment satisfactorily. This example is the same as Example 1 except that the σ_{x_i}'s are not assumed to be negligible. They are assumed to be constant for all the data points (i.e., $\sigma_{x_i} = K_{cx}$). For K_{cx} equal to 0.1, the investment required for equipment is assumed to be $2000, and is inversely proportional to K_{cx}^2. In other words, if K_{cx} is to be reduced to 0.05, then the investment would have to be increased to $8000.

The equation for the total cost C is similar to Equation 4.10.1, but an extra term is required for the investment in equipment to measure the values of x_i:

$$C = C_1 \cdot n + C_2 \cdot \frac{1.0}{K_{cy}} + C_3 \cdot \left(\frac{0.1}{K_{cx}}\right)^2, \qquad (4.10.14)$$

where

$$C_1 = \$100, \quad C_2 = \$4000, \quad \text{and} \quad C_3 = \$2000.$$

Equation 4.10.2 is still valid; however, Equation 4.5.8 must be used in place of Equation 4.10.3. Solving for $K_{cx}{}^2$, we obtain:

$$K_{cx}{}^2 = \frac{\sigma_{a_2}{}^2 \cdot n \cdot (x_b - x_a)^2}{12 \cdot a_2{}^2} - \frac{K_{cy}{}^2}{a_2{}^2}. \qquad (4.10.15)$$

Equation 4.10.15 could be substituted into Equation 4.10.14 and then equations for the partial derivatives $\partial C / \partial n$ and $\partial C / \partial K_{cy}$ could be obtained. Optimum values of n and K_{cy} could then be obtained by setting the partial derivatives equal to zero and solving the two equations. The required value for K_{cx} could then be obtained by substituting the values of n and K_{cy} into Equation 4.10.15.

Although this procedure for determining the optimum values of n, K_{cx}, and K_{cy} is theoretically valid, the equations are cumbersome. For this problem the simplest procedure is to locate the optimum operating conditions by calculating C for several combinations of n, K_{cx}, and K_{cy}. For each combination of K_{cx} and K_{cy}, the most conservative estimate of n is made by assuming that a_2 is at its maximum value (i.e., $a_2 = 10.0$). From Equations 4.5.8 and 4.5.24:

$$n = \frac{12}{\hat{\sigma}_{a_2}{}^2 \cdot (x_b - x_a)^2} \cdot (K_{cy}{}^2 + 100 K_{cx}{}^2). \qquad (4.10.16)$$

If we substitute 0.05 for $\hat{\sigma}_{a_2}$ and assume that $x_b - x_a \approx 16.0$, then:

$$n \approx 18.8 \cdot (K_{cy}{}^2 + 100 \cdot K_{cx}{}^2). \qquad (4.10.17)$$

Using Equations 4.10.17 and 4.10.14, C can be determined for any combination of values of K_{cy} and K_{cx}. Results are tabulated in Table 4.9.

From Table 4.9 it can be seen that the optimum operating conditions are approximately $K_{cy} = 1.0$, $K_{cx} = 0.1$, and $n = 38$. Equipment should therefore be selected with this optimum in mind.

In each of the examples considered in this section the values of x_1 and x_n were assumed to be fixed. In addition, the uncertainty of the resulting value of a_1 was not considered in the analyses. It can be seen that if these quantities are considered, the problem of optimizing an experiment can

TABLE 4.9. Total cost[a] of the proposed experiment as a function of K_{cy} and K_{cx}

K_{cy}	K_{cx}	n	C
0.8	0.08	24	$10 520
	0.09	28	10 270
	0.10	31	10 100
	0.11	35	10 150
	0.12	40	10 390
1.0	0.08	31	$10 220
	0.09	34	9 870
	0.10	38	9 800
	0.11	42	9 850
	0.12	46	9 990
1.2	0.08	40	$10 450
	0.09	43	10 100
	0.10	46	9 930
	0.11	50	9 980
	0.12	55	10 220

[a] The values of n were determined by increasing the value of n computed using Equation 4.10.17 to the next integer. The values of C were determined using Equation 4.10.14 and the constants $C_1 = 100, $C_2 = 4000, and $C_3 = 2000.

become extremely complex, even for the simple straight-line experiments. For very complicated experiments for which it is essential to locate the optimum operating conditions, such modern techniques as linear or non-linear programing must be used.

4.11 Summary

In this chapter the formalism of prediction analysis has been developed. The notation $\hat{\sigma}_{a_k}$ is used to denote the "predicted value" of σ_{a_k}. Under the assumption that "systematic errors" are negligible, the values of $\hat{\sigma}_{a_k}$ are shown to be equal to the "mean values" of the "unbiased estimates" of σ_{a_k} if the experiment is repeated many times.

Assuming a "direct approach" for solution, the procedure for computing the values of $\hat{\sigma}_{a_k}$ can be summarized:

1. Select the function f (i.e. Equation 3.3.1 or 3.3.2) which will be used to relate the dependent and independent variables.
2. List the function F (Equation 3.6.1) and all its derivatives (Equation 3.6.3).
3. Assume values for all of the experimental variables and parameters including n (the number of data points); x_{ji}, $\sigma_{x_{ji}}$, $i = 1, 2, \cdots, n$ and

$j = 1, 2, \cdots, m$ (all independent variables); $a_k, k = 1, 2, \cdots, p$ (all un-known parameters). The values of $\sigma_{x_{ji}}$ are usually determined by assuming a relationship with the values of x_{ji} (see Equations 4.5.2, 4.5.3, and 4.5.4 for examples). The values of x_{ji} can often be specified by a simple equation. For example, if the values are evenly spaced, and $m = 1$, Equation 4.7.1 can be used.

4. Compute the values of y_i by assuming that they are equal to the values of the function f as determined using the assumed values of x_{ji} and a_k.

5. Compute the values of σ_{y_i} using the values of y_i and an assumed model which relates these quantities (see Equations 4.5.2, 4.5.3, and 4.5.4 for examples).

6. Using the expressions for the derivatives of F with respect to the dependent and independent variables (i.e., $F_y{}^i$ and $F_{x_j}{}^i$ of Equation 3.6.3), and the assumed and computed values of the variables and parameters, compute the numerical values of $\hat{F}_y{}^i$ and $\hat{F}_{x_j}{}^i$.

7. Using Equation 3.6.15, compute the values of $\hat{L}_i$.

8. Using the expressions for the derivatives of F with respect to the unknown parameters (i.e., $F_{a_k}{}^i$ of Equation 3.6.3) and the assumed and computed values of the variables and parameters, compute the numerical values of $\hat{F}_{a_k}{}^i$.

9. Using Equations 3.6.21 and 3.6.22, compute all the elements $\hat{C}_{kl}$ of the "predicted coefficient matrix" $\hat{C}$.

10. Invert the C matrix. A matrix inversion procedure is discussed in Section 4.8.

11. Using Equations 4.3.3 and 4.3.4, compute the values of $\hat{\sigma}_{a_k}$ and $\hat{\sigma}_{lk}$.

12. If "confidence intervals" are to be specified for the proposed experiment, the intervals are determined using Equation 4.3.5 for the unknown parameters and Equation 4.3.6 for the function f. For comparison of the predicted accuracy of the results on the basis of confidence intervals, the products $\hat{\sigma}_{a_k} \cdot t_{(\beta/2, n-p)}$ are determined. Example 4 of Section 4.5 shows that for experiments in which the values of $n - p$ are small, this product is a more realistic measure of accuracy than just $\hat{\sigma}_{a_k}$.

13. Repeat the procedure from Step 3 for as many combinations of variables and parameters as is required to understand the functional dependence of the values of $\hat{\sigma}_{a_k}$ and to plan the proposed experiment.

If an "analytical" solution is sought, the procedure outlined above is altered. Analytical expressions are used to obtain estimates of the elements $\hat{C}_{kl}{}^{-1}$. In this manner, the calculations required in Steps 4 through 10 are avoided. Advantages and disadvantages of the "analytical approach" are discussed in Section 4.6.

Design of prediction and least squares analysis computer programs are discussed in Section 4.7. Treatment of systematic error in least squares, as well as prediction analyses, is considered in Section 4.9.

Application of prediction analysis to the problem of optimizing the design of experiments is discussed in Section 4.10. The method for locating the optimum operating conditions is dependent upon the complexity of the experiment and the stated objectives.

Problems

1. The dependent and independent variables for a proposed experiment are to be related by Equation 4.5.1. The minimum value of x is 2.0, and the values of x are to be evenly spaced. The spacing between points is 0.2. The uncertainties for the values of y are each equal to 1.0 and the uncertainties for the values of x are each equal to 0.1. The possible range of values for a_1 is 50. to 100., and the possible range of values for a_2 is 5.0 to 10. How many data points are required so that a_1 is determined to no more than ± 0.5 and a_2 is determined to no more than ± 0.1?

2. The experiment proposed in Problem 1 is modified so that each value of y is determined to an accuracy of 1%. How many data points are required to satisfy the experimental objectives?

3. Derive equations for $\hat{\sigma}_{12}$ for Cases I, II, and III of Section 4.5. (Note—These equations are derived in Section 5.2.)

4. Derive equations for $\hat{\sigma}_{a1}$, $\hat{\sigma}_{a2}$, and $\hat{\sigma}_{12}$ for the following class of experiments:

$$y = a_1 \exp\,(-a_2 x)$$

$$\hat{\sigma}_{y_i} = K_{fy} \cdot y_i,$$

$$\hat{\sigma}_{x_i} = K_{cx}.$$

The values of x_i are to be evenly spaced. (Note—The required equations are derived in Case II of Section 6.3.)

5. A "straight-line experiment" is to be performed for which the values of σ_{y_i} will be proportional to the values of y_i. The uncertainties σ_{x_i} will be negligible. The purpose of the experiment is to measure a_2 to at least 1% accuracy. The possible range of values of a_1 is 50. to 100. and the possible range of values of a_2 is 5.0 to 10. Assume that the values of x_i must fall within the range 4.0 to 20.0 and are to be evenly spaced.

It is estimated that each measurement of y will cost \$100. To measure y to 1% accuracy, equipment costing \$4000 must be acquired. The accuracy can be improved by an additional investment in funds. Assume that the total cost of the equipment is inversely proportional to the percent accuracy and the equipment cannot be reused once the experiment has been completed. To minimize the total cost of the experiment, how many data points should be included in the experiment, and how much should be invested in equipment?

Computer Projects

1. For a proposed experiment, the dependent and independent variables are to be related by a straight line. The expected values of a_1 and a_2 are 20. and 2., respectively. The values of σ_{y_i} are proportional to y_i and the values of σ_{x_i} are negligible. Using Equations 4.5.56 and 4.5.57, compute the asymptotic values of the following dimensionless groups:

$$\Theta_1 \equiv \frac{\hat{\sigma}_{a_1} \cdot n^{1/2}}{a_1 \cdot K_{fy}},$$

$$\Theta_2 \equiv \frac{\hat{\sigma}_{a_2} \cdot n^{1/2}}{a_2 \cdot K_{fy}}.$$

The values of x are to be evenly spaced within the region $x_a = 0.0$ to $x_b = 10.0$. Using the direct approach, compute the values of Θ_1 and Θ_2 as functions of n. (Note—The minimum value of n is equal to the number of unknown parameters plus one which for straight-line experiments is three.) How many points are required so that the asymptotic solutions for the Θ's are within 1% and 0.1% of the direct solutions?

2. Using the direct approach, perform a prediction analysis for the following class of experiments:

$$y = a_1 \exp\left(-a_2 x\right),$$

$$\hat{\sigma}_{y_i} = K_{fy} \cdot y_i,$$

$$\sigma_{x_i} = K_{cx}.$$

The values of x_i are to be evenly spaced. Analytical solutions for this problem are included in Case II of section 6.3. Compare the results from the direct approach with the results included in Section 6.3.

Part II   Applications

The second part of this book is limited to applications of prediction analysis. Experiments can be classified according to the function used for "fitting" the data. The functions considered in Part II include:

Chapter 5—The Polynomial
Chapter 6—The Exponential
Chapter 7—The Sine Series
Chapter 8—The Gaussian

These functions have been selected because they represent some of the most important classes of experiments. They are useful in many diverse areas of science and technology.

An additional function is considered in Chapter 9. Analysis of this function, although of only limited applicability, is included because it illustrates the potential usefulness of prediction analysis for analyzing complicated experiments. In addition, the analysis includes treatment of a systematic error which affects the experiment in a complicated manner. (In Chapters 5 through 8, it is assumed that systematic errors are negligible.)

Numerical results are included in each chapter; however, no attempt is made to include results for all possible combinations of the experimental variables. An attempt has been made to include results for the most interesting combinations of the experimental variables, and to include enough data to yield an over-all picture for each function. Examples of how the results might be used to plan an experiment are included in the summaries of the following chapters.

For each function analyzed, equations for $\hat{F}_{a_k}{}^i$ and $\hat{L}_i$ are included. Using these equations in the framework of a *prediction analysis computer program*, the reader should be able to extend the results to other combinations of the experimental variables if the need arises. Wherever feasible, analytical solutions for the $\hat{\sigma}_{a_k}$'s and $\hat{\sigma}_{jk}$'s are derived. If the derivation is extremely complicated or impossible, only numerical results obtained using the *direct approach* (described in Section 4.4) are included.

In Chapters 5 through 8 the emphasis is on experiments in which the values of the independent variable are evenly spaced. It should be realized, however, that the method of prediction analysis is not limited to just this special case. The main reasons for making this assumption are: (1) most experiments are performed in this manner, and (2) the complexity of an analytical analysis of the experiment is reduced. In Chapter 9 an experiment is analyzed for which the values of the independent variables are not evenly spaced.

Chapter 5

The Polynomial

5.1 Introduction

The polynomial is an important function which is useful for analysis of many different types of experiments. The general form of the polynomial function is[1]:

$$y = \sum_{k=1}^{p} a_k x^{k-1}. \tag{5.1.1}$$

The index $p - 1$ is often called the *order* of the polynomial. For example, a *first-order* polynomial is characterized by a value of p equal to 2:

$$y = a_1 + a_2 x, \tag{5.1.2}$$

and a *second-order* polynomial is characterized by a value of p equal to 3:

$$y = a_1 + a_2 x + a_3 x^2. \tag{5.1.3}$$

For many experiments, use of the polynomial function is dictated by the physical laws governing the experiment. For example, use of a first-order polynomial (or straight line) is required in Example 4 of Section 4.5 because the behavior of a spring under variable load conditions necessitates this choice.

For other experiments, however, the physical laws governing the experiment are extremely complicated or unknown. For such experiments all that is required is a curve to fit the experimental data. The resulting curve might then be used for such purposes as interpolation, integration, differentiation, etc. Choice of the polynomial function is often made for this type of experiment because the polynomial is a *linear function*. As explained in Section 3.9, least squares analysis of linear functions can be

---

[1] No attempt is made in this chapter to consider orthogonal polynomials; however, a brief discussion of orthogonality follows Equation 7.5.4.

made without resorting to iterative solutions if the uncertainties associated with the independent variable are negligible. In addition, such operations as interpolation, integration, and differentiation are relatively simple if the function is a polynomial.

Prediction analysis of the polynomial function is dependent upon the order of the polynomial, the magnitudes of the uncertainties for the x and y variables, and the other experimental variables and parameters. Analysis of three cases of the first-order polynomial were included in Section 4.5. Extension of these cases are included in Section 5.2. The second-order polynomial is considered in Section 5.3 and higher-order polynomials are discussed in Section 5.4.

It is assumed in this chapter that systematic errors are negligible. If systematic errors are considered as an important factor in a proposed experiment, the prediction analysis should be amended using the techniques discussed in Section 4.9.

5.2 The First-Order Polynomial

The first-order polynomial is the function which characterizes an important class of experiments referred to as the "straight-line experiment" in Section 4.5. Equation 5.1.2 is used to express the relationship between the dependent and independent variables for all experiments within this class:

$$y = a_1 + a_2x. \tag{5.1.2}$$

The cases within this class can be characterized by the choice of the functions used to determine the uncertainties σ_{x_i} and σ_{y_i}:

$$\sigma_{x_i} = g(x_i), \tag{1.3.2}$$

$$\sigma_{y_i} = h(y_i). \tag{1.3.3}$$

The most important choices for the functions g and h are listed as Equations 4.5.2, 4.5.3, and 4.5.4:

Constant uncertainty:
$$\sigma_{x_i} = K_{cx},$$
$$\sigma_{y_i} = K_{cy}. \tag{4.5.2}$$

Constant fractional uncertainty:
$$\sigma_{x_i} = K_{fx} \cdot | x_i |,$$
$$\sigma_{y_i} = K_{fy} \cdot | y_i |. \tag{4.5.3}$$

Counting statistics:
$$\sigma_{x_i} = K_{sx} \cdot (x_i)^{1/2},$$
$$\sigma_{y_i} = K_{sy} \cdot (y_i)^{1/2}. \tag{4.5.4}$$

Three cases were discussed in Section 4.5. In this section the analyses

TABLE 5.1. Summary of cases analyzed for the *first-order polynomial*

Case	σ_{y_i}	σ_{x_i}	Equations		
			$\hat{\sigma}_{a_1}$	$\hat{\sigma}_{a_2}$	$\hat{\sigma}_{12}$
I	K_{cy}	K_{cx}	5.2.12	5.2.13	5.2.14
II	$K_{fy} \cdot \lvert y_i \rvert$	K_{cx}	5.2.52	5.2.53	5.2.54
III	$K_{sy} \cdot (y_i)^{1/2}$	K_{cx}	5.2.89	5.2.90	5.2.91

included in Section 4.5 are extended and the results are presented in a more general form. The cases considered in this section are summarized in Table 5.1.

Case I.–*Constant uncertainty for both the x and y variables.*

For this case, the values of σ_{y_i} and σ_{x_i} are:

$$\sigma_{y_i} = K_{cy},$$
$$\sigma_{x_i} = K_{cx}. \tag{4.5.5}$$

This case is the same as Case I of Section 4.5. Equations 4.5.8 through 4.5.10 are the appropriate expressions for $\hat{L}_i$, $\hat{F}_{a_1}{}^i$, and $\hat{F}_{a_2}{}^i$:

$$\hat{L}_i = K_{cy}^2 + (a_2 \cdot K_{cx})^2 \equiv L, \tag{4.5.8}$$

$$\hat{F}_{a_1}{}^i = -1, \tag{4.5.9}$$

$$\hat{F}_{a_2}{}^i = -x_i. \tag{4.5.10}$$

Substituting Equations 4.5.8 through 4.5.10 into 4.4.1 and proceeding in the manner outlined in Section 4.4, it is shown in Section 4.5 that the correct expressions for the variances $\hat{\sigma}_{a_1}^2$ and $\hat{\sigma}_{a_2}^2$ are:

$$\hat{\sigma}_{a_1}^2 = \frac{L}{n} \cdot \frac{(x^2)_{av}}{(x^2)_{av} - (x_{av})^2}, \tag{4.5.16}$$

$$\hat{\sigma}_{a_2}^2 = \frac{L}{n} \cdot \frac{1}{(x^2)_{av} - (x_{av})^2}. \tag{4.5.17}$$

The analysis of Case I of Section 4.5 can be extended to include the covariance $\hat{\sigma}_{12}$. From Equations 3.7.18 and 4.5.11 through 4.5.13:

$$\hat{\sigma}_{12} = \hat{C}_{12}^{-1} = \frac{L}{n} \cdot \frac{-x_{av}}{(x^2)_{av} - (x_{av})^2}. \tag{5.2.1}$$

If the values of x are evenly spaced (i.e., Equation 4.4.3 is valid), then equations for the values of x_{av} and $(x^2)_{av}$ can be derived. In Section 4.5 it is shown that:

$$x_{av} = \tfrac{1}{2}(x_b + x_a),$$
(4.5.21)

and as n approaches infinity:

$$(x^2)_{av} \to \frac{1}{3} \frac{x_b^3 - x_a^3}{x_b - x_a}.$$
(4.5.22)

The limits x_a and x_b are defined by Equations 4.4.7 and 4.4.8.

A more useful method of treating the problem is to derive the equations in terms of dimensionless groups. Defining r_x and r_a as:

$$r_x \equiv \frac{x_b + x_a}{2 \cdot (x_b - x_a)} = \frac{x_{av}}{n \cdot \Delta x},$$
(5.2.2)

$$r_a \equiv \frac{a_2 \cdot (x_b - x_a)}{a_1} = \frac{a_2 \cdot n \cdot \Delta x}{a_1},$$
(5.2.3)

and substituting Equations 5.2.2 and 5.2.3 into Equations 4.5.23 and 4.5.24, as n approaches infinity, it is seen that:

$$\hat{\sigma}_{a_1}^2 \to \frac{4L}{n} \cdot (3r_x^2 + \tfrac{1}{4}),$$
(5.2.4)

and

$$\hat{\sigma}_{a_2}^2 \to \frac{12L}{n} \cdot \left(\frac{a_2}{a_1 r_a}\right)^2.$$
(5.2.5)

It should be noticed that if $a_1 = 0$, $r_a \to \infty$ and therefore Equation 5.2.5 is indeterminate. An alternative form of this equation can be derived by substituting Equation 5.2.3 into Equation 5.2.5:

$$\hat{\sigma}_{a_2}^2 \to \frac{12L}{n^3 \Delta x^2}.$$
(5.2.6)

Substituting Equations 4.5.21 and 4.5.22 into Equation 5.2.1 and then substituting 5.2.2 and 5.2.3 into the resulting equation, an expression for $\hat{\sigma}_{12}$ can be derived;

$$\hat{\sigma}_{12} \to -\frac{12L}{n} \cdot \frac{a_2 r_x}{a_1 r_a},$$
(5.2.7)

or, alternatively,

$$\hat{\sigma}_{12} \to \frac{12L r_x}{n^2 \Delta x}.$$
(5.2.8)

The most useful manner for presenting the results is in terms of dimensionless groups. Defining Φ_1, Φ_2, and Φ_{12},

$$\Phi_1 \equiv \hat{\sigma}_{a_1} \cdot \left(\frac{n}{L}\right)^{1/2}, \tag{5.2.9}$$

$$\Phi_2 \equiv \hat{\sigma}_{a_2} \cdot \left(\frac{n}{L}\right)^{1/2} \cdot \frac{a_1}{a_2}, \tag{5.2.10}$$

$$\Phi_{12} \equiv \hat{\sigma}_{12} \cdot \frac{n}{L} \cdot \frac{a_1}{a_2}, \tag{5.2.11}$$

from Equations 5.2.4, 5.2.5, and 5.2.7 it can be seen that as n approaches infinity,

$$\Phi_1 \rightarrow (12r_x^2 + 1)^{1/2}, \tag{5.2.12}$$

$$\Phi_2 \rightarrow \frac{(12)^{1/2}}{|r_a|}, \tag{5.2.13}$$

$$\Phi_{12} \rightarrow - \frac{12r_x}{r_a}. \tag{5.2.14}$$

The Φ's are plotted in Figures 5.1 and 5.2 as functions of r_x and r_a. Equations 5.2.12 through 5.2.14 are only the asymptotic solutions for the Φ's (i.e., the solutions for infinite n). Figures 5.1 and 5.2 are therefore exact only for n equal to infinity. The Φ's can be computed for finite values of n by using the *direct approach* discussed in Section 4.4. The minimum permissible value of n is equal to the number of unknown parameters plus one. For a first-order polynomial the number of unknown parameters is two (i.e., a_1 and a_2) and therefore the minimum value of n for a least squares or prediction analysis of the first-order polynomial is three. Ratios of the Φ's computed for $n = 3$ and $n \rightarrow \infty$ are included in Table 5.2 for several combinations of r_a and r_x.

From Table 5.2 it can be seen that the Φ's are not strongly dependent upon n. As n is increased from 3, the Φ's rapidly approach the asymptotic solutions. At $n = 3$ and $r_x = 1.5$, Φ_{12} is about 12% above the asymptotic solution, but at $n = 8$, it is only 1% above the asymptotic solution. It should be noticed that the discrepancy between finite and infinite values of n for Φ_1 and Φ_{12} is dependent upon r_x. For Φ_2, however, the discrepancy is independent of r_x. It can be concluded from Table 5.2 that Equations 5.2.12 through 5.2.14 and Figures 5.1 and 5.2 are good approximations for the Φ's even for small values of $n - 2$.

Example 1—Assume that the values of y can be measured to an uncertainty of 1.0 (i.e., $K_{cy} = 1.0$), the values of x can be measured to an

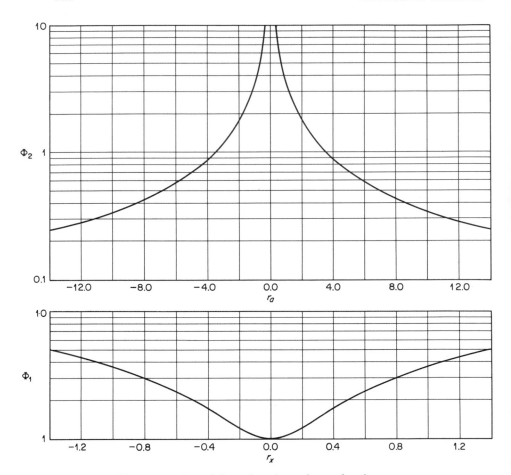

Figure 5.1. Φ_1 and Φ_2 as functions of r_x and r_a for $n \to \infty$.

TABLE 5.2. Ratios of the Φ's computed at $n = 3$ and $n \to \infty$ for several combinations of r_x and r_a

r_x	r_a	$\dfrac{(\Phi_1)_{n=3}}{(\Phi_1)_{n\to\infty}}$	$\dfrac{(\Phi_2)_{n=3}}{(\Phi_2)_{n\to\infty}}$	$\dfrac{(\Phi_{12})_{n=3}}{(\Phi_{12})_{n\to\infty}}$
0.0	0.0	1.00	1.06	1.00
1.5	0.0	1.06	1.06	1.123
0.0	50.0	1.00	1.06	1.00
1.5	50.0	1.06	1.06	1.123

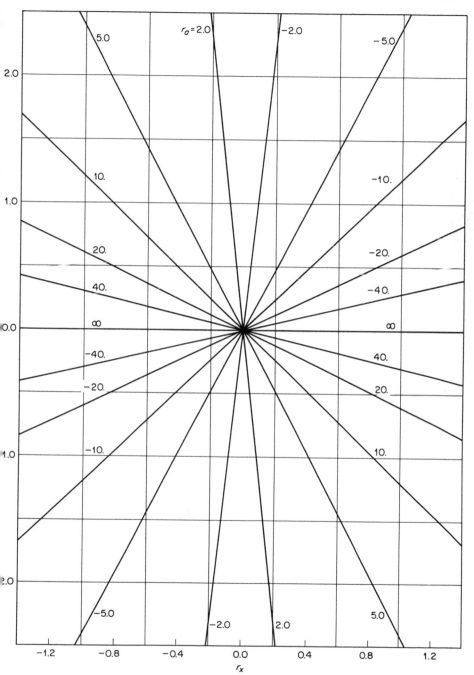

Figure 5.2. Φ_{12} as a function of r_x and r_a for $n \rightarrow \infty$.

uncertainty of 0.1 (i.e., $K_{cx} = 0.1$), and the maximum possible value of a_2 is 10. Also, assume that the values of x must fall within the range 4.0 to 20. If the purpose of the experiment is to measure a_1 to ± 0.5, how many data points are required if the values of x are to be evenly spaced within the given range?

The most conservative estimate is made if the maximum possible value of r_x is assumed. From Equation 5.2.2, the maximum value of r_x corresponds to the minimum value of $n \cdot \Delta x$ which is $20 - 4 = 16$. Since $x_{av} = (20 + 4)/2 = 12$, $r_x = 12/16 = 0.75$, and from Equation 5.2.12 (or Figure 5.1), $\Phi_1 = 2.78$. From Equation 4.5.8, $L = 1.0^2 + (10.0 \times 0.1)^2 = 2.0$, and therefore from Equation 5.2.9, $n = 2.0 \times 2.78^2/0.5^2 = 62$ data points. In other words, if the experiment is planned to include about 62 evenly spaced data points, the estimated value of σ_{a_1} determined from a least squares analysis of the experimental data should be about 0.5.

Example 2—Assume that the experiment discussed in Example 1 is performed using 62 data points which are evenly spaced within the given range. If the resulting values of a_1 and a_2 are 1.0 and 10.0, and if these values are used to extrapolate to a value of $x = 30$, what will be the predicted uncertainty for the value of y determined by extrapolation?

In Example 1, performing the experiment using 62 data points was shown to lead to a value of $\hat{\sigma}_{a_1} = 0.5$. From Equation 5.2.3, $r_a \approx 10.0 \times 16.0/1 = 160.$; and from Equations 5.2.13 and 5.2.14, $\Phi_2 \approx 0.0216$, and $\Phi_{12} \approx -12.0 \times 0.75/160. = -0.0562$. From Equations 5.2.10 and 5.2.11, $\hat{\sigma}_{a_2} = 0.0216 \times (10.0/1.0)/(62/2.0)^{1/2} = 0.0388$, and $\hat{\sigma}_{12} = -0.0562 \times (10.0/1.0)/(62/2.0) = -0.0181$.

To predict the uncertainty in the extrapolated value of y corresponding to $x = 30$; Equation 4.3.7 must be used:

$$\hat{\sigma}_f{}^2 = \hat{F}_{a_1} \cdot \hat{F}_{a_1} \cdot \hat{C}_{11}{}^{-1} + \hat{F}_{a_2} \cdot \hat{F}_{a_2} \cdot \hat{C}_{22}{}^{-1} + 2F_{a_1} \cdot F_{a_2} \cdot \hat{C}_{12}{}^{-1}, \quad (5.2.15)$$

where

$$y = f(x; a_1, a_2) = f(30; 1, 10) = 301, \quad (5.2.16)$$

$$\hat{F}_{a_1} = -1, \quad (5.2.17)$$

$$\hat{F}_{a_2} = -x = -30, \quad (5.2.18)$$

$$\hat{C}_{11}{}^{-1} = \hat{\sigma}_{a_1}{}^2 = 0.5^2, \quad (5.2.19)$$

$$\hat{C}_{22}{}^{-1} = \hat{\sigma}_{a_2}{}^2 = 0.0388^2, \quad (5.2.20)$$

and

$$\hat{C}_{12}{}^{-1} = \hat{\sigma}_{12} = -0.0181. \quad (5.2.21)$$

Substituting Equations 5.2.17 through 5.2.21 into Equation 5.2.15 and then taking the square root, we obtain:

$$\hat{\sigma}_f = (0.5^2 + 900 \times 0.0388^2 - 60 \times 0.0181)^{1/2} = 0.72. \quad (5.2.22)$$

That is, the extrapolated value of y corresponding to $x = 30$ should be about 301.0 ± 0.7. It is interesting to note that the uncertainty for the extrapolated value of y is less than the uncertainty corresponding to the individual measurements of y_i (i.e., $\sigma_{y_i} = 1.0$). This result is explained by the fact that the uncertainty for the extrapolated value of y is obtained by analysis of all the experimental data (i.e., 62 separate data points).

An interesting point is left to the reader to prove. Assume that the coordinate system is changed so that the range is no longer 4 to 20 but, instead, is x_1 to $x_1 + 16$. A change in the coordinate system will result in a new value of a_1. From Equations 5.2.4, 5.2.5, and 5.2.7, it can be shown that the value of $\hat{\sigma}_{a_2}$ will not change but $\hat{\sigma}_{a_1}$ and σ_{12} will change. In the example, $\hat{\sigma}_f$ was computed at $x = 30$. In the new coordinate system, the point $x = 30$ is equivalent to $x = x_1 + 26$. It is left to the reader to show that the value of $\hat{\sigma}_f$ computed at $x = x_1 + 26$ is the same as the value determined at $x = 30$ in the preceding example. That is, even if the x and y axes are changed, the value of σ_f will not change for Case I.

Case II. *Constant fractional uncertainty for the variable y and constant uncertainty for the variable x.*

For this case, the values of σ_{y_i} and σ_{x_i} are:

$$\sigma_{y_i} = K_{fy} \cdot | y_i |,$$

$$\sigma_{x_i} = K_{cx}. \tag{5.2.23}$$

The absolute value notation is necessary because there is no physical significance for negative values of σ_{y_i}. The difference between this case and Case II of Section 4.5 should be noted. For this case, the analysis is for the more general problem in which K_{cx} is not necessarily equal to zero.

The equation for $\hat{L}_i$ is determined by substituting Equation 5.2.23 into Equation 3.6.16:

$$\hat{L}_i = K_{fy}^2 y_i^2 + (\hat{F}_x^i)^2 K_{cx}^2. \tag{5.2.24}$$

The symbol $\hat{F}_x^i$ denotes the derivative of $\hat{F}^i$ with respect to x:

$$\hat{F}^i = y_i - a_1 - a_2 x_i, \tag{5.2.25}$$

and therefore

$$\hat{F}_x^i = -a_2. \tag{5.2.26}$$

Substituting Equation 5.2.26 into Equation 5.2.24 gives:

$$\hat{L}_i = K_{fy}^2 y_i^2 + a_2^2 K_{cx}^2. \tag{5.2.27}$$

The derivatives $\hat{F}_{a_1}^i$ and $\hat{F}_{a_2}^i$ are also required in the analysis and are expressed by Equations 4.5.9 and 4.5.10:

$$\hat{F}_{a_1}^i = -1, \tag{4.5.9}$$

$$\hat{F}_{a_2}^i = -x_i. \tag{4.5.10}$$

Substituting Equations 4.5.9, 4.5.10, and 5.2.27 into Equation 4.4.1 gives

$$\hat{C}_{11} = \sum_{i=1}^{n} \frac{1}{K_{fy}^2 y_i^2 + a_2^2 K_{cx}^2} = \frac{n}{K_{fy}^2} \cdot \left(\frac{1}{y^2 + \lambda^2}\right)_{av}, \qquad (5.2.28)$$

$$\hat{C}_{12} = \sum_{i=1}^{n} \frac{x_i}{K_{fy}^2 y_i^2 + a_2^2 K_{cx}^2} = \frac{n}{K_{fy}^2} \cdot \left(\frac{x}{y^2 + \lambda^2}\right)_{av}, \qquad (5.2.29)$$

$$\hat{C}_{22} = \sum_{i=1}^{n} \frac{x_i^2}{K_{fy}^2 y_i^2 + a_2^2 K_{cx}^2} = \frac{n}{K_{fy}^2} \cdot \left(\frac{x^2}{y^2 + \lambda^2}\right)_{av}, \qquad (5.2.30)$$

where

$$\lambda \equiv a_2 K_{cx}/K_{fy}, \qquad (5.2.31)$$

and the subscript av refers to the average values of the quantities within the parentheses.

The predicted variances $\hat{\sigma}_{a_1}^2$ and $\hat{\sigma}_{a_2}^2$ are determined from Equations 4.5.14 and 4.5.15:

$$\hat{\sigma}_{a_1}^2 = \hat{C}_{11}^{-1} = \frac{\hat{C}_{22}}{\hat{C}_{22} \cdot \hat{C}_{11} - \hat{C}_{12}^2}, \qquad (4.5.14)$$

$$\hat{\sigma}_{a_2}^2 = \hat{C}_{22}^{-1} = \frac{\hat{C}_{11}}{\hat{C}_{22} \cdot \hat{C}_{11} - \hat{C}_{12}^2}. \qquad (4.5.15)$$

Similarly, the predicted value of the covariance can be determined:

$$\hat{\sigma}_{12} = \hat{C}_{12}^{-1} = \frac{-\hat{C}_{12}}{\hat{C}_{22} \cdot \hat{C}_{11} - \hat{C}_{12}^2}. \qquad (5.2.32)$$

Substituting Equations 5.2.28, 5.2.29, and 5.2.30 into Equations 4.5.14, 4.5.15, and 5.2.32, we obtain:

$$\hat{\sigma}_{a_1}^2 = \frac{K_{fy}^2}{n} \cdot \left\{ \left(\frac{x^2}{y^2 + \lambda^2}\right)_{av} \Big/ \left[\left(\frac{x^2}{y^2 + \lambda^2}\right)_{av} \cdot \left(\frac{1}{y^2 + \lambda^2}\right)_{av} - \left(\frac{x}{y^2 + \lambda^2}\right)_{av}^2 \right] \right\},$$

$$(5.2.33)$$

$$\hat{\sigma}_{a_2}^2 = \frac{K_{fy}^2}{n} \cdot \left\{ \left(\frac{1}{y^2 + \lambda^2}\right)_{av} \Big/ \left[\left(\frac{x^2}{y^2 + \lambda^2}\right)_{av} \cdot \left(\frac{1}{y^2 + \lambda^2}\right)_{av} - \left(\frac{x}{y^2 + \lambda^2}\right)_{av}^2 \right] \right\},$$

$$(5.2.34)$$

$$\hat{\sigma}_{12} = \frac{K_{fy}^2}{n} \cdot \left\{ -\left(\frac{x}{y^2 + \lambda^2}\right)_{av} \Big/ \left[\left(\frac{x^2}{y^2 + \lambda^2}\right)_{av} \cdot \left(\frac{1}{y^2 + \lambda^2}\right)_{av} - \left(\frac{x}{y^2 + \lambda^2}\right)_{av}^2 \right] \right\}.$$

$$(5.2.35)$$

If it is assumed that the values of x are evenly spaced (i.e., Equation 4.4.3 is valid), integral expressions for the required average values can be derived. As n approaches infinity,

$$\left(\frac{1}{y^2 + \lambda^2}\right)_{av} \rightarrow \int_{x_a}^{x_b} \frac{dx}{y^2 + \lambda^2} \Big/ (x_b - x_a) = \int_{y_a}^{y_b} \frac{dy}{y^2 + \lambda^2} \Big/ (y_b - y_a,), \quad (5.2.36)$$

because dy equals $a_2 dx$, and $a_2(x_b - x_a)$ equals $y_b - y_a$. Integrating the numerator of Equation 5.2.36:

$$\left(\frac{1}{y^2 + \lambda^2}\right)_{av} \rightarrow \frac{\tan^{-1}(y_b/\lambda) - \tan^{-1}(y_a/\lambda)}{\lambda \cdot (y_b - y_a)} . \quad (5.2.37)$$

Similarly, as n approaches infinity,

$$\left(\frac{x}{y^2 + \lambda^2}\right)_{av} \rightarrow \int_{y_a}^{y_b} \frac{y - a_1}{y^2 + \lambda^2} \, dy \Big/ (y_b - y_a), \quad (5.2.38)$$

$$\left(\frac{x}{y^2 + \lambda^2}\right)_{av} \rightarrow \left[\tfrac{1}{2} \ln\left(\frac{\lambda^2 + y_b^2}{\lambda^2 + y_a^2}\right) - \frac{a_1}{\lambda}\left(\tan^{-1}\frac{y_b}{\lambda} - \tan^{-1}\frac{y_a}{\lambda}\right)\right] \Big/ a_2(y_b - y_a) ,$$

$$(5.2.39)$$

$$\left(\frac{x^2}{y^2 + \lambda^2}\right)_{av} \rightarrow \int_{y_a}^{y_b} \frac{(y - a_1)^2}{y^2 + \lambda^2} \, dy \Big/ (y_b - y_a) , \quad (5.2.40)$$

$$\left(\frac{x^2}{y^2 + \lambda^2}\right)_{av} \rightarrow \left[(y_b - y_a) + \left(\frac{a_1^2 - \lambda^2}{\lambda}\right)\right.$$

$$\times \left. \left\{\tan^{-1}\frac{y_b}{\lambda} - \tan^{-1}\frac{y_a}{\lambda}\right\} - a_1 \ln\left(\frac{\lambda^2 + y_b^2}{\lambda^2 + y_a^2}\right)\right] \Big/ a_2^2(y_b - y_a) . \quad (5.2.41)$$

It is interesting to note that Equations 5.2.37, 5.2.39, and 5.2.41 reduce to Equations 4.5.39, 4.5.42, and 4.5.44 as λ approaches zero *if the signs of y_a and y_b are the same*. Consider Figure 5.3. As $\lambda \rightarrow 0$, $\phi = (\pi/2) - \theta \rightarrow 0$,

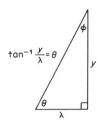

Figure 5.3. Triangle representing $\tan^{-1} y/\lambda$.

and therefore $\phi \rightarrow \tan \phi = \lambda/y$. The angle θ therefore approaches $(\pi/2) - (\lambda/y)$ as $\lambda \rightarrow 0$. Substituting this limit into Equation 5.2.37, as $\lambda \rightarrow 0$,

$$\frac{\tan^{-1}(y_b/\lambda) - \tan^{-1}(y_a/\lambda)}{\lambda(y_b - y_a)} \rightarrow \frac{\frac{1}{2}\pi - (\lambda/y_b) - \frac{1}{2}\pi + (\lambda/y_a)}{\lambda(y_b - y_a)} = \frac{1}{y_a y_b}.$$

$$(5.2.42)$$

Substituting Equation 5.2.42 into 5.2.37 yields Equation 4.5.39. In addition, as $\lambda \rightarrow 0$,

$$\frac{1}{2}\ln\left(\frac{\lambda^2 + y_b^2}{\lambda^2 + y_a^2}\right) \rightarrow \frac{1}{2}\ln\left(\frac{y_b^2}{y_a^2}\right) = \ln\left(\frac{y_b}{y_a}\right). \qquad (5.2.43)$$

Using Equations 5.2.42 and 5.2.43, it can be seen that Equations 5.2.39 and 5.2.41 reduce to Equations 4.5.42 and 4.5.44 as $\lambda \rightarrow 0$.

If the signs of y_a and y_b are different, Equation 5.2.37 reduces to Equation 5.2.44 as $\lambda \rightarrow 0$:

$$\frac{\tan^{-1}(y_b/\lambda) - \tan^{-1}(y_a/\lambda)}{\lambda(y_b - y_a)} \rightarrow \frac{\frac{1}{2}\pi - (\lambda/y_b) - [-\frac{1}{2}\pi + (\lambda/y_a)]}{\lambda(y_b - y_a)}$$

$$= \frac{(\pi/\lambda) - y_b^{-1} - y_a^{-1}}{y_b - y_a}. \qquad (5.2.44)$$

It is more convenient if the results are presented in terms of the independent rather than the dependent variable. The following equations may be substituted into 5.2.37, 5.2.39, and 5.2.41:

$$y_a = a_1 + a_2 x_a,$$
$$y_b = a_1 + a_2 x_b. \qquad (5.2.45)$$

If the dimensionless groups r_x, r_a, and ω_f are substituted into the resulting equations, then as $n \rightarrow \infty$,

$$\left(\frac{1}{y^2 + \lambda^2}\right)_{av} \rightarrow \frac{R_1}{a_1^2}, \qquad (5.2.46)$$

$$\left(\frac{x}{y^2 + \lambda^2}\right)_{av} \rightarrow \frac{(2r_a)^{-1}R_2 - R_1}{a_2 \cdot a_1}, \qquad (5.2.47)$$

$$\left(\frac{x^2}{y^2 + \lambda^2}\right) \rightarrow \frac{1 + (1 - \omega_f^2)R_1 - r_a^{-1}R_2}{a_2^2}, \qquad (5.2.48)$$

where

$$\omega_f \equiv \frac{\lambda}{a_1} = \frac{a_2 \cdot K_{cx}}{a_1 \cdot K_{fy}}, \tag{5.2.49}$$

$$R_1 \equiv \frac{1}{r_a \omega_f} \left[\tan^{-1} \left(\frac{1 + r_x r_a + \frac{1}{2} r_a}{\omega_f} \right) - \tan^{-1} \left(\frac{1 + r_x r_a - \frac{1}{2} r_a}{\omega_f} \right) \right], \tag{5.2.50}$$

and

$$R_2 \equiv \ln \left(\frac{\omega_f^2 + (1 + r_x r_a + \frac{1}{2} r_a)^2}{\omega_f^2 + (1 + r_x r_a - \frac{1}{2} r_a)^2} \right). \tag{5.2.51}$$

These equations are valid for all values of r_x, r_a, and ω_f.

Substituting Equations 5.2.46 through 5.2.48 into Equations 5.2.33 through 5.2.35, analytical expressions can be derived for $\hat{\sigma}_{a_1}{}^2$, $\hat{\sigma}_{a_2}{}^2$, and $\hat{\sigma}_{12}$. These equations can then be expressed in terms of the dimensionless groups Θ_1, Θ_2, and Θ_{12}. As n approaches infinity:

$$\Theta_1 \equiv \frac{\hat{\sigma}_{a_1} \cdot n^{1/2}}{a_1 \cdot K_{fy}} \rightarrow \left(\frac{1 - r_a^{-1} R_2 + R_1}{R_1 - \omega_f^2 R_1^2 - (4 r_a^2)^{-1} R_2^2} \right)^{1/2}, \tag{5.2.52}$$

$$\Theta_2 \equiv \frac{\hat{\sigma}_{a_2} \cdot n^{1/2}}{a_2 \cdot K_{fy}} \rightarrow \left(\frac{R_1}{R_1 - \omega_f^2 R_1^2 - (4 r_a^2)^{-1} R_2^2} \right)^{1/2}, \tag{5.2.53}$$

and

$$\Theta_{12} \equiv \frac{\hat{\sigma}_{12} \cdot n}{a_1 \cdot a_2 \cdot K_{fy}^2} \rightarrow \frac{-(2 r_a)^{-1} R_2 + R_1}{R_1 - \omega_f^2 R_1^2 - (4 r_a^2)^{-1} R_2^2}. \tag{5.2.54}$$

Outside the region $-(1/r_a) - \frac{1}{2} < r_x < -(1/r_a) + \frac{1}{2}$, as $\omega_f \rightarrow 0$, it can be seen from Equations 5.2.50 and 5.2.51 that:

$$R_1 \rightarrow [(1 + r_x r_a)^2 - r_a^2/4]^{-1} \equiv R_{10}, \tag{5.2.55}$$

and

$$R_2 \rightarrow 2 \cdot \ln \left(\frac{1 + r_x r_a + r_a/2}{1 + r_x r_a - r_a/2} \right) \equiv R_{20}. \tag{5.2.56}$$

If Equations 5.2.55 and 5.2.56 are substituted into Equations 5.2.52 through 5.2.54, as $\omega_f \rightarrow 0$,

$$\Theta_1 \rightarrow \left(\frac{1 - r_a^{-1} R_{20} + R_{10}}{R_{10} - (4 r_a^2)^{-1} R_{20}^2} \right)^{1/2}, \tag{5.2.57}$$

$$\Theta_2 \rightarrow \left(\frac{R_{10}}{R_{10} - (4 r_a^2)^{-1} R_{20}^2} \right)^{1/2}, \tag{5.2.58}$$

and

$$\Theta_{12} \to \frac{-(2r_a)^{-1}R_{20} + R_{10}}{R_{10} - (4r_a^2)^{-1}R_{20}^2} .$$
(5.2.59)

Substituting Equations 5.2.45, 5.2.2, 5.2.3, 5.2.55 and 5.2.56 into Equations 4.5.56 and 4.5.57 and then into Equations 5.2.52 and 5.2.53, it can be shown that the results for Case II of Section 4.5 can be expressed in dimensionless form by Equations 5.2.57 and 5.2.58. It should be remembered that these results are valid only outside the range $-(1/r_a) - \frac{1}{2} < r_x < -(1/r_a) + \frac{1}{2}$. Within this range the signs of y_a and y_b are different so as $\omega_f \to 0$, $R_1 \to \infty$ (see Equation 5.2.44), and therefore all the Θ's approach one.

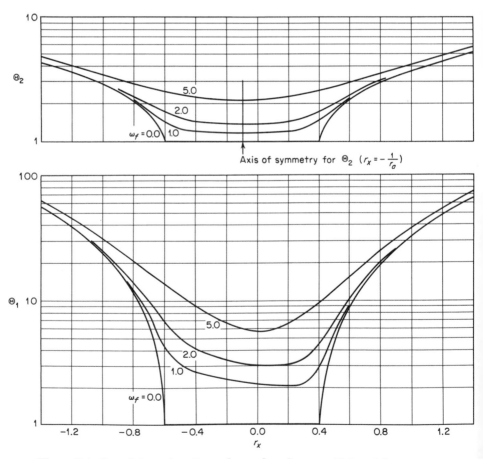

Figure 5.4. Θ_1 and Θ_2 as functions of r_x and ω_f for $r_a = 10.0$ and for $n \to \infty$.

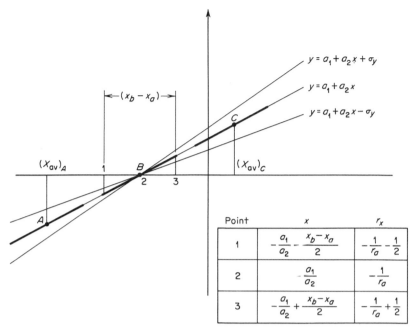

Point	x	r_x
1	$-\dfrac{a_1}{a_2} - \dfrac{x_b - x_a}{2}$	$-\dfrac{1}{r_a} - \dfrac{1}{2}$
2	$-\dfrac{a_1}{a_2}$	$-\dfrac{1}{r_a}$
3	$-\dfrac{a_1}{a_2} + \dfrac{x_b - x_a}{2}$	$-\dfrac{1}{r_a} + \dfrac{1}{2}$

Figure 5.5. Three-line segments for which r_a, ω_f, and $x_b - x_a = n \cdot \Delta x$ are the same, but x_{av} (and therefore r_x) is different.

The dimensionless groups Θ_1 and Θ_2 are plotted as functions of r_x and ω_f in Figure 5.4. It is observed that Θ_2 is symmetrical about a value of $r_x = -1/r_a$. At this value of r_x, the corresponding value of x_{av} is $-a_1/a_2$ and the corresponding value of y is zero. This can be seen schematically in Figure 5.5. The line segments A and C are equidistant from B and the magnitudes of y and σ_y are the same for both segments. One might therefore expect symmetry about $r_x = -1/r_a$ for Θ_2. Proof of this symmetry can be derived from Equations 5.2.50 and 5.2.53.

In Figure 5.6, Θ_1 and Θ_2 are plotted as functions of r_x and r_a for $\omega_f = 0$. For the three values of r_a considered (i.e., 1, 10, and 100), symmetry about the points $-1/r_a$ is observed for Θ_2. It is also noted that the shapes of the curves are the same. In other words, if the abscissa was $r_x + (1/r_a)$ rather than r_x, the three curves would be identical (if ω_f was equal to zero).

The behavior of Θ_1 is more complicated. There is no symmetry about the point $r_x = -1/r_a$, even for $\omega_f = 0$. For $r_a = 1.0$ and $\omega_f = 0.0$, as r_x increases beyond $-(1/r_a) + \frac{1}{2}$ (i.e., -0.5), Θ_1 is observed to increase, then decrease to 1.0 and then increase monotonically. This behavior can be explained by examining Equation 5.2.57. A value of $\Theta_1 = 1.0$ should be observed when the numerator and denominator of the equation are identical. This criterion is satisfied outside the region $r_x = -(1/r_a) \pm \frac{1}{2}$

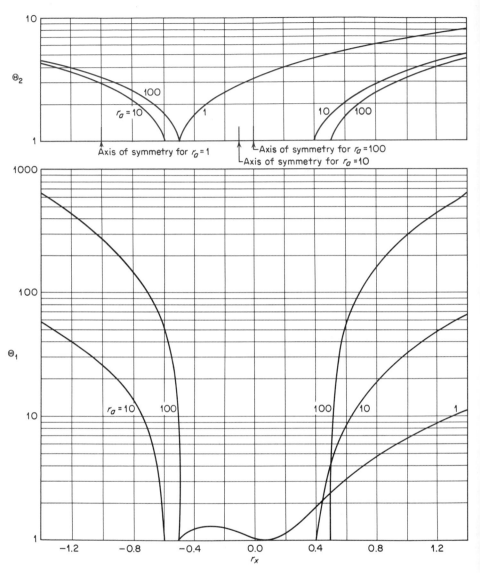

Figure 5.6. Θ_1 and Θ_2 as functions of r_x and r_a for $\omega_f = 0.0$ and for $n \to \infty$.

only when:

$$1 - r_a^{-1}R_{20} = -[(2r_a)^{-1}R_{20}]^2. \qquad (5.2.60)$$

Solving for R_{20} gives:

$$R_{20} - 4r_aR_{20} + 4r_a^2 = 0, \qquad (5.2.61)$$

$$R_{20} = 2r_a \pm \tfrac{1}{2}(16r_a^2 - 16r_a^2)^{1/2} = 2r_a, \qquad (5.2.62)$$

and from Equation 5.2.56,

$$\frac{1 + r_x r_a + (r_a/2)}{1 + r_x r_a - (r_a/2)} = e^{r_a}. \tag{5.2.63}$$

Solving for r_x gives:

$$(r_x)_{\Theta_1 = 1} = \frac{1}{2} - \frac{1}{r_a} \cdot \frac{e^{r_a} - 1}{e^{r_a} + 1}. \tag{5.2.64}$$

From Equation 5.2.64 it is seen that for $e^{r_a} \gg 1$, $(r_x)_{\Theta_1=1} \to \frac{1}{2} - (1/r_a)$ which is the upper limit of the region $-1/r_a - \frac{1}{2} < r_x < -1/r_a + \frac{1}{2}$. Thus minima of Θ_1 as a function of r_x also exist for $r_a = 10.$ and $100.$; however, they are so close to $-1/r_a + \frac{1}{2}$ that they are not observed in Figure 5.6.

The region $-1/r_a - \frac{1}{2} < r_x < -1/r_a + \frac{1}{2}$ is seen from the graphs to be the most favorable region for performing an experiment if Case II (i.e., Equation 5.2.23) is applicable. In this region the values of Θ_1 and Θ_2 and also Θ_{12} are smallest.

If $|1 + r_x r_a + r_a/2|$ and $|1 + r_x r_a - r_a/2|$ are large compared to $|\omega_f|$, then it can be shown (using Equations 5.2.46 and 5.2.48), that:

$$\frac{\Theta_1}{\Theta_2} = \left(\frac{a_2^2 [x^2/(y^2 + \lambda^2)]_{av}}{a_1^2 [1/(y^2 + \lambda^2)]_{av}}\right)^{1/2} \approx \left(\frac{1 + (1 - \omega_f^2) R_{10} - r_a^{-1} R_{20}}{R_{10}}\right)^{1/2}. \tag{5.2.65}$$

Using the following equality:

$$\frac{1 + r_x r_a + \frac{1}{2} r_a}{1 + r_x r_a - \frac{1}{2} r_a} = 1 + \frac{r_a}{1 + r_x r_a - \frac{1}{2} r_a}, \tag{5.2.66}$$

it can be seen (from Equation 5.2.56) that if $|1 + r_x r_a - r_a/2| \gg |r_a|$, then,

$$R_{20} \approx \frac{2 r_a}{1 + r_x r_a - \frac{1}{2} r_a}, \tag{5.2.67}$$

because $2 \log (1 + x) \approx 2x$ for small values of x. Substituting Equations 5.2.55 and 5.2.67 into Equation 5.2.65 gives:

$$\frac{\Theta_1}{\Theta_2} \approx [(1 + r_x r_a)^2 - \frac{1}{4} r_a^2 + (1 - \omega_f^2) - 2(1 + r_x r_a + \frac{1}{2} r_a)]^{1/2}$$

$$= [r_x^2 r_a^2 - (r_a + \frac{1}{4} r_a^2 + \omega_f^2)]^{1/2}. \tag{5.2.68}$$

If $r_x^2 r_a^2 \gg r_a + r_a^2/4 + \omega_f^2$, then

$$\frac{\Theta_1}{\Theta_2} = \frac{\hat{\sigma}_{a1}}{\hat{\sigma}_{a2}} \cdot \frac{a_2}{a_1} \approx r_x r_a = \frac{a_2 \cdot x_{av}}{a_1}. \tag{5.2.69}$$

From Equation 5.2.69 it is seen that as $|r_x|$ increases, the ratio $\hat{\sigma}_{a1}/\hat{\sigma}_{a2}$ approaches x_{av}.

TABLE 5.3. Ratios[a] of the Θ's computed at $n = 3$ and $n \to \infty$ for several combinations of r_x, r_a and ω_f.

r_x	r_a	ω_f	$\dfrac{(\Theta_1)_{n=3}}{(\Theta_1)_{n\to\infty}}$	$\dfrac{(\Theta_2)_{n=3}}{(\Theta_2)_{n\to\infty}}$	$\dfrac{(\Theta_{12})_{n=3}}{(\Theta_{12})_{n\to\infty}}$
0.5	0.0	0.0	1.045	1.06	1.12
1.5	0.0	0.0	1.06	1.06	1.125
0.5	10.0	0.0	1.60	1.29	2.66
0.5	10.0	1.0	1.42	1.23	2.10
1.5	10.0	0.0	1.09	1.08	1.17
1.5	10.0	1.0	1.08	1.08	1.17
0.5	50.0	0.0	3.19	1.53	8.85
1.5	50.0	0.0	1.09	1.08	1.17

[a] These results were determined using Equations 5.2.52 through 5.2.54 for $n \to \infty$ and the direct approach for $n = 3$.

If the slope a_2 is zero, the dimensionless group Θ_2 is meaningless (see Equation 5.2.53). To estimate $\hat{\sigma}_{a_2}$, one must know the value of the product $a_2 \cdot \Theta_2$. This product is a constant and is independent of r_x and ω_f. In fact,

$$\hat{\sigma}_{a_2} = \frac{(12)^{1/2}}{n \cdot \Delta x} \cdot \frac{K_{fy}}{n^{1/2}} \quad \text{if} \quad a_2 = 0. \tag{5.2.70}$$

Ratios of the Θ's computed for $n = 3$ (which is the minimum permissible value of n) and $n \to \infty$ are included in Table 5.3 for several combinations of r_a, r_x, and ω_f.

From Table 5.3 it can be seen that the Θ's are not strongly dependent upon n except for the combination of small values of r_x and large values of r_a. Even for these combinations, as n increases, the ratios rapidly approach one. For example, for $n = 15$, the value of Θ_1 is only about 1.25 times the asymptotic values as compared to 3.19 for $n = 3$, $r_x = 0.5$, and $r_a = 50$. The values of the Θ's for finite values of n were computed using the direct approach described in Section 4.4.

Case III. *Counting statistics for the variable y and constant uncertainty for the variable x.*

For this case the values of σ_{y_i} and σ_{x_i} are:

$$\begin{aligned} \sigma_{y_i} &= K_{sy} \cdot (y_i)^{1/2}, \\ \sigma_{x_i} &= K_{cx}. \end{aligned} \tag{5.2.71}$$

For experiments of this type a negative number of counts is not possible, so all values of y_i must be greater than or equal to zero. The difference between this case and Case III of Section 4.5 should be noted. For this case, the analysis is for the more general problem in which K_{cx} is not necessarily equal to zero.

The appropriate expression for $\hat{L}_i$ is determined by substituting Equation 5.2.71 into Equation 3.6.16:

$$\hat{L}_i = K_{sy}{}^2 \cdot y_i + a_2{}^2 \cdot K_{cx}{}^2. \tag{5.2.72}$$

The elements of the $\hat{C}$ matrix are determined by substituting Equations 4.5.9, 4.5.10 and 5.2.72 into Equation 4.4.1:

$$\hat{C}_{11} = \sum_{i=1}^{n} \frac{1}{K_{sy}{}^2 y_i + a_2{}^2 K_{cx}{}^2} = \frac{n}{K_{sy}{}^2} \cdot \left(\frac{1}{y + \delta}\right)_{av}, \tag{5.2.73}$$

$$\hat{C}_{12} = \sum_{i=1}^{n} \frac{x_i}{K_{sy}{}^2 y_i + a_2{}^2 K_{cx}{}^2} = \frac{n}{K_{sy}{}^2} \cdot \left(\frac{x}{y + \delta}\right)_{av}, \tag{5.2.74}$$

$$\hat{C}_{22} = \sum_{i=1}^{n} \frac{x_i{}^2}{K_{sy}{}^2 y_i + a_2{}^2 K_{cx}{}^2} = \frac{n}{K_{sy}{}^2} \cdot \left(\frac{x^2}{y + \delta}\right)_{av}, \tag{5.2.75}$$

where

$$\delta \equiv a_2{}^2 K_{cx}{}^2 / K_{sy}{}^2, \tag{5.2.76}$$

and the subscript av refers to average values of the quantities inside the parentheses.

Following the procedure used in Cases I and II for determining the predicted variances $\hat{\sigma}_{a1}{}^2$ and $\hat{\sigma}_{a2}{}^2$, and the predicted covariance $\hat{\sigma}_{12}$:

$$\hat{\sigma}_{a1}{}^2 = \frac{K_{sy}{}^2}{n} \cdot \left\{\left(\frac{x^2}{y + \delta}\right)_{av} \Big/ \left[\left(\frac{x^2}{y + \delta}\right)_{av} \cdot \left(\frac{1}{y + \delta}\right)_{av} - \left(\frac{x}{y + \delta}\right)_{av}^2\right]\right\},$$
$$\tag{5.2.77}$$

$$\hat{\sigma}_{a2}{}^2 = \frac{K_{sy}{}^2}{n} \cdot \left\{\left(\frac{1}{y + \delta}\right)_{av} \Big/ \left[\left(\frac{x^2}{y + \delta}\right)_{av} \cdot \left(\frac{1}{y + \delta}\right)_{av} - \left(\frac{x}{y + \delta}\right)_{av}^2\right]\right\},$$
$$\tag{5.2.78}$$

$$\hat{\sigma}_{12} = \frac{K_{sy}{}^2}{n} \cdot \left\{-\left(\frac{x}{y + \delta}\right)_{av} \Big/ \left[\left(\frac{x^2}{y + \delta}\right)_{av} \cdot \left(\frac{1}{y + \delta}\right)_{av} - \left(\frac{x}{y + \delta}\right)_{av}^2\right]\right\}.$$
$$\tag{5.2.79}$$

If it is assumed that the values of x are evenly spaced (i.e., Equation

4.4.3 is valid), then integral expressions for the required average values can be developed. As n approaches infinity,

$$\left(\frac{1}{y+\delta}\right)_{av} \rightarrow \int_{x_a}^{x_b} \frac{dx}{y+\delta} \bigg/ (x_b - x_a) = \int_{x_a}^{x_b} \frac{dx}{a_2 x + a_1 + \delta} \bigg/ (x_b - x_a) ,$$

(5.2.80)

$$\left(\frac{1}{y+\delta}\right)_{av} \rightarrow \frac{1}{a_2(x_b - x_a)} \cdot \ln\left(\frac{a_2 x_b + a_1 + \delta}{a_2 x_a + a_1 + \delta}\right).$$

(5.2.81)

Defining a new dimensionless ratio:

$$\omega_s \equiv \frac{\delta}{a_1} = \frac{a_2{}^2 \cdot K_{cx}{}^2}{a_1 \cdot K_{sy}{}^2} ,$$

(5.2.82)

and substituting Equations 5.2.2., 5.2.3, and 5.2.82 into Equation 5.2.81, we obtain:

$$\left(\frac{1}{y+\delta}\right)_{av} \rightarrow \frac{R_3}{a_1 r_a} ,$$

(5.2.83)

where

$$R_3 \equiv \ln\left(\frac{r_x r_a + \frac{1}{2} r_a + 1 + \omega_s}{r_x r_a - \frac{1}{2} r_a + 1 + \omega_s}\right).$$

(5.2.84)

Similarly,

$$\left(\frac{x}{y+\delta}\right)_{av} \rightarrow \frac{1}{a_2{}^2(x_b - x_a)} \cdot \left[a_2(x_b - x_a) - (\delta + a_1)\ln\left(\frac{a_2 x_b + a_1 + \delta}{a_2 x_a + a_1 + \delta}\right)\right],$$

(5.2.85)

$$\left(\frac{x}{y+\delta}\right)_{av} \rightarrow \frac{1}{a_2 r_a} \cdot [r_a - (\omega_s + 1)R_3],$$

(5.2.86)

$$\left(\frac{x^2}{y+\delta}\right)_{av} \rightarrow \frac{1}{a_2{}^3(x_b - x_a)} \cdot \left[\frac{1}{2}[(a_1 + a_2 x_b)^2 - (a_1 + a_2 x_a)^2] - (\delta + 2a_1)\right.$$

$$\left. \times (a_2 x_b - a_2 x_a) + (a_1 + \delta)^2 \ln\left(\frac{a_2 x_b + a_1 + \delta}{a_2 x_a + a_1 + \delta}\right)\right], \quad (5.2.87)$$

$$\left(\frac{x^2}{y+\delta}\right)_{av} \rightarrow \frac{a_1}{a_2{}^2 r_a} \cdot [(1 + r_x r_a)r_a - (\omega_s + 2)r_a + (1 + \omega_s)^2 R_3].$$

(5.2.88)

Substituting Equations 5.2.83, 5.2.86, and 5.2.88 into Equations 5.2.77 through 5.2.79, and then defining the dimensionless groups Ψ_k, we obtain:

$$\Psi_1 \equiv \frac{\hat{\sigma}_{a_1} \cdot n^{1/2}}{K_{sy} \cdot (a_1)^{1/2}} \rightarrow \left(\frac{r_x r_a - \omega_s - 1 + [(1+\omega_s)^2/r_a] \cdot R_3}{[r_a^{-1} + r_x + (\omega_s/r_a)] \cdot R_3 - 1}\right)^{1/2}, \qquad (5.2.89)$$

$$\Psi_2 \equiv \frac{\hat{\sigma}_{a_2} \cdot n^{1/2}(a_1)^{1/2}}{K_{sy} \cdot a_2} \rightarrow \left(\frac{r_a^{-1} \cdot R_3}{[r_a^{-1} + r_x + (\omega_s/r_a)] \cdot R_3 - 1}\right)^{1/2}, \qquad (5.2.90)$$

$$\Psi_{12} \equiv \frac{\hat{\sigma}_{12} \cdot n}{K_{sy}^2 \cdot a_2} \rightarrow \frac{-1 + [(\omega_s+1)/r_a] \cdot R_3}{[r_a^{-1} + r_x + (\omega_s/r_a)] \cdot R_3 - 1}. \qquad (5.2.91)$$

It should be remembered that this case is limited to only those problems in which all the y_i's are positive. If the slope a_2 is positive, this condition is satisfied if:

$$r_x > -r_a^{-1} + \tfrac{1}{2}, \qquad a_2 > 0, \qquad (5.2.92)$$

or if a_2 is negative,

$$r_x < -r_a^{-1} - \tfrac{1}{2}, \qquad a_2 < 0. \qquad (5.2.93)$$

Equations 5.2.89 through 5.2.91 can be simplified as $\omega_s \rightarrow 0$:

$$R_3 \rightarrow R_{30} \equiv \ln\left(\frac{r_x r_a + 1 + \tfrac{1}{2}r_a}{r_x r_a + 1 - \tfrac{1}{2}r_a}\right), \qquad (5.2.94)$$

and

$$\Psi_1 \rightarrow \left(\frac{r_x r_a - 1 + r_a^{-1}R_{30}}{(r_a^{-1} + r_x)R_{30} - 1}\right)^{1/2}, \qquad (5.2.95)$$

$$\Psi_2 \rightarrow \left(\frac{r_a^{-1}R_{30}}{(r_a^{-1} + r_x)R_{30} - 1}\right)^{1/2}, \qquad (5.2.96)$$

$$\Psi_{12} \rightarrow \left(\frac{-1 + r_a^{-1}R_{30}}{(r_a^{-1} + r_x)R_{30} - 1}\right). \qquad (5.2.97)$$

Using a procedure similar to the procedure outlined for Case II, it can be shown that the results for Case III of Section 4.5 can be expressed in dimensionless form by Equations 5.2.95 and 5.2.96.

The Ψ's are plotted as functions of r_x, ω_s, and r_a in Figures 5.7 and 5.8. These curves are good approximations even for small values of n. Ratios of the Ψ's computed for $n = 3$ (which is the minimum permissible value of n) and $n \rightarrow \infty$ are included in Table 5.4 for several combinations of r_x, r_a, and ω_s.

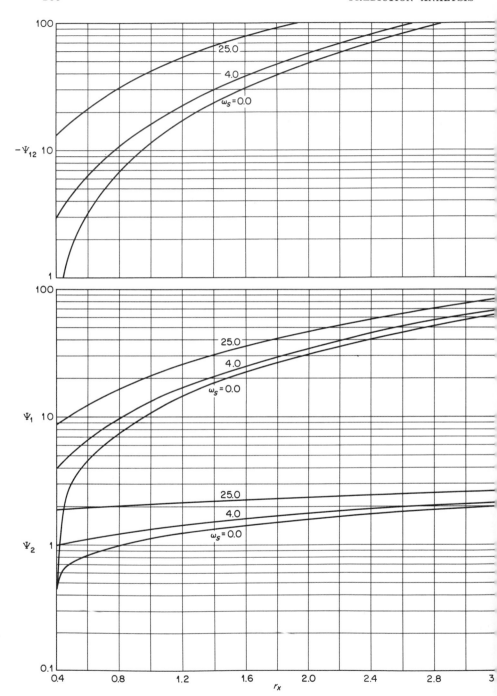

Figure 5.7. Ψ_1, Ψ_2, and Ψ_{12} as functions of r_x and ω_s for $r_a = 10.0$ and for $n \to \infty$.

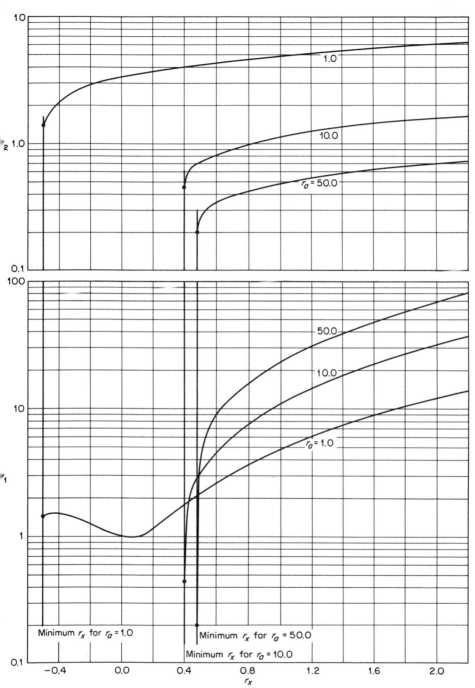

Figure 5.8. Ψ_1 and Ψ_2 as functions of r_x and r_a for $\omega_s = 0.0$ and for $n \to \infty$.

TABLE 5.4. Ratios of the Ψ's computed at $n = 3$ and $n \rightarrow \infty$ for several combinations of r_x, r_a, and ω_s

r_x	r_a	ω_s	$\dfrac{(\Psi_1)_{n=3}}{(\Psi_1)_{n\to\infty}}$	$\dfrac{(\Psi_2)_{n=3}}{(\Psi_2)_{n\to\infty}}$	$\dfrac{(\Psi_{12})_{n=3}}{(\Psi_{12})_{n\to\infty}}$
0.4	10.0	0.0	5.56	1.69	6.50
0.4	10.0	1.0	1.19	1.15	1.61
0.5	10.0	0.0	1.21	1.15	1.52
1.0	10.0	0.0	1.08	1.08	1.17
1.0	10.0	1.0	1.08	1.07	1.14
1.5	10.0	0.0	1.07	1.07	1.15
2.0	10.0	0.0	1.065	1.065	1.14

From Table 5.4 it can be seen that the Ψ's are not strongly dependent upon n. As r_x approaches its limiting value (i.e., Equation 5.2.92 or 5.2.93), the ratios of the Ψ's computed at $n = 3$ and $n \rightarrow \infty$ increase. For $r_a = 10$. and $\omega_s = 0.0$ the limiting value of r_x is 0.4 if a_2 is positive. As n increases, the ratios rapidly approach one. For example, for $n = 24$ and $r_x = 0.5$, the value of Ψ_1 is less than 1% above the asymptotic value as compared to 21% for $n = 3$. Although not shown in the table, as r_a increases, the ratios increase. For ω_s not equal to zero, the ratios are decreased. This effect is particularly important as r_x approaches its limiting value. The values of the Ψ's were computed using the direct approach (described in Section 4.4) for $n = 3$.

If the slope a_2 is zero, Ψ_2 is meaningless (see Equation 5.2.90). To estimate $\hat{\sigma}_{a_2}$ one must know the value of the product $a_2 \cdot \psi_2$. This product is a constant and is independent of r_x and ω_s. The correct expression for $\hat{\sigma}_{a_2}$ is similar to Equation 5.2.70:

$$\hat{\sigma}_{a_2} = \frac{(12)^{1/2}}{n \cdot \Delta x} \cdot \frac{K_{sy}}{n^{1/2}} \quad \text{if} \quad a_2 = 0. \tag{5.2.98}$$

5.3 The Second-Order Polynomial

Equation 5.1.3 is the relationship between the dependent and independent variables for all experiments within this class:

$$y = a_1 + a_2 x + a_3 x^2. \tag{5.1.3}$$

The cases that are considered in this section are summarized in Table 5.5.

Equations for the $\hat{\sigma}_{a_k}$'s are not derived; however, equations are included for the elements of the $\hat{C}$ matrix. Results from direct calculations of the $\hat{\sigma}_{a_k}$'s are presented in graphical and tabular forms.

TABLE 5.5. Summary of cases analyzed for the *second-order polynomial* class of experiments

Case	σ_{y_i}	σ_{x_i}	Figure	Table
I	K_{cy}	K_{cx}	5.9	—
II	$K_{fy} \cdot \|y_i\|$	K_{cx}	5.10	5.6, 5.7
III	$K_{sy} \cdot (y_i)^{1/2}$	K_{cx}	5.11	5.8, 5.9

Case I. *Constant uncertainty for both the x and y variables.*

For this case the values of σ_{y_i} and σ_{x_i} are:

$$\sigma_{y_i} = K_{cy},$$

$$\sigma_{x_i} = K_{cx}. \tag{5.3.1}$$

The only difference between this case and Case I of Section 5.2 is that the function includes a third parameter (i.e., a_3). The analysis is therefore for a parabola rather than a straight line. The equations for $\hat{F}_{a_1}{}^i$, and $\hat{F}_{a_2}{}^i$ are the same for the two cases, but the equation for $\hat{L}_i$ must be modified:

$$L_i = K_{cy}{}^2 + (a_2 + 2a_3x_i)^2 K_{cx}{}^2, \tag{5.3.2}$$

$$\hat{F}_{a_1}{}^i = -1, \tag{4.5.9}$$

$$\hat{F}_{a_2}{}^i = -x_i. \tag{4.5.10}$$

An additional equation for $\hat{F}_{a_3}{}^i$ is required for this class of experiments:

$$\hat{F}_{a_3}{}^i = -x_i{}^2. \tag{5.3.3}$$

Equations for the elements of the matrix $\hat{C}$ are obtained by substituting Equations 4.5.9, 4.5.10, 5.3.2, and 5.3.3 into Equation 4.4.1:

$$\hat{C}_{11} = n \cdot \left(\frac{1}{K_{cy}{}^2 + (a_2 + 2a_3x)^2 K_{cx}{}^2} \right)_{av}, \tag{5.3.4}$$

$$\hat{C}_{12} = n \cdot \left(\frac{x}{K_{cy}{}^2 + (a_2 + 2a_3x)^2 K_{cx}{}^2} \right)_{av}, \tag{5.3.5}$$

$$\hat{C}_{13} = n \cdot \left(\frac{x^2}{K_{cy}{}^2 + (a_2 + 2a_3x)^2 K_{cx}{}^2} \right)_{av}, \tag{5.3.6}$$

$$\hat{C}_{22} = n \cdot \left(\frac{x^2}{K_{cy}{}^2 + (a_2 + 2a_3x)^2 K_{cx}{}^2} \right)_{av}, \tag{5.3.7}$$

$$\hat{C}_{23} = n \cdot \left(\frac{x^3}{K_{cy}{}^2 + (a_2 + 2a_3x)^2 K_{cx}{}^2} \right)_{av}, \tag{5.3.8}$$

$$\hat{C}_{33} = n \cdot \left(\frac{x^4}{K_{cy}{}^2 + (a_2 + 2a_3x)^2 K_{cx}{}^2} \right)_{av}. \tag{5.3.9}$$

Expressions for the $\hat{\sigma}$'s require inversion of a 3×3 symmetrical matrix:

$$\hat{\sigma}_{a_1}^2 = \hat{C}_{11}^{-1} = \frac{\hat{C}_{22} \cdot \hat{C}_{33} - (\hat{C}_{23})^2}{D}, \qquad (5.3.10)$$

$$\hat{\sigma}_{a_2}^2 = \hat{C}_{22}^{-1} = \frac{\hat{C}_{11} \cdot \hat{C}_{33} - (\hat{C}_{13})^2}{D}, \qquad (5.3.11)$$

$$\hat{\sigma}_{a_3}^2 = \hat{C}_{33}^{-1} = \frac{\hat{C}_{11} \cdot \hat{C}_{22} - (\hat{C}_{12})^2}{D}, \qquad (5.3.12)$$

$$\hat{\sigma}_{12} = \hat{C}_{12}^{-1} = \frac{-\hat{C}_{12} \cdot \hat{C}_{33} + \hat{C}_{13} \cdot \hat{C}_{23}}{D}, \qquad (5.3.13)$$

$$\hat{\sigma}_{13} = \hat{C}_{13}^{-1} = \frac{\hat{C}_{12} \cdot \hat{C}_{23} - \hat{C}_{13} \cdot \hat{C}_{22}}{D}, \qquad (5.3.14)$$

$$\hat{\sigma}_{23} = \hat{C}_{23}^{-1} = \frac{-\hat{C}_{11} \cdot \hat{C}_{23} + \hat{C}_{12} \cdot \hat{C}_{13}}{D}, \qquad (5.3.15)$$

where D is the determinant of the matrix $\hat{C}$:

$$D = \hat{C}_{11} \cdot (\hat{C}_{22} \cdot \hat{C}_{33} - \hat{C}_{12} \cdot \hat{C}_{12}) - \hat{C}_{12} \cdot (\hat{C}_{12} \cdot \hat{C}_{33} - \hat{C}_{13} \cdot \hat{C}_{23})$$
$$+ \hat{C}_{13} \cdot (\hat{C}_{12} \cdot \hat{C}_{23} - \hat{C}_{13} \cdot \hat{C}_{22}). \qquad (5.3.16)$$

Assuming that the values of x are evenly spaced (i.e., Equation 4.4.3 is valid), and then following the procedure of Section 5.2, analytical expressions for the $\hat{\sigma}$'s can be derived. The expressions are, however, extremely complicated, even if $K_{cx} = 0$. Fortunately, the method of prediction analysis is not limited only to those problems that can be treated analytically. In Section 4.4 the *direct approach* is discussed. Using this approach, experiments of any degree of complexity can be analyzed. Results were obtained in this manner and are plotted in Figure 5.9 in terms of the following dimensionless groups:

$$\Omega_1 \equiv \frac{\hat{\sigma}_{a_1} \cdot n^{1/2}}{K_{cy} \cdot (1 + \omega_c^2)^{1/2}}, \qquad (5.3.17)$$

$$\Omega_2 \equiv \frac{\hat{\sigma}_{a_2} \cdot n^{1/2} \cdot (n \cdot \Delta x)}{K_{cy} \cdot (1 + \omega_c^2)^{1/2}}, \qquad (5.3.18)$$

$$\Omega_3 \equiv \frac{\hat{\sigma}_{a_3} \cdot n^{1/2} \cdot (n \cdot \Delta x)^2}{K_{cy} \cdot (1 + \omega_c^2)^{1/2}}, \qquad (5.3.19)$$

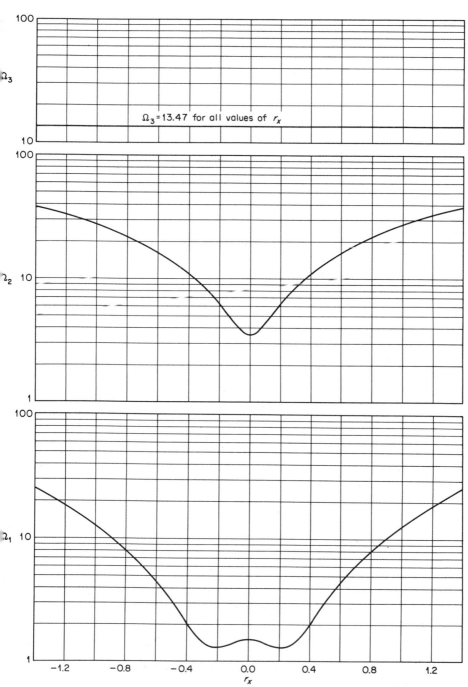

Figure 5.9. Ω_1, Ω_2, and Ω_3 as functions of r_x for all values of r_a and r_p if $K_{cx} = 0$, and for all values of r_a and ω_e if $a_3 = 0$.

where ω_c is defined as:

$$\omega_c \equiv \frac{a_2 \cdot K_{cx}}{K_{cy}}. \qquad (5.3.20)$$

It should be noted that ω_c is also dimensionless.

Four dimensionless groups are required to completely specify the problem. The choice of the groups is, of course, arbitrary. An important criterion for selection is that the groups should facilitate presentation of the results. The following groups satisfy this criterion—r_x (Equation 5.2.2), r_a (Equation 5.2.3), ω_c (Equation 5.3.20), and r_p:

$$r_p \equiv \frac{a_1 \cdot a_3}{a_2{}^2}. \qquad (5.3.21)$$

The values of Ω_1, Ω_2, and Ω_3 are fixed for a given combination of r_x, r_a, r_p, and ω_c.

An examination of Equations 5.3.4 through 5.3.16 shows that the Ω's are independent of a_1, a_2, and a_3 and therefore r_a and r_p if $K_{cx} = 0$. If a_2 is large compared to $2a_3 x_{max}$, then the Ω's are insensitive to r_a, r_p, and ω_c. If $a_3 = 0$, the Ω's are independent of r_a and ω_c. For $K_{cx} = 0$, $\Omega_3 = 13.47$ and is independent of r_x as well as r_a and r_p.

The Ω's were computed using the direct approach with a value of $n = 24$. The results are within 1% of the asymptotic values (i.e., $n \to \infty$) and are also good approximations for small values of n. The asymptotes were determined using the direct approach by increasing n until the Ω's converged. For experiments of this class, the minimum possible number of data points is four.

Case II. *Constant fractional uncertainty for the variable y and constant uncertainty for the variable x.*

For this case the values of σ_{y_i} and σ_{x_i} are:

$$\sigma_{y_i} = K_{fy} \cdot |y_i|,$$
$$\sigma_{x_i} = K_{cx}. \qquad (5.3.22)$$

The only difference between this case and Case II of Section 5.2 is that the function includes a third parameter (i.e., a_3). The equation for $\hat{L}_i$ must be modified:

$$\hat{L}_i = K_{fy}{}^2 y_i{}^2 + (a_2 + 2a_3 x_i)^2 K_{cx}{}^2. \qquad (5.3.23)$$

The equations for $\hat{F}_{a_1}{}^i$, $\hat{F}_{a_2}{}^i$, and $\hat{F}_{a_3}{}^i$ are the same for this case and Case I of this section.

Equations for the elements of the matrix $\hat{C}$ are obtained by substituting

Equations 5.3.23, 4.5.9, 4.5.10, and 5.3.3 into Equation 4.4.1:

$$\hat{C}_{11} = \frac{n}{K_{fy}{}^2} \cdot \left(\frac{1}{y^2 + \lambda^2 [1 + (2a_3x/a_2)]^2} \right)_{av}, \tag{5.3.24}$$

$$\hat{C}_{12} = \frac{n}{K_{fy}{}^2} \cdot \left(\frac{x}{y^2 + \lambda^2 [1 + (2a_3x/a_2)]^2} \right)_{av}, \tag{5.3.25}$$

$$\hat{C}_{13} = \frac{n}{K_{fy}{}^2} \cdot \left(\frac{x^2}{y^2 + \lambda^2 [1 + (2a_3x/a_2)]^2} \right)_{av}, \tag{5.3.26}$$

$$\hat{C}_{22} = \frac{n}{K_{fy}{}^2} \cdot \left(\frac{x^2}{y^2 + \lambda^2 [1 + (2a_3x/a_2)]^2} \right)_{av}, \tag{5.3.27}$$

$$\hat{C}_{23} = \frac{n}{K_{fy}{}^2} \cdot \left(\frac{x^3}{y^2 + \lambda^2 [1 + (2a_3x/a_2)]^2} \right)_{av}, \tag{5.3.28}$$

$$\hat{C}_{33} = \frac{n}{K_{fy}{}^2} \cdot \left(\frac{x^4}{y^2 + \lambda^2 [1 + (2a_3x/a_2)]^2} \right)_{av}, \tag{5.3.29}$$

where λ is defined by Equation 5.2.31. Equations for the $\hat{\sigma}$'s can be derived by substituting Equations 5.3.24 through 5.3.29 into Equation 5.3.10 through 5.3.16.

If it is assumed that the values of x are evenly spaced (i.e., Equation 4.4.3 is valid), approximate expressions for the required average values can be derived in a manner similar to that used for derivation of Equations 5.2.52 through 5.2.54. The resulting equations are extremely complicated and therefore derivation of equations for the $\hat{\sigma}$'s are not included. However, results from direct computations of the $\hat{\sigma}_{a_k}$'s (but not the $\hat{\sigma}_{jk}$'s) are presented in graphical and tabular forms.

Results are plotted in Figure 5.10 in terms of dimensionless groups Θ_k:

$$\Theta_k \equiv \frac{\hat{\sigma}_{a_k} \cdot n^{1/2}}{K_{fy} \cdot a_k} \tag{5.3.30}$$

and as functions of r_x and r_p (defined by Equations 5.2.2 and 5.3.21) for $\omega_f = 0.0$ (defined by Equation 5.2.49), $r_a = 10.0$ (defined by Equation 5.2.3), and $n = 24$.

Although the results are exact only for $n = 24$, they are good approximations for small as well as large values of n. It may be concluded from Figure 5.10 that the region $(-1/r_a - \frac{1}{2}) < r_x < (-1/r_a + \frac{1}{2})$ is the best region for performing an experiment because the Θ_k's are smallest in this region. It is also noted that if r_p is increased, Θ_1 and Θ_2 are also increased, and Θ_3 approaches a symmetric distribution.

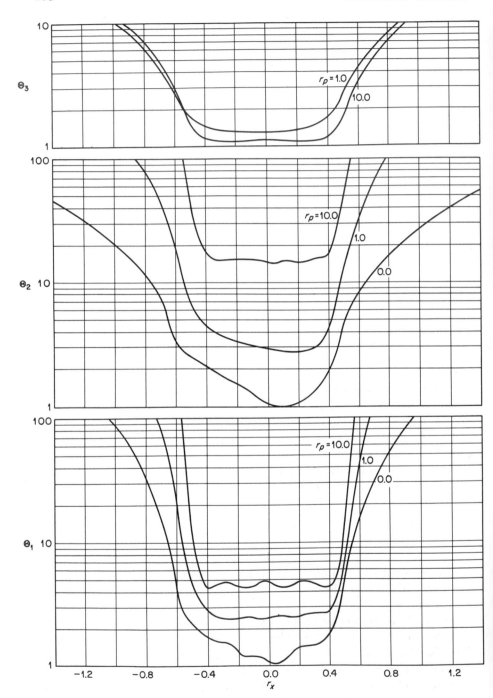

Figure 5.10. Θ_1, Θ_2, and Θ_3 as functions of r_x and r_p for $r_a = 10.0$, $\omega_f = 0$, and $n = 24$.

TABLE 5.6. Values[a] of Θ_1, Θ_2, and Θ_3 as functions of r_x and ω_f
for $r_a = 10.0$ and $r_p = 1.0$

r_x	$\omega_f = 0.0$			$\omega_f = 2.0$		
	Θ_1	Θ_2	Θ_3	Θ_1	Θ_2	Θ_3
−1.0	729.6	183.5	10.66	844.2	208.8	11.93
−0.8	201.1	74.92	6.081	268.6	95.46	7.404
−0.6	21.05	17.51	2.662	39.02	29.11	3.775
−0.4	2.747	4.412	1.462	5.114	9.378	2.219
−0.2	2.461	3.283	1.344	4.186	6.465	2.016
0.0	2.484	2.896	1.330	4.293	5.257	2.003
0.2	2.585	2.787	1.380	4.661	5.430	2.095
0.4	2.762	4.334	1.717	4.997	8.377	2.404
0.6	44.30	32.39	4.208	72.94	46.97	5.509
0.8	296.5	105.9	8.232	365.9	126.4	9.535
1.0	946.8	233.9	13.35	1061.9	259.1	14.61

[a] These values were computed using the direct approach with $n = 24$.

The curves in Figure 5.10 are for $\omega_f = 0.0$. As ω_f is increased, the Θ_k's are increased. As an example, values of the Θ_k's are compared in Table 5.6 for $\omega_f = 0.0$ and $\omega_f = 2.0$. For both sets of results, $r_a = 10.0$ and $r_p = 1.0$.

The influence of r_a upon Θ_1 and Θ_2 is seen in Figure 5.6 to be complicated even for the *first-order polynomial*. For the *second-order polynomial* the complexity is greater. In Table 5.7, values of Θ_k are compared for several combinations of r_a, r_x, and ω_f. The value of r_p is 0.0 for all the combinations.

Several points concerning the results in Table 5.7 require explanation:

1. If $a_3 = 0$, from Equation 5.3.30 it is seen that $\Theta_3 \to \infty$; however, the product $a_3 \cdot \Theta_3$ is finite. For this reason, $a_3 \cdot \Theta_3$ is tabulated rather than Θ_3. It is interesting to note that $a_3 \cdot \Theta_3$ is approximately proportional to r_a. It should be noted that even if $a_3 = 0$, the results do not reduce to the case of the first-order polynomial.

2. For $\omega_f = 0.0$ and $r_x = 0.0$ and 0.3, the values of Θ_1 are observed to oscillate in an apparently random manner. This phenomenon is explained by the fact that there is a *singularity* in the region $-1/r_a - \frac{1}{2} < r_x < -1/r_a + \frac{1}{2}$ if r_p and ω_f are equal to zero. If $\omega_f = 0$, $\lambda = 0$ and therefore there is a value of x within the region for which $y^2 + \lambda^2(1 + 2a_3x/a_2)^2 = 0$. From Equations 5.3.24 through 5.3.29, it can be seen that the values of the elements of the matrix $\hat{C}$ will be strongly influenced by the data point closest to this singularity. For this reason one can expect erratic results for a prediction analysis in which the region of interest contains a singu-

TABLE 5.7. Values[a] of Θ_1, Θ_2, and $a_3 \cdot \Theta_3$ as functions of r_x, ω_f, and r_a

r_x	ω_f	r_a	Θ_1	Θ_2	$a_3 \cdot \Theta_3$
0.0	0.0	2.0	1.077	2.075	0.0844[b]
		5.0	1.005	1.236	0.174
		10.0	1.076	1.068	0.350
		20.0	1.562	1.018	0.708
		50.0	1.083	1.024	1.734
0.0	2.0	2.0	1.413	2.250	0.130
		5.0	2.138	1.552	0.259
		10.0	3.816	1.426	0.508
		20.0	7.388	1.393	1.012
		50.0	18.30	1.383	2.526
0.3	0.0	2.0	1.492	4.709	0.192
		5.0	1.622	1.388	0.211
		10.0	1.480	1.297	0.360
		20.0	1.689	1.385	0.720
		50.0	2.036	1.440	1.749
0.6	0.0	2.0	6.883	15.48	0.281
		5.0	10.18	10.08	0.481
		10.0	15.43	8.158	0.805
		20.0	25.71	7.145	1.444
		50.0	56.30	6.511	3.351

[a] For all combinations, $r_p = 0$ and $n = 24$.
[b] Units of a_3.

larity. If, however, ω_f is unequal to zero, λ is unequal to zero and no singularity can exist.

3. The values of Θ_2 approach infinity as r_a approaches zero. This behavior can be explained by considering Equation 5.3.30. The product $a_2 \cdot \Theta_2$ approaches a limiting value as a_2 approaches zero.

Case III. *Counting statistics for the variable y and constant uncertainty for the variable x.*

For this case the values of σ_{y_i} and σ_{x_i} are:

$$\sigma_{y_i} = K_{sy} \cdot y_i^{1/2},$$
$$\sigma_{x_i} = K_{cx}. \tag{5.3.31}$$

All the values of y_i must be equal to or greater than zero.

The only difference between this case and Case III of Section 5.2 is that the function includes a third parameter (i.e., a_3). The equation for $\hat{L}_i$ must be modified:

$$\hat{L}_i = K_{sy}^2 y_i + (a_2 + 2a_3 x_i)^2 K_{cx}^2. \tag{5.3.32}$$

The equations for $\hat{F}_{a_1}{}^i$, $\hat{F}_{a_2}{}^i$, and $\hat{F}_{a_3}{}^i$ are the same for this case and Case I of this section.

Equations for the elements of the matrix $\hat{C}$ are obtained by substituting Equations 5.3.32, 4.5.9, 4.5.10, and 5.3.3 into Equation 4.4.1:

$$\hat{C}_{11} = \frac{n}{K_{sy}{}^2} \cdot \left(\frac{1}{y + \delta[1 + (2a_3x/a_2)]^2} \right)_{av} , \tag{5.3.33}$$

$$\hat{C}_{12} = \frac{n}{K_{sy}{}^2} \cdot \left(\frac{x}{y + \delta[1 + (2a_3x/a_2)]^2} \right)_{av} , \tag{5.3.34}$$

$$\hat{C}_{13} = \frac{n}{K_{sy}{}^2} \cdot \left(\frac{x^2}{y + \delta[1 + (2a_3x/a_2)]^2} \right)_{av} , \tag{5.3.35}$$

$$\hat{C}_{22} = \frac{n}{K_{sy}{}^2} \cdot \left(\frac{x^2}{y + \delta[1 + (2a_3x/a_2)]^2} \right)_{av} , \tag{5.3.36}$$

$$\hat{C}_{23} = \frac{n}{K_{sy}{}^2} \cdot \left(\frac{x^3}{y + \delta[1 + (2a_3x/a_2)]^2} \right)_{av} , \tag{5.3.37}$$

$$\hat{C}_{33} = \frac{n}{K_{sy}{}^2} \cdot \left(\frac{x^4}{y + \delta[1 + (2a_3x/a_2)]^2} \right)_{av} , \tag{5.3.38}$$

where δ is defined by Equation 5.2.76. Equations for the $\hat{\sigma}$'s can be derived by substituting Equations 5.3.33 through 5.3.38 into Equations 5.3.10 through 5.3.16.

If it is assumed that the values of x are evenly spaced (i.e., Equation 4.4.3 is valid), approximate expressions for the required average values can be derived in a manner similar to that used for derivation of Equations 5.2.88, 5.2.91, and 5.2.93. The resulting equations are extremely cumbersome. Although derivations of equations for the $\hat{\sigma}$'s are not included, results obtained using the direct approach are presented in graphical and tabular forms.

The results are plotted in Figure 5.11 in terms of dimensionless groups Ψ_k:

$$\Psi_k \equiv \frac{\hat{\sigma}_{a_k} \cdot n^{1/2} \cdot a_1{}^{1/2}}{a_k \cdot K_{sy}} , \tag{5.3.39}$$

as functions of r_x and r_p (defined by Equations 5.2.2 and 5.3.21) for $\omega_s = 0.0$ (defined by Equation 5.2.82), $r_a = 10.0$ (defined by Equation 5.2.3), and $n = 24$. Since all the values of y_i must be equal to or greater than zero for this case, there is a limitation on the values of r_x. To determine the limiting values of r_x, the value of x corresponding to the minimum value

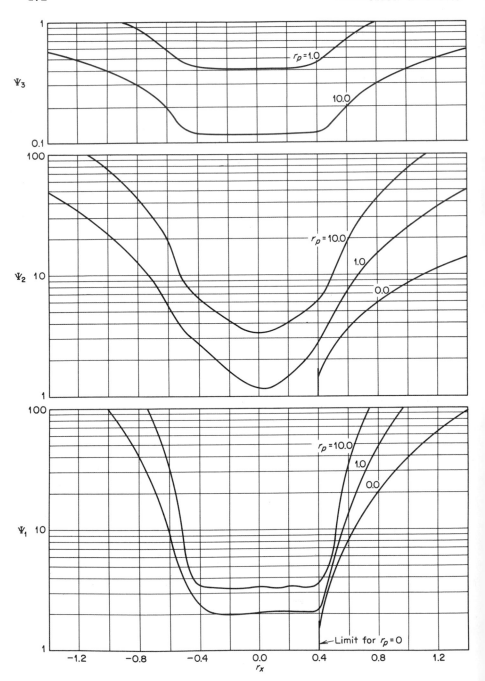

Figure 5.11. Ψ_1, Ψ_2, and Ψ_3 as functions of r_x and r_p for $r_a = 10.0$, $\omega_s = 0.0$, and $n = 24$.

of y is first located. To simplify the following analysis, it is assumed that a_3 is positive. Setting the derivative dy/dx equal to zero gives:

$$\frac{dy}{dx} = a_2 + 2a_3 x_{min} = 0,$$

therefore,

$$x_{min} = -\frac{a_2}{2a_3}, \qquad (5.3.40)$$

and

$$y_{min} = a_1 + a_2 x_{min} + a_3 x_{min}^2$$

$$= a_1 - \frac{a_2^2}{4a_3}. \qquad (5.3.41)$$

If the minimum value of y is greater than zero, there is no limitation on the value of r_x. Dividing Equation 5.3.41 by a_1 and substituting $r_p \equiv a_3 \cdot a_1/a_2^2$ into the equation, it can be seen that:

$$y_{min} > 0 \quad \text{if} \quad r_p > 0.25. \qquad (5.3.42)$$

If $r_p < 0.25$, the limiting values of r_x can be determined by solving for the roots of the following equation:

$$y = a_1 + a_2 x + a_3 x^2 = 0, \qquad (5.3.43)$$

$$x_{limits} = -\frac{a_2}{2a_3} \pm \left(\frac{a_2^2}{4a_3^2} - \frac{a_1}{a_3}\right)^{1/2}. \qquad (5.3.44)$$

For values of x outside this range, all values of y_i should be positive. If, however, a_3 is negative, y is positive for all values within the range. The limiting values of r_x can be determined by dividing Equation 5.3.44 by $n \cdot \Delta x$:

$$(r_x)_{limits} = \frac{x_{limits}}{n \cdot \Delta x} = -\frac{a_2}{2a_3(n \cdot \Delta x)} \pm \left(\frac{a_2^2}{4a_3^2(n \cdot \Delta x)^2} - \frac{a_1}{a_3(n \cdot \Delta x)^2}\right)^{1/2}.$$

$$(5.3.45)$$

Substituting $r_a \equiv a_2(n \cdot \Delta x)/a_1$ and $r_p \equiv a_1 \cdot a_3/a_2^2$ into the equation, Equation 5.3.45 can be reduced to a more useful form:

$$(r_x)_{limits} = (r_p r_a)^{-1}(-\tfrac{1}{2} \pm (\tfrac{1}{4} - r_p)^{1/2}]. \qquad (5.3.46)$$

Equation 5.3.46 verifies that for $r_p > 0.25$ there are no real roots of Equation 5.3.43 and therefore there are no limits on the value of r_x. From Equation 5.3.46 it can be shown that if r_p and a_3 are positive and if r_p is less than 0.25, then r_x must be:

$$r_x > \tfrac{1}{2} + (r_p r_a)^{-1}[-\tfrac{1}{2} + (\tfrac{1}{4} - r_p)^{1/2}],$$

or   (5.3.47)

$$r_x < -\tfrac{1}{2} - (r_p r_a)^{-1}[-\tfrac{1}{2} + (\tfrac{1}{4} - r_p)^{1/2}],$$

if all values of y_i are to be greater than zero. Equation 5.3.47 is indeterminate if $a_3 = 0$. For this special case, the limit for the first-order polynomial (i.e., Equations 5.2.92 and 5.2.93 are applicable):

$$r_x > -r_a^{-1} + \tfrac{1}{2} \quad \text{if} \quad r_p = 0 \quad \text{and} \quad a_2 > 0, \qquad (5.2.92)$$

$$r_x < -r_a^{-1} - \tfrac{1}{2} \quad \text{if} \quad r_p = 0 \quad \text{and} \quad a_2 < 0. \qquad (5.2.93)$$

From Figure 5.11 it can be seen that as r_p increases, Ψ_1 and Ψ_2 increase. As a_3 (and therefore r_p) approaches zero, from Equation 5.3.39 it is seen that Ψ_3 approaches infinity. However, the product $a_3 \cdot \Psi_3$ approaches a finite value. The region $(-1/r_a - \tfrac{1}{2}) < r_x < (-1/r_a + \tfrac{1}{2})$ is seen to be the best region for performing an experiment because the Ψ_k's are smallest in this region.

The curves in Figure 5.11 are for $\omega_s = 0$. As ω_s is increased, the Ψ_k's are increased. As an example, values of the Ψ_k's are compared in Table 5.8 for $\omega_s = 0.0$ and $\omega_s = 4.0$. For both sets of results, $r_a = 10.0$ and $r_p = 1.0$.

TABLE 5.8. Values[a] of Ψ_1, Ψ_2, and Ψ_3 as functions of r_x and ω_s
for $r_a = 10.0$, and $r_p = 1.0$

r_x	$\omega_s = 0.0$			$\omega_s = 4.0$		
	Ψ_1	Ψ_2	Ψ_3	Ψ_1	Ψ_2	Ψ_3
-1.0	96.28	22.32	1.194	393.7	91.37	4.892
-0.8	39.73	12.83	0.8982	160.0	51.95	3.650
-0.6	9.342	5.584	0.5876	30.67	20.40	2.231
-0.4	2.405	2.571	0.4330	5.095	8.395	1.519
-0.2	1.976	1.619	0.4121	4.130	5.729	1.486
0.0	2.069	1.162	0.4094	4.203	4.314	1.492
0.2	2.127	1.428	0.4186	4.623	4.889	1.496
0.4	2.179	2.612	0.4720	4.711	7.523	1.561
0.6	13.78	7.448	0.7441	53.24	29.54	2.983
0.8	48.45	15.24	1.048	197.0	62.16	4.281
1.0	109.9	25.21	1.337	450.5	103.4	5.491

[a] These values were computed using the direct approach with $n = 24$.

TABLE 5.9. Values[a] of Ψ_1, Ψ_2, and Ψ_3 as functions of r_x and r_a

r_x	r_a	Ψ_1	Ψ_2	Ψ_3
0.0	10.0	2.069	1.162	0.4094
	20.0	2.643	1.075	0.1883
	50.0	3.971	1.028	0.0718
	100.0	6.904	1.010	0.0356
0.3	10.0	2.092	1.850	0.4330
	20.0	2.674	1.614	0.1928
	50.0	3.685	1.520	0.0728
	100.0	5.321	1.497	0.0360
0.6	10.0	13.78	7.448	0.7441
	20.0	23.64	6.713	0.3429
	50.0	53.85	6.311	0.1307
	100.0	104.6	6.188	0.0644

[a] For all combinations $r_p = 1.0$, $\omega_s = 0.0$, and $n = 24$.

The influence of r_a upon the Ψ_k's is complicated. In Table 5.9, values of Ψ_k are compared for several combinations of r_a and r_x. The values of r_p and ω_s are 1.0, and 0.0 for all the combinations.

From Table 5.9 it is seen that as r_a increases, Ψ_2 and Ψ_3 decrease, and Ψ_1 increases. It is interesting to note that Ψ_3 is approximately inversely proportional to r_a. The values of Ψ_2 decrease at a much slower rate than the values of Ψ_3.

5.4 Higher-Order Polynomials

Equation 5.1.1 is the general form of the polynomial function:

$$y = \sum_{k=1}^{p} a_k x^{k-1}. \tag{5.1.1}$$

The index $p - 1$ is called the *order* of the polynomial. As the order increases, analytical solutions for the $\hat{\sigma}_{a_k}$'s and the $\hat{\sigma}_{jk}$'s become more complicated.

A general equation for the elements of the coefficient matrix $\hat{C}$ can be expressed as:

$$\hat{C}_{jk} = n \cdot \left(\frac{x^{j+k-2}}{\hat{L}_i}\right)_{av}, \tag{5.4.1}$$

where $\hat{L}_i$ is defined by Equation 3.6.16. For a polynomial of order $p - 1$:

$$\hat{L}_i = \sigma_{y_i}^2 + [a_2 + 2a_3x + \cdots + (p - 1) \cdot a_p x^{p-2}]\sigma_{x_i}^2. \tag{5.4.2}$$

The simplest procedure for predicting the σ's associated with the parameters of higher-order polynomials is to use the direct approach described in Section 4.4. Inversion of the matrix $\hat{C}$ leads to solutions for the predicted variances $\hat{\sigma}_{a_k}{}^2$ and covariances $\hat{\sigma}_{jk}$:

$$\hat{\sigma}_{a_k}{}^2 = \hat{C}_{kk}{}^{-1}, \tag{5.4.3}$$

$$\hat{\sigma}_{jk} = \hat{C}_{jk}{}^{-1}. \tag{5.4.4}$$

Often the polynomial is to be fitted to experimental data, but the correct order of the polynomial is unknown. Intuitively, one realizes that the uncertainties $\hat{\sigma}_{a_k}$, as well as the actual values a_k of the unknown parameters, will be dependent upon the order of the polynomial used to fit the data.[2] The tendency is to use a polynomial of higher order than might be required in order to be on the "safe-side."

The problem can best be illustrated if constant errors are assumed for both the dependent and independent variables:

$$\sigma_{y_i} = K_{cy},$$

$$\sigma_{x_i} = K_{cx}. \tag{5.4.5}$$

For the proposed experiment a first-order polynomial is sufficient to relate the dependent and independent variables. If data are obtained and a first-order polynomial is fit to the data, as K_{cy} and K_{cx} approach zero, and if there are no systematic errors:

$$a_1 \rightarrow \alpha_1,$$

$$a_2 \rightarrow \alpha_2. \tag{5.4.6}$$

That is, the parameters a_1 and a_2, determined by a least squares analysis, approach the true values α_1 and α_2 as the uncertainty in the data approaches zero.

If a second-order polynomial is fit to the data, as K_{cy} and K_{cx} approach zero:

$$a_1 \rightarrow \alpha_1,$$

$$a_2 \rightarrow \alpha_2, \tag{5.4.7}$$

$$a_3 \rightarrow 0.$$

Similarly, if a third-order polynomial is fit to the data, as K_{cy} and K_{cx} approach zero:

$$a_1 \rightarrow \alpha_1,$$

$$a_2 \rightarrow \alpha_2,$$

$$a_3 \rightarrow 0, \tag{5.4.8}$$

$$a_4 \rightarrow 0.$$

[2] The dependence on order can be reduced by using "orthogonal polynomials." The subject of orthogonal polynomials is discussed by F. B. Hildebrand, "Introduction to Numerical Analysis," McGraw-Hill Book Co., New York, 1956.

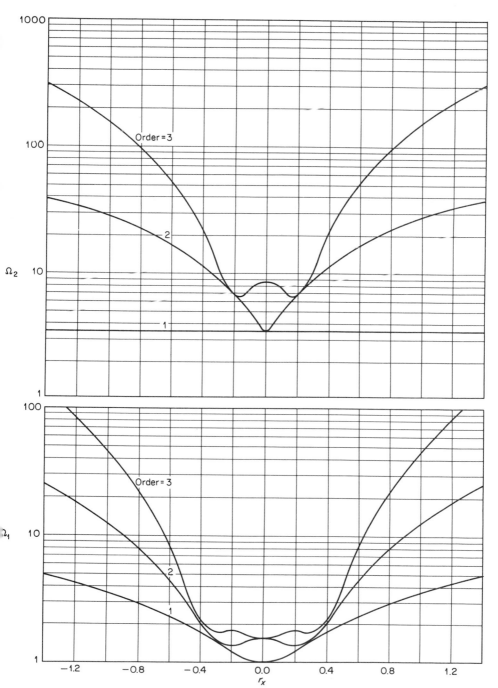

Figure 5.12. Ω_1 and Ω_2 as functions of r_x and the order of the polynomial for $n = 24$ and $K_{cx} = 0$.

A comparison of Equations 5.4.6, 5.4.7, and 5.4.8 suggests that as long as the order of the polynomial is greater than the order required by the realities of the problem, correct values of the parameters will be obtained. For those cases in which the uncertainties of the individual data points approach zero, this is true. However, for real experiments with finite uncertainties the values of the σ_{a_k}'s as well as the a_k's must be considered. By increasing the order of the polynomial beyond the required order, the uncertainties σ_{a_1} and σ_{a_2} are increased. Prediction analysis is next applied to the problem of estimating how large this loss of accuracy will be.

Results for Ω_1 and Ω_2 (defined by Equations 5.3.17 and 5.3.18) are plotted in Figure 5.12 as a function of r_x and the order of the polynomial. The results were obtained by direct computation with $n = 24$. The results are, however, good approximations for small as well as large values of n. From the curves it is seen that the loss of accuracy is dependent upon the value of r_x. If the absolute value of r_x is large, the loss of accuracy can be considerable.

As an example, consider a set of data with values of x evenly spaced from 2.5 to 11.5. If the total number of data points is 10, then:

$$n \cdot \Delta x = n \cdot \frac{x_n - x_1}{n - 1} = \tfrac{10}{9} \cdot (11.5 - 2.5) = 10.0. \qquad (5.4.9)$$

The value of r_x is:

$$r_x = \frac{x_{\mathrm{av}}}{n \cdot \Delta x} = \frac{(11.5 + 2.5)/2}{10.0} = 0.70. \qquad (5.4.10)$$

From Figure 5.12, the values of Ω_1 and Ω_2 can be determined for $r_x = 0.70$ for orders of one, two, and three:

TABLE 5.10. Ω_1 and Ω_2 as functions of the order of the polynomial for $r_x = 0.70$

Order	Ω_1	Ω_2
1	2.94	3.47
2	6.08	19.2
3	14.2	73.4

From Equations 5.3.17 and 5.3.18, equations for $\hat{\sigma}_{a_1}$ and $\hat{\sigma}_{a_2}$ can be derived:

$$\hat{\sigma}_{a_1} = \frac{\Omega_1 \cdot K_{cy} \cdot (1 + \omega_c^2)^{1/2}}{n^{1/2}}, \qquad (5.4.11)$$

$$\hat{\sigma}_{a_2} = \frac{\Omega_2 \cdot K_{cy} \cdot (1 + \omega_c^2)^{1/2}}{n^{1/2} \cdot (n \cdot \Delta x)}. \qquad (5.4.12)$$

TABLE 5.11. $\hat{\sigma}_{a_1}$ and $\hat{\sigma}_{a_2}$ as functions of the order of the polynomial for $r_x = 0.70$, $K_{cy} = 1.0$, $\omega_c = 0.0$, and $n = 24$

Order	$\hat{\sigma}_{a1}$	$\hat{\sigma}_{a2}$
1	0.933	0.110
2	1.93	0.610
3	4.50	2.33

Using Equations 5.4.11 and 5.4.12, the values of Ω listed in Table 5.10, $\omega_c = 0.0$ and a value of $K_{cy} = 1.0$, values of $\hat{\sigma}_{a_1}$ and $\hat{\sigma}_{a_2}$ can be determined as functions of the order of the polynomial and are listed in Table 5.11 for orders of one, two, and three.

A comparison of the values of $\hat{\sigma}_{a_1}$ and $\hat{\sigma}_{a_2}$ in Table 5.11 suggests that there is a considerable loss of accuracy if the order is increased beyond the correct order for the experiment (which for this example is one). For $r_x = 0.7$, increasing the order from one to two results in an increase in $\hat{\sigma}_{a_1}$ by a factor of more than two, and an increase in $\hat{\sigma}_{a_2}$ by a factor of more than five. It should be emphasized, however, that using an order *less* than the correct order will also lead to considerable inaccuracies.

It should be concluded that it is inadvisable to analyze data using a polynomial of a higher-order than is required. For experiments in which the order of the polynomial is unknown, it is worthwhile to re-analyze the data using several different values for the order. It might be possible to ascertain the correct order by examining the results to see at which order the a_k's "stabilize."

5.5 Summary

In this chapter experiments for which the polynomial function is used to relate the dependent and independent variables have been considered. The polynomials discussed in the chapter are listed in Table 5.12.

In Sections 5.2 and 5.3 the cases considered were *constant error, constant fractional error, and counting statistics* for the dependent variable. For the independent variable, only *constant or zero error* was considered.

Several generalizations can be made about the results.

1. The predicted uncertainties are inversely proportional to the square root of the number of data points.

TABLE 5.12. Summary of the functions discussed in Chapter 5

Function	Section
$y = a_1 + a_2 x$	5.2
$y = a_1 + a_2 x + a_3 x^2$	5.3
$y = \sum_{k=1}^{p} a_k x^{k-1}$	5.4

2. The predicted uncertainties are proportional to the constant K_{cy}, K_{fy}, or K_{sy}.

3. If all other factors are held constant, increasing the order of the polynomial beyond the correct order increases the predicted uncertainties.

As an example of how the results in this chapter might be used, consider the following problem. The dependent and independent variables are to be related by a second-order polynomial. All values of the dependent variable y are to be measured to 1% accuracy, and the uncertainties for the measured values of x are assumed to be zero. The range of values of x is 1.0 to 10.0, and the expected values of the unknown parameters are $a_1 = 1.0$, $a_2 = 1.0$, and $a_3 = 10.0$. If the number of data points is to be ten, and if the points are evenly spaced over the entire range, what are the predicted uncertainties for the unknown parameters a_1, a_2, and a_3? Spacing ten points over the range 1.0 to 10.0 implies that $\Delta x = 1.0$. From Equations 5.2.2, 5.2.3, 5.3.21, and 5.2.49, the values of r_x, r_a, r_p, and ω_f are 0.55, 10.0, 10.0, and 0.0, respectively. For this problem, Case II of Section 5.3 is applicable, and from Figure 5.10, $\Theta_1 = 80$, $\Theta_2 = 100$, and $\Theta_3 = 2.4$. From Equation 5.3.30, $\hat{\sigma}_{a_1} = 80 \times 0.01 \times 1.0/(10)^{1/2} = 0.25$, $\hat{\sigma}_{a_2} = 100 \times 0.01 \times 1.0/(10)^{1/2} = 0.32$, and $\hat{\sigma}_{a_3} = 2.4 \times 0.01 \times 10/(10)^{1/2} = 0.076$. It is therefore seen that only a_3 can be determined to high precision in the proposed experiment (i.e., 0.76%). The parameters a_1 and a_2 can be determined to accuracies of about 25% and 32%.

It should be noted that Figure 5.10 is applicable to this problem only because $r_a = 10.0$ and $\omega_f = 0.0$. If data are required for combinations of the dimensionless groups not included in this chapter, it might be possible to determine the required information by interpolation or extrapolation. An alternative procedure is to extend the results to the required combinations using the techniques and equations included in Chapter 4 and in this chapter.

Problems

1. The dependent and independent variables for a proposed experiment are to be related by a first-order polynomial. The values of y_i are to be determined to 1% accuracy, and the values of x are to be determined to ± 0.1. The expected values of a_1 and a_2 are 10.0 and 2.0, respectively. The values of x are to be evenly spaced with the minimum value equal to 5.0, the maximum value equal to 15.0, and the spacing between points equal to 0.5. Predict the values of σ_{a_1} and σ_{a_2} for the proposed experiment. Ans.: $\hat{\sigma}_{a_1} = 0.36$, $\hat{\sigma}_{a_2} = 0.026$.

2. The predicted uncertainties for a_1 and a_2 of Problem I exceed the objectives of the proposed experiment. If both a_1 and a_2 are to be determined to at least 1% accuracy, to what value should the number of data points be increased? Assume that the minimum and maximum values of x remain the same, and only the spacing between the points is changed.

3. The dependent and independent variables for a proposed experiment are to be related by a second-order polynomial. The uncertainties for the values of y_i are to be estimated by "counting statistics," and the uncertainties for the values of x are negligible. The product $n \cdot \Delta x$ is equal to 10.0, and the expected values of a_1, a_2, and a_3 are each equal to 2.0. What choice of values of x_a and x_b minimize the predicted values of σ_{a_1}, σ_{a_2}, and σ_{a_3}? Using these values for x_a and x_b, what are the 95% confidence interval half-widths for a_1, a_2, and a_3 if n is 4, 10, or 100?

4. The time requirement for the experiment described in Problem 3 is 10 minutes per data point. If the number of data points is 10, and if all the a_n's are to be determined to at least 1% accuracy, is 10 minutes per data point sufficient? If it is not, what value should be chosen? (Note—Assume that the values of y_i are proportional to the time per data point. The expected values of the a_n's are also dependent upon the time per data point.)

5. The dependent and independent variables for a proposed experiment are to be related by a first-order polynomial. The values of y_i are to be determined to ±2.0, the values of x are to be determined to ±0.5, and the expected values of a_1 and a_2 are 8.0 and 2.0. The values of x are to be evenly spaced with the minimum value equal to 0.0, the maximum value equal to 20.0, and the spacing between points equal to 2.0. The curve resulting from a least squares analysis of the experimental data is to be used for extrapolation to a value of x equal to 30. Predict the uncertainty and 95% confidence interval half-width for the value of y corresponding to $x = 30$.

Computer Projects

1. Recompute the results presented in Figure 5.10 for $r_a = 5.0$ and 20.0, as well as for 10.0.

2. Recompute the results presented in Figure 5.11 for $r_a = 5.0$ and 20.0 as well as for 10.0.

3. Predict the values of σ_{a_1}, σ_{a_2}, and σ_{12} as functions of the experimental variables for the following cases of the "first-order polynomial" class of experiments:

(A) $\sigma_{y_i} = K_{fy} \cdot | y_i |$ $\sigma_{x_i} = K_{fy} \cdot | x_i |$,
(B) $\sigma_{y_i}^2 = (K_{fy} \cdot y_i)^2 + K_{cy}^2$ $\sigma_{x_i} = 0$,
(C) $\sigma_{y_i}^2 = K_{sy}^2 \cdot y_i + K_{cy}^2$ $\sigma_{x_i} = 0, y \geq 0$.
For all cases, the values of x are to be evenly spaced.

4. Predict the values of σ_{a_1}, σ_{a_2}, and σ_{a_3} as functions of the experimental variables for the same cases as in the previous problem but for the "second-order polynomial" class of experiments.

5. Predict the values of σ_{a_1}, σ_{a_2}, and σ_{12} as functions of the experimental variables for the following cases of the "first-order polynomial" class of experiments:

(A) $\sigma_{y_i} = K_{cy}$; $\sigma_{x_i} = K_{cy}$; $x_{i+1} = x_i + i \cdot \Delta x$,
(B) Same as A except $x_{i+1} = x_i + \Delta x/i$.
Compare the results for these cases to the results from Case I of Section 5.2.

Chapter 6

The Exponential

6.1 Introduction

The exponential function is required for analysis of many types of experiments. The simplest experiments to analyze are those in which the dependent and independent variables are related by the *simple exponential* function:

$$y = a_1 \exp (-a_2 x). \qquad (6.1.1)$$

Often, however, an additional term must be included:

$$y = a_1 \exp (-a_2 x) + a_3. \qquad (6.1.2)$$

Equation 6.1.2 is referred to as the *exponential-plus-background* function. The parameter a_3 is the *background* for the variable y.

Equations 6.1.1 and 6.1.2 can be generalized to the *sum-of-exponentials* and *sum-of-exponentials-plus-background* functions:

$$y = a_1 \exp (-a_2 x) + a_3 \exp (-a_4 x) + \cdots + a_{p-1} \exp (-a_p x), \qquad (6.1.3)$$

and

$$y = a_1 \exp (-a_2 x) + a_3 \exp (-a_4 x) + \cdots + a_{p-2} \exp (-a_{p-1} x) + a_p, \qquad (6.1.4)$$

where p is the number of unknown parameters.

Analysis of these experiments is, of course, independent of the physical nature of the variables. For many of the experiments that fall within the general category of *exponential experiments*, the independent variable x represents time. For such experiments, y is referred to as a *time dependent variable* and is said to "decay" or "build up" as a function of time.

The notation system used throughout the book denotes the unknown parameters as a_k ($k = 1, 2, \cdots, p$). Therefore an implication of Equations

6.1.1 through 6.1.4 is that the amplitudes and background (i.e., a_1, a_3, $\cdots$), and the decay constants (i.e., a_2, a_4, $\cdots$) are all unknown parameters. For many experiments, the decay constants are known and only the amplitudes and background are unknown (for example, an experiment in which the unknown proportions of several known radioactive isotopes are to be determined). For these experiments, the notation must be modified. If the kth decay constant is denoted as λ_k, Equations 6.1.1 through 6.1.4 are replaced by Equations 6.1.5 through 6.1.8:

$$y = a_1 \exp\left(-\lambda_1 x\right), \tag{6.1.5}$$

$$y = a_1 \exp\left(-\lambda_1 x\right) + a_2, \tag{6.1.6}$$

$$y = a_1 \exp\left(-\lambda_1 x\right) + a_2 \exp\left(-\lambda_2 x\right) + \cdots + a_p \exp\left(-\lambda_p x\right), \tag{6.1.7}$$

and

$$y = a_1 \exp\left(-\lambda_1 x\right) + a_2 \exp\left(-\lambda_2 x\right) + \cdots + a_{p-1} \exp\left(-\lambda_{p-1} x\right) + a_p. \tag{6.1.8}$$

The decay constants λ_k are known and can be considered as input quantities to the analysis.

Experiments of the types represented by Equations 6.1.1 through 6.1.4 require nonlinear least squares analysis of the experimental data. If the number of unknown parameters (i.e., p) is large, the convergence problems may be severe. Even if p is only four or five, if the decay constants (a_2 and a_4) are close in value, or if one of the amplitudes (a_1 or a_3) is close to zero, then it will be difficult to achieve convergence. (See the paragraph following Equation 4.8.14).

Least squares analyses of experiments of the types represented by Equations 6.1.5 through 6.1.8 are not usually plagued by convergence problems. In fact, if the uncertainties in the values of x (i.e., the $\hat{\sigma}_{x_i}$'s) are negligible, the least squares analyses are linear.

It is assumed in this chapter that systematic errors are negligible. If systematic errors are considered as an important factor in a proposed experiment, the prediction analysis should be amended using the techniques discussed in Section 4.9.

6.2 The Simple Exponential

Equation 6.1.1 is the relationship between the dependent and independent variables for experiments in this class:

$$y = a_1 \exp\left(-a_2 x\right). \tag{6.1.1}$$

The cases within this class that are considered in this section are summarized in Table 6.1.

TABLE 6.1. Summary of cases analyzed for the *simple exponential* class of experiments

Case	σ_{y_i}	σ_{x_i}	Equations		
			$\hat{\sigma}_{a_1}$	$\hat{\sigma}_{a_2}$	$\hat{\sigma}_{12}$
I	K_{cy}	K_{cx}	6.2.23	6.2.24	6.2.25
II	$K_{fy} \cdot y_i$	K_{cx}	6.2.50	6.2.51	6.2.52
III	$K_{sy} \cdot (y_i)^{1/2}$	K_{cx}	6.2.70	6.2.71	6.2.72

Case I. *Constant uncertainty for both the x and y variables.*

For this case the values of σ_{y_i} and σ_{x_i} are:

$$\sigma_{y_i} = K_{cy},$$
$$\sigma_{x_i} = K_{cx}. \qquad (6.2.1)$$

The equation for $\hat{L}_i$ is determined by substituting Equation 6.2.1 into Equation 3.6.16:

$$\hat{L}_i = K_{cy}^2 + (\hat{F}_x{}^i)^2 K_{cx}^2. \qquad (6.2.2)$$

The symbol $\hat{F}_x{}^i$ denotes the derivative of $\hat{F}^i$ with respect to x_i:

$$\hat{F}^i = y_i - a_1 \exp(-a_2 x_i), \qquad (6.2.3)$$

and therefore

$$\hat{F}_x{}^i = a_1 a_2 \exp(-a_2 x_i). \qquad (6.2.4)$$

Substituting Equation 6.2.4 into Equation 6.2.2 gives:

$$\hat{L}_i = K_{cy}^2 + [a_1 a_2 \exp(-a_2 x_i)]^2 K_{cx}^2. \qquad (6.2.5)$$

The derivatives $\hat{F}_{a_1}{}^i$ and $\hat{F}_{a_2}{}^i$ are also required:

$$\hat{F}_{a_1}{}^i = -\exp(-a_2 x_i), \qquad (6.2.6)$$

$$\hat{F}_{a_2}{}^i = a_1 x_i \exp(-a_2 x_i). \qquad (6.2.7)$$

Substituting Equations 6.2.5, 6.2.6, and 6.2.7 into Equation 4.4.1 gives:

$$\hat{C}_{11} = \sum_{i=1}^{n} \frac{\exp(-2a_2 x_i)}{K_{cy}^2 + [a_1 a_2 \exp(-a_2 x_i)]^2 K_{cx}^2}, \qquad (6.2.8)$$

$$\hat{C}_{12} = \sum_{i=1}^{n} \frac{-a_1 x_i \exp(-2a_2 x_i)}{K_{cy}^2 + [a_1 a_2 \exp(-a_2 x_i)]^2 K_{cx}^2}, \qquad (6.2.9)$$

and

$$\hat{C}_{12} = \sum_{i=1}^{n} \frac{a_1^2 x_i^2 \exp{(-2a_2 x_i)}}{K_{cy}^2 + [a_1 a_2 \exp{(-a_2 x_i)}]^2 K_{cx}^2} \cdot \qquad (6.2.10)$$

If K_{cx} is negligibly small, Equations 6.2.8 through 6.2.10 reduce to much simpler forms:

$$\hat{C}_{11} = \sum_{i=1}^{n} \frac{\exp{(-2a_2 x_i)}}{K_{cy}^2} = \frac{n}{K_{cy}^2} \cdot [\exp{(-2a_2 x)}]_{av}, \qquad (6.2.11)$$

$$\hat{C}_{12} = \sum_{i=1}^{n} \frac{-a_1 x_i \exp{(-2a_2 x_i)}}{K_{cy}^2} = \frac{n}{K_{cy}^2} \cdot [-a_1 x \exp{(-2a_2 x)}]_{av}, (6.2.12)$$

$$\hat{C}_{22} = \sum_{i=1}^{n} \frac{a_1^2 x_i^2 \exp{(-2a_2 x_i)}}{K_{cy}^2} = \frac{n}{K_{cy}^2} \cdot [a_1^2 x^2 \exp{(-2a_2 x)}]_{av}. \qquad (6.2.13)$$

The subscript av denotes the average values of the quantities within the parentheses.

If it is assumed that the values of x are evenly spaced (i.e., Equation 4.4.3 is valid), then integral expressions for the required average values can be developed. As n approaches infinity:

$$[\exp{(-2a_2 x)}]_{av} \rightarrow \int_{x_a}^{x_b} \exp{(-2a_2 x)}\, dx \Big/ (x_b - x_a)$$

$$= \frac{\exp{(-2a_2 x_a)} - \exp{(-2a_2 x_b)}}{2 \cdot a_2 \cdot (x_b - x_a)}, \qquad (6.2.14)$$

where x_a and x_b are defined by Equations 4.4.7 and 4.4.8:

$$x_a = x_1 - \frac{\Delta x}{2}, \qquad (4.4.7)$$

$$x_b = x_n + \frac{\Delta x}{2}. \qquad (4.4.8)$$

Defining the dimensionless groups z_a and z_b:

$$z_a \equiv a_2 \cdot x_a, \qquad (6.2.15)$$

$$z_b \equiv a_2 \cdot x_b, \qquad (6.2.16)$$

and substituting these groups into Equation 6.2.14, we obtain:

$$[\exp{(-2a_2 x)}]_{av} \rightarrow \frac{\exp{(-2z_a)} - \exp{(-2z_b)}}{2 \cdot (z_b - z_a)}. \qquad (6.2.17)$$

Similarly:

$$[-a_1 x \exp(-2a_2 x)]_{av} \rightarrow$$

$$-\frac{a_1}{a_2} \cdot \frac{(z_a + \frac{1}{2}) \exp(-2z_a) - (z_b + \frac{1}{2}) \exp(-2z_b)}{2 \cdot (z_b - z_a)}, \quad (6.2.18)$$

$$[a_1^2 x^2 \exp(-2a_2 x)]_{av} \rightarrow$$

$$\frac{a_1^2}{a_2^2} \cdot \frac{(z_a^2 + z_a + \frac{1}{2}) \exp(-2z_a) - (z_b^2 + z_b + \frac{1}{2}) \exp(-2z_b)}{2 \cdot (z_b - z_a)}. \quad (6.2.19)$$

The predicted variances $\hat{\sigma}_{a1}^2$ and $\hat{\sigma}_{a2}^2$ and the predicted covariance $\hat{\sigma}_{12}$ are related to the $\hat{C}_{jk}$'s:

$$\hat{\sigma}_{a1}^2 = \hat{C}_{11}^{-1} = \frac{\hat{C}_{22}}{\hat{C}_{11} \cdot \hat{C}_{22} - \hat{C}_{12}^2}, \quad (6.2.20)$$

$$\hat{\sigma}_{a2}^2 = \hat{C}_{22}^{-1} = \frac{\hat{C}_{11}}{\hat{C}_{11} \cdot \hat{C}_{22} - \hat{C}_{12}^2}, \quad (6.2.21)$$

$$\hat{\sigma}_{12} = \hat{C}_{12}^{-1} = \frac{-\hat{C}_{12}}{\hat{C}_{11} \cdot \hat{C}_{22} - \hat{C}_{12}^2}. \quad (6.2.22)$$

Equations for the $\hat{\sigma}$'s can be derived by substituting Equations 6.2.17 through 6.2.19 into Equations 6.2.11 through 6.2.13 and then into Equations 6.2.20 through 6.2.22. After rearranging terms, a new set of dimensionless groups are defined, and as n approaches infinity:

$$\Lambda_1 \equiv \frac{\hat{\sigma}_{a1} \cdot n^{1/2}}{K_{cy} \cdot (z_b - z_a)^{1/2}} \rightarrow$$

$$\left(\frac{2 \cdot \{(z_a^2 + z_a + \frac{1}{2}) - (z_b^2 + z_b + \frac{1}{2}) \exp[-2(z_b - z_a)]\}}{D_1} \right)^{1/2}, \quad (6.2.23)$$

$$\Lambda_2 \equiv \frac{\hat{\sigma}_{a2} \cdot a_1 \cdot n^{1/2}}{K_{cy} \cdot a_2 \cdot (z_b - z_a)^{1/2}} \rightarrow \left(\frac{2 \cdot \{1 - \exp[-2(z_b - z_a)]\}}{D_1} \right)^{1/2}, \quad (6.2.24)$$

$$\Lambda_{12} \equiv \frac{\hat{\sigma}_{12} \cdot a_1 \cdot n}{K_{cy}^2 \cdot a_2 \cdot (z_b - z_a)} \rightarrow$$

$$\frac{2 \cdot \{(z_a + \frac{1}{2}) - (z_b + \frac{1}{2}) \exp[-2(z_b - z_a)]\}}{D_1}, \quad (6.2.25)$$

where D_1 is defined as:

$$D_1 \equiv \exp\left(-2z_a\right) \cdot \left(\{1 - \exp\left[-2(z_b - z_a)\right]\}\right.$$
$$\times \{(z_a^2 + \tfrac{1}{2}z_a + \tfrac{1}{2}) - (z_b^2 + \tfrac{1}{2}z_b + \tfrac{1}{2}) \exp\left[-2(z_b - z_a)\right]\}$$
$$\left. - \{(z_a + \tfrac{1}{2}) - (z_b + \tfrac{1}{2}) \exp\left[-2(z_b - z_a)\right]\}^2\right). \quad (6.2.26)$$

Equation 6.2.26 can be simplified:

$$D_1 = \exp\left(-2z_a\right) \cdot \{\tfrac{1}{4} - (2z_az_b - z_b^2 - z_a^2 - \tfrac{1}{2})$$
$$\times \exp\left[-2(z_b - z_a)\right] + \tfrac{1}{4} \exp\left[-4(z_b - z_a)\right]\}. \quad (6.2.27)$$

The Λ's are plotted in Figure 6.1 as functions of z_a and $z_b - z_a$. The curves are good approximations for most values of n. For example, for $z_a = 0.0$ and $z_b - z_a = 5.0$, Λ_1 for $n = 3$ is 2.56 times the asymptotic value, but for $n = 13$, it is only 1.05 times the asymptotic value.

From Figure 6.1 it is seen that the curves approach asymptotes as the value of $z_b - z_a$ increases. In fact, the curve for $z_b - z_a$ equal to infinity is very close to the curve for $z_b - z_a$ equal to five.

If K_{cx} is not neglected, the derivation of equations for the Λ's is extremely complicated. A simpler procedure is to use the direct approach which is discussed in Section 4.4. Using a value of $n = 24$, values of Λ_1 and Λ_2 were determined for several values of ω_c which is defined as follows:

$$\omega_c \equiv \frac{a_1 \cdot a_2 \cdot K_{cx}}{K_{cy}}. \quad (6.2.28)$$

The results are plotted in Figure 6.2. It is seen that the deviation between the curves is greatest for $z_a = 0$. The curves are exact for $n = 24$ and are good approximations for most values of n. For example, for $z_a = 0.0$, $z_b - z_a = 5.0$, and $\omega_c = 1.0$, the values of Λ_1 for $n = 3, 13, 24$ and infinity are 5.26, 2.58, 2.53, and 2.51, respectively.

Case II. *Constant fractional uncertainty for the variable y and constant uncertainty for the variable x.*

For this case the values of σ_{y_i} and σ_{x_i} are:

$$\sigma_{y_i} = K_{fy} \cdot y_i, \quad (6.2.29)$$

$$\sigma_{x_i} = K_{cx}.$$

The equation for $\hat{L}_i$ is determined by substituting Equations 6.2.29 and 6.2.4 into Equation 3.6.16:

$$\hat{L}_i = K_{fy}^2 \cdot y_i^2 + [a_1a_2 \exp\left(-a_2x_i\right)]^2 K_{cx}^2. \quad (6.2.30)$$

Equations 6.2.6 and 6.2.7 for $\hat{F}_{a_1}{}^i$ and $\hat{F}_{a_2}{}^i$ are valid for Case II as well as

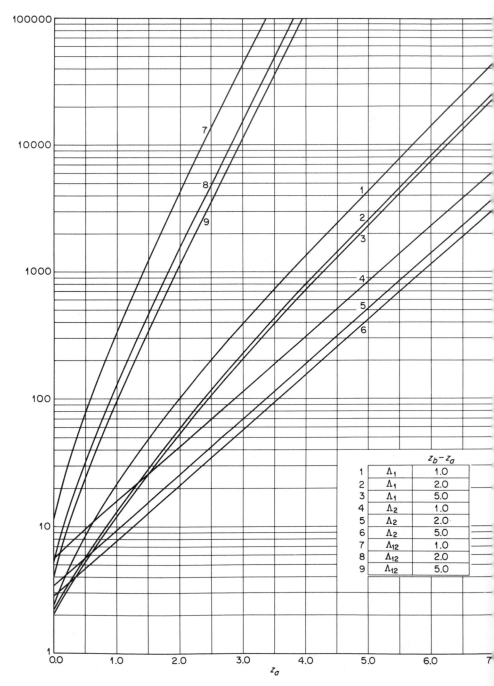

Figure 6.1. Λ_1, Λ_2, and Λ_{12} as function of z_a and $z_b - z_a$ for $K_{cx} = 0$ and $n \to \infty$.

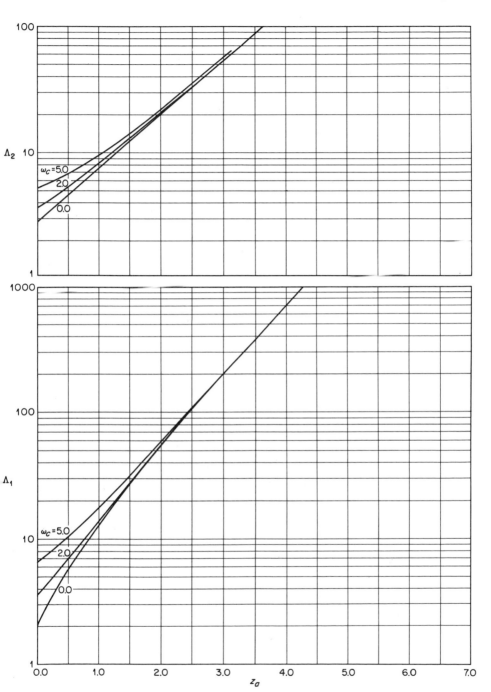

Figure 6.2. Λ_1 and Λ_2 as functions of z_a and ω_c for $z_b - z_a = 5.0$, and $n = 24$.

for Case I. Substituting Equations 6.2.6, 6.2.7, and 6.2.30 into Equation 4.4.1 gives:

$$\hat{C}_{11} = \sum_{i=1}^{n} \frac{\exp\left(-2a_2 x_i\right)}{K_{fy}^2 y_i^2 + [a_1 a_2 \exp\left(-a_2 x_i\right)]^2 K_{cx}^2}, \tag{6.2.31}$$

$$\hat{C}_{12} = \sum_{i=1}^{n} \frac{-a_1 x_i \exp\left(-2a_2 x_i\right)}{K_{fy}^2 y_i^2 + [a_1 a_2 \exp\left(-a_2 x_i\right)]^2 K_{cx}^2}, \tag{6.2.32}$$

$$\hat{C}_{22} = \sum_{i=1}^{n} \frac{a_1^2 x_i^2 \exp\left(-2a_2 x_i\right)}{K_{fy}^2 y_i^2 + [a_1 a_2 \exp\left(-a_2 x_i\right)]^2 K_{cx}^2}. \tag{6.2.33}$$

Substituting Equation 6.1.1 into Equations 6.2.31 through 6.2.33, defining the dimensionless group ω_f:

$$\omega_f \equiv \frac{a_2 \cdot K_{cx}}{K_{fy}}, \tag{6.2.34}$$

and canceling the factor $\exp\left(-2a_2 x_i\right)$ from the numerators and denominators, it can be seen that:

$$\hat{C}_{11} = \frac{n}{K_{fy}^2 a_1^2} \cdot \left(\frac{1}{1 + \omega_f^2}\right), \tag{6.2.35}$$

$$\hat{C}_{12} = \frac{n}{K_{fy}^2 a_1^2} \cdot \left(\frac{-a_1 x}{1 + \omega_f^2}\right)_{av}, \tag{6.2.36}$$

$$\hat{C}_{22} = \frac{n}{K_{fy} a_1^2} \cdot \left(\frac{a_1^2 x^2}{1 + \omega_f^2}\right)_{av}. \tag{6.2.37}$$

If it is assumed that the values of x are evenly spaced (i.e., Equation 4.4.3 is valid), then the required average values can be estimated by integration over the applicable range. As n approaches infinity:

$$\frac{1}{1 + \omega_f^2} = \left(\frac{1}{1 + \omega_f^2}\right) \cdot \frac{z_b - z_a}{z_b - z_a}, \tag{6.2.38}$$

$$\left(\frac{-a_1 x}{1 + \omega_f^2}\right)_{av} \rightarrow \left(\frac{-a_1/a_2}{1 + \omega_f^2}\right) \cdot \frac{\frac{1}{2}(z_b^2 - z_a^2)}{z_t - z_a}, \tag{6.2.39}$$

and

$$\left(\frac{a_1^2 x^2}{1 + \omega_f^2}\right)_{av} \rightarrow \left(\frac{a_1^2/a_2^2}{1 + \omega_f^2}\right) \cdot \frac{\frac{1}{3}(z_b^3 - z_a^3)}{z_b - z_a}, \tag{6.2.40}$$

where z_a and z_b are defined by Equations 6.2.15 and 6.2.16. Substituting

Equations 6.2.35 through 6.2.40 into Equations 6.2.20 through 6.2.22, as n approaches infinity, gives:

$$\hat{\sigma}_{a_1}^2 = \hat{C}_{11}^{-1} \rightarrow \frac{K_{fy}^2}{n} \cdot a_1^2 \cdot (1 + \omega_f^2) \cdot (z_b - z_a) \cdot \frac{\frac{1}{3}(z_b^3 - z_a^3)}{D_2}, \qquad (6.2.41)$$

$$\hat{\sigma}_{a_2}^2 = \hat{C}_{22}^{-1} \rightarrow \frac{K_{fy}^2}{n} \cdot a_2^2 \cdot (1 + \omega_f^2) \cdot (z_b - z_a) \cdot \frac{z_b - z_a}{D_2}, \qquad (6.2.42)$$

$$\hat{\sigma}_{12} = \hat{C}_{12}^{-1} \rightarrow \frac{K_{fy}^2}{n} \cdot a_1 \cdot a_2 \cdot (1 + \omega_f^2) \cdot (z_b - z_a) \cdot \frac{\frac{1}{2}(z_b^2 - z_a^2)}{D_2}, \qquad (6.2.43)$$

where D_2 is defined as:

$$D_2 \equiv \tfrac{1}{3}(z_b^3 - z_a^3) \cdot (z_b - z_a) - \tfrac{1}{4}(z_b^2 - z_a^2)^2. \qquad (6.2.44)$$

Rearranging terms, and defining a new set of dimensionless groups, as n approaches infinity, gives:

$$\vartheta_1 \equiv \frac{\hat{\sigma}_{a_1} \cdot n^{1/2}}{K_{fy} a_1 \cdot (1 + \omega_f^2)^{1/2}} \rightarrow \left(\frac{\frac{1}{3}(z_b^3 - z_a^3) \cdot (z_b - z_a)}{D_2} \right)^{1/2}, \qquad (6.2.45)$$

$$\vartheta_2 \equiv \frac{\hat{\sigma}_{a_2} \cdot n^{1/2}}{K_{fy} a_2 \cdot (1 + \omega_f^2)^{1/2}} \rightarrow \left(\frac{(z_b - z_a)^2}{D_2} \right)^{1/2}, \qquad (6.2.46)$$

$$\vartheta_{12} \equiv \frac{\hat{\sigma}_{12} \cdot n}{K_{fy}^2 a_1 \cdot a_2 \cdot (1 + \omega_f^2)} \rightarrow \frac{\frac{1}{2}(z_b^2 - z_a^2)(z_b - z_a)}{D_2}. \qquad (6.2.47)$$

Equations 6.2.45 through 6.2.47 can be simplified considerably. Using the following equalities:

$$z_b^3 - z_a^3 = (z_b^2 + z_b z_a + z_a^2) \cdot (z_b - z_a), \qquad (6.2.48)$$

and

$$z_b^2 - z_a^2 = (z_b + z_a) \cdot (z_b - z_a), \qquad (6.2.49)$$

it can be shown that the equations reduce to:

$$\vartheta_1 \rightarrow \left(4 + \frac{12 z_b z_a}{(z_b - z_a)^2} \right)^{1/2}, \qquad (6.2.50)$$

$$\vartheta_2 \rightarrow \frac{(12)^{1/2}}{z_b - z_a}, \qquad (6.2.51)$$

and

$$\vartheta_{12} \rightarrow \frac{6(z_b + z_a)}{(z_b - z_a)^2}. \qquad (6.2.52)$$

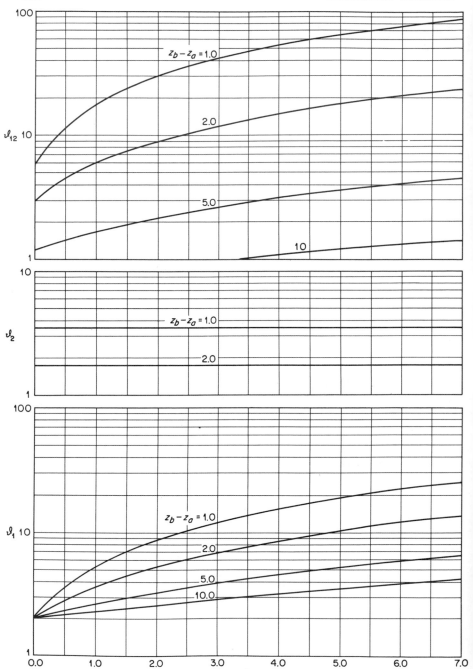

Figure 6.3. ϑ_1, ϑ_2, and ϑ_{12} as functions of z_a and $z_b - z_a$ for $n \rightarrow \infty$, and for all values of ω_f.

The ϑ's are plotted in Figure 6.3 as functions of z_a and $z_b - z_a$. Although the curves are only exact for n equal to infinity, they are good approximations for all possible values of n. For example, for $z_a = 0.0$, $z_b - z_a = 5$, and $n = 3, 13, 23$ and infinity, $\vartheta_1 = 2.0916, 2.0045, 2.0014$, and 2.000, respectively.

Case III. *Counting statistics for the variable y and constant uncertainty for the variable x.*

For this case the values of σ_{y_i} and σ_{x_i} are:

$$\sigma_{y_i} = K_{sy} \cdot (y_i)^{1/2},$$
$$\sigma_{x_i} = K_{cx}.$$
(6.2.53)

The equation for $\hat{L}_i$ is determined by substituting Equations 6.2.53 and 6.2.4 into Equation 3.6.16:

$$\hat{L}_i = K_{sy}^2 y_i + [a_1 a_2 \exp(-a_2 x_i)]^2 K_{cx}^2.$$
(6.2.54)

Equations 6.2.6 and 6.2.7 are valid for Case III as well as Case I. Substituting Equations 6.2.6, 6.2.7, and 6.2.54 into Equation 4.4.1:

$$\hat{C}_{11} = \sum_{i=1}^{n} \frac{\exp(-2a_2 x_i)}{K_{sy}^2 y_i + [a_1 a_2 \exp(-a_2 x_i)]^2 K_{cx}^2},$$
(6.2.55)

$$\hat{C}_{12} = \sum_{i=1}^{n} \frac{-a_1 x_i \exp(-2a_2 x_i)}{K_{sy}^2 y_i + [a_1 a_2 \exp(-a_2 x_i)]^2 K_{cx}^2},$$
(6.2.56)

$$\hat{C}_{22} = \sum_{i=1}^{n} \frac{a_1^2 x_i^2 \exp(-2a_2 x_i)}{K_{sy}^2 y_i + [a_1 a_2 \exp(-a_2 x_i)]^2 K_{cx}^2}.$$
(6.2.57)

Substituting Equation 6.1.1 into Equations 6.2.31 through 6.2.33, defining the dimensionless group ω_s:

$$\omega_s \equiv \frac{a_2 \cdot (a_1)^{1/2} \cdot K_{cx}}{K_{sy}},$$
(6.2.58)

and canceling the factor $\exp(-a_2 x_i)$ from the numerators and denominators, it can be seen that:

$$\hat{C}_{11} = \frac{n}{K_{sy}^2 a_1} \cdot \left(\frac{\exp(-a_2 x)}{1 + \omega_s^2 \exp(-a_2 x)} \right)_{av},$$
(6.2.59)

$$\hat{C}_{12} = \frac{n}{K_{sy}^2 a_1} \cdot \left(\frac{-a_1 x \exp(-a_2 x)}{1 + \omega_s^2 \exp(-a_2 x)} \right)_{av},$$
(6.2.60)

$$\hat{C}_{22} = \frac{n}{K_{sy}^2 a_1} \cdot \left(\frac{a_1^2 x^2 \exp(-a_2 x)}{1 + \omega_s^2 \exp(-a_2 x)} \right)_{av}.$$
(6.2.61)

If it is assumed that $\omega_s = 0$ and that the values of x are evenly spaced (i.e., Equation 4.4.3 is valid), then the required average values can be estimated by integration over the applicable range. As n approaches infinity:

$$[\exp{(-a_2x)}]_{av} \rightarrow \frac{\exp{(-z_a)} - \exp{(-z_b)}}{z_b - z_a}, \tag{6.2.62}$$

$$[-a_1x \exp{(-a_2x)}]_{av} \rightarrow$$

$$-\frac{a_1}{a_2} \cdot \frac{(z_a + 1) \exp{(-z_a)} - (z_b + 1) \exp{(-z_b)}}{z_b - z_a}, \tag{6.2.63}$$

$$[a_1^2x^2 \exp{(-a_2x)}]_{av} \rightarrow$$

$$\frac{a_1^2}{a_2^2} \cdot \frac{(z_a^2 + 2z_a + 2) \exp{(-z_a)} - (z_b^2 + 2z_b + 2) \exp{(-z_b)}}{z_b - z_a}, \tag{6.2.64}$$

where z_a and z_b are defined by Equations 6.2.15 and 6.2.16. Substituting Equations 6.2.59 through 6.2.64 into Equations 6.2.20 through 6.2.22, as n approaches infinity, gives:

$$\hat{\sigma}_{a_1}^2 = \hat{C}_{11}^{-1} \rightarrow \frac{K_{sy}^2}{n} a_1(z_b - z_a)$$

$$\times \frac{(z_a^2 + 2z_a + 2) - (z_b^2 + 2z_b + 2) \exp{[-(z_b - z_a)]}}{D_3}, \tag{6.2.65}$$

$$\hat{\sigma}_{a_2}^2 = \hat{C}_{22}^{-1} \rightarrow \frac{K_{sy}^2}{n} \frac{a_2^2}{a_1} (z_b - z_a) \frac{1 - \exp{[-(z_b - z_a)]}}{D_3}, \tag{6.2.66}$$

$$\hat{\sigma}_{12} = \hat{C}_{12}^{-1} \rightarrow \frac{K_{sy}^2}{n} a_2(z_b - z_a) \frac{(z_a + 1) - (z_b + 1) \exp{[-(z_b - z_a)]}}{D_3}, \tag{6.2.67}$$

where D_3 is defined as:

$$D_3 \equiv \exp{(z_a)} \cdot [\{(1 - \exp{[-(z_b - z_a)]}\}$$

$$\times \{z_a^2 + 2z_a + 2 - \exp{[-(z_b - z_a)]} \cdot (z_b^2 + 2z_b + 2)\}$$

$$- \{z_a + 1 - \exp{[-(z_b - z_a)]}(z_b + 1)\}^2]. \tag{6.2.68}$$

Equation 6.2.68 can be simplified:

$$D_3 = \exp{(-z_a)} \cdot \{1 + \exp{[-(z_b - z_a)]} \cdot (2z_az_b - z_b^2 - z_a^2 - 2)$$

$$+ \exp{[-2(z_b - z_a)]}\}. \tag{6.2.69}$$

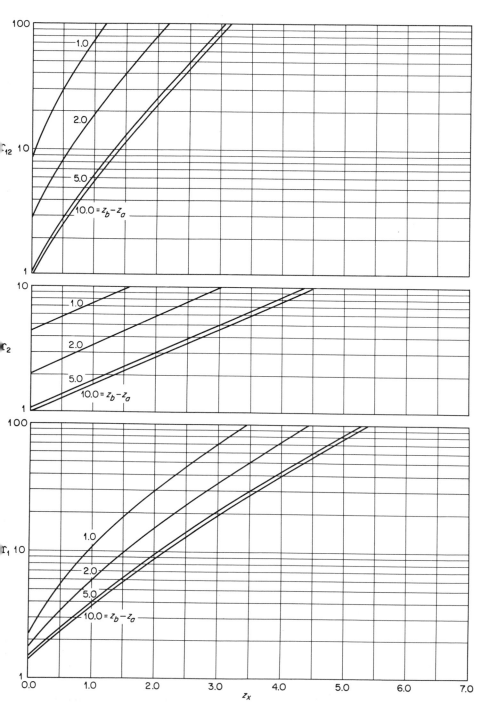

Figure 6.4. Υ_1, Υ_2, and Υ_{12} as functions of z_a and $z_b - z_a$ for $\omega_s = 0.0$ and $n \to \infty$.

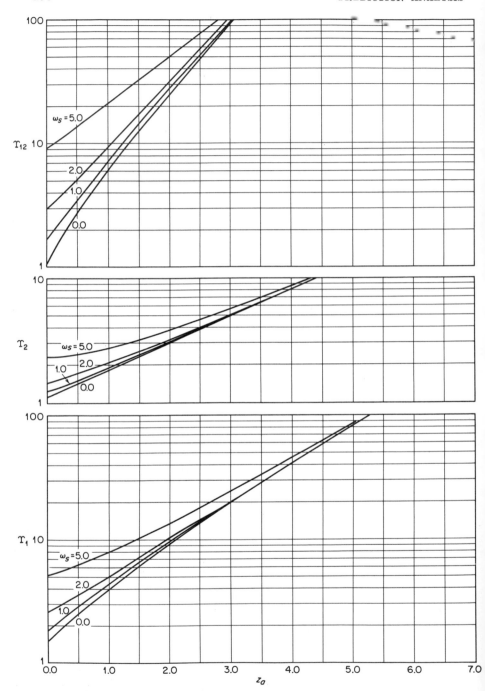

Figure 6.5. Υ_1, Υ_2, and Υ_{12} as functions of z_a and ω_s for $z_b - z_a = 5.0$ and $n = 24$.

If we rearrange terms and define a new set of dimensionless groups, as n approaches infinity,

$$\Upsilon_1 \equiv \frac{\hat{\sigma}_{a1} \cdot n^{1/2}}{K_{sy} \cdot (a_1)^{1/2} \cdot (z_b - z_a)^{1/2}} \rightarrow$$

$$\left(\frac{(z_a^2 + 2z_a + 2) - (z_b + 2z_b + 2) \exp\left[-(z_b - z_a)\right]}{D_3} \right)^{1/2}, \quad (6.2.70)$$

$$\Upsilon_2 \equiv \frac{\hat{\sigma}_{a2} \cdot (a_1)^{1/2} \cdot n^{1/2}}{K_{sy} \cdot a_2 \cdot (z_b - z_a)^{1/2}} \rightarrow \left(\frac{1 - \exp\left[-(z_b - z_a)\right]}{D_3} \right)^{1/2}, \quad (6.2.71)$$

$$\Upsilon_{12} \equiv \frac{\hat{\sigma}_{12} \cdot n}{K_{sy}^2 \cdot a_2 \cdot (z_b - z_a)} \rightarrow \frac{(z_a + 1) - (z_b + 1) \exp\left[-(z_b - z_a)\right]}{D_3}.$$

$$(6.2.72)$$

It should be remembered that Equations 6.2.70 through 6.2.72 are valid only when $\omega_s = 0$.

The Υ's are plotted in Figure 6.4 as functions of z_a and $z_b - z_a$. Although the curves are only exact for n equal to infinity, they are good approximations for most values of n. For example, for $z_a = 0.0$, $z_b - z_a = 5.0$, and $n = 3, 13, 23,$ and infinity, $\Upsilon_1 = 1.920, 1.482, 1.470,$ and 1.464, respectively.

Values of the Υ's computed using the direct approach (discussed in Section 4.4) are plotted in Figure 6.5 for values of ω_s other than zero. Although the curves are only exact for $n = 24$, they are good approximations for most values of n. For example, for $z_a = 0.0$, $z_b - z_a = 5.0$, $\omega_s = 1.0$, and for $n = 3, 13, 24,$ and infinity, $\Upsilon_1 = 2.20, 1.86, 1.85,$ and 1.84, respectively. The value of Υ_1 for $n \to \infty$ was determined by increasing n until Υ_1 approached an asymptote.

6.3 The Exponential-Plus-Background

Equation 6.1.2 is the relationship between the dependent and independent variables for all experiments within this class:

$$y = a_1 \exp(-a_2 x) + a_3. \quad (6.1.2)$$

The cases that are considered in this section are summarized in Table 6.2.

TABLE 6.2. Summary of cases analyzed for the *exponential-plus-background* class of experiments

Case	σ_{yi}	σ_{xi}	Figure
I	K_{cy}	K_{cx}	6.6
II	$K_{fy} \cdot y_i$	K_{cx}	6.7
III	$K_{sy} \cdot (y_i)^{1/2}$	K_{cx}	6.8

Case I. *Constant uncertainty for both the x and y variables.*

For this case the values of σ_{y_i} and σ_{x_i} are:

$$\sigma_{y_i} = K_{cy},$$

$$\sigma_{x_i} = K_{cx}. \tag{6.3.1}$$

The only difference between this case and Case I of Section 6.2 is that the function includes a third parameter (i.e., a_3). The equations for $\hat{L}_i$, $\hat{F}_{a_1}{}^i$, and $\hat{F}_{a_2}{}^i$ are the same for the two cases:

$$\hat{L}_i = K_{cy}{}^2 + [a_1 a_2 \exp{(-a_2 x_i)}]^2 K_{cx}{}^2, \tag{6.2.5}$$

$$\hat{F}_{a_1}{}^i = -\exp{(-a_2 x_i)}, \tag{6.2.6}$$

$$\hat{F}_{a_2}{}^i = a_1 x_i \exp{(-a_2 x_i)}, \tag{6.2.7}$$

but for this case, an additional equation for $\hat{F}_{a_3}{}^i$ is required:

$$\hat{F}_{a_3}{}^i = -1. \tag{6.3.2}$$

Equations for the elements of the matrix $\hat{C}$ are obtained by substituting Equations 6.2.5, 6.2.6, 6.2.7 and 6.3.2 into Equation 4.4.1:

$$\hat{C}_{11} = \frac{n}{K_{cy}{}^2} \cdot \left(\frac{\exp{(-2a_2 x)}}{1 + \omega_c{}^2 \exp{(-2a_2 x)}} \right)_{av}, \tag{6.3.3}$$

$$\hat{C}_{12} = \frac{n}{K_{cy}{}^2} \cdot \left(\frac{-a_1 x \exp{(-2a_2 x)}}{1 + \omega_c{}^2 \exp{(-2a_2 x)}} \right)_{av}, \tag{6.3.4}$$

$$\hat{C}_{13} = \frac{n}{K_{cy}{}^2} \cdot \left(\frac{\exp{(-a_2 x)}}{1 + \omega_c{}^2 \exp{(-2a_2 x)}} \right)_{av}, \tag{6.3.5}$$

$$\hat{C}_{22} = \frac{n}{K_{cy}{}^2} \cdot \left(\frac{a_1{}^2 x^2 \exp{(-2a_2 x)}}{1 + \omega_c{}^2 \exp{(-2a_2 x)}} \right)_{av}, \tag{6.3.6}$$

$$\hat{C}_{23} = \frac{n}{K_{cy}{}^2} \cdot \left(\frac{-a_1 x \exp{(-a_2 x)}}{1 + \omega_c{}^2 \exp{(-2a_2 x)}} \right)_{av}, \tag{6.3.7}$$

$$\hat{C}_{33} = \frac{n}{K_{cy}{}^2} \cdot \left(\frac{1}{1 + \omega_c{}^2 \exp{(-2a_2 x)}} \right)_{av}, \tag{6.3.8}$$

where ω_c is defined by Equation 6.2.28 to be $a_1 \cdot a_2 \cdot K_{cx}/K_{cy}$.

Equations for the $\hat{\sigma}$'s can be obtained by substituting Equations 6.3.3 through 6.3.8 into Equations 5.3.10 through 5.3.16. Even if ω_c is assumed to be zero, the resulting equations are extremely complicated. Although the derivation of the equation is not presented for this case, an examination

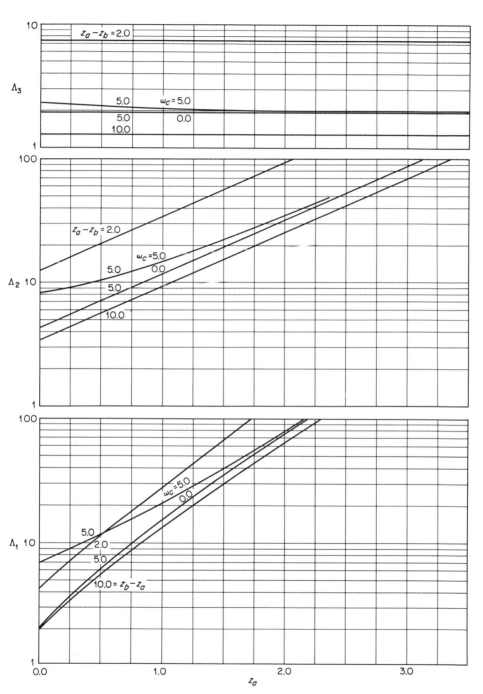

Figure 6.6. Λ_1, Λ_2, and Λ_3 as functions of z_a and $z_b - z_a$, for $n = 24$, and for all values of a_3/a_1. $\omega_c = 0.0$ except as noted.

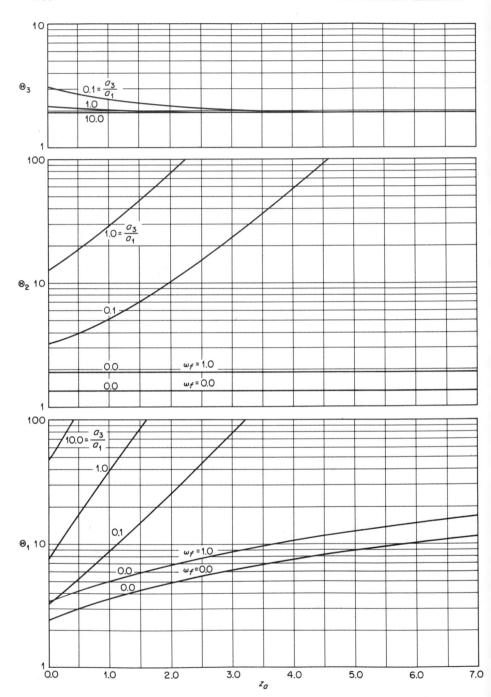

Figure 6.7. Θ_1, Θ_2, and Θ_3 as functions of z_a and a_3/a_1 for $z_b - z_a = 5.0$, $n = 24$, and $\omega_f = 0.0$ except as noted.

of Equations 6.3.3 through 6.3.8 reveals that the $\hat{\sigma}$'s are independent of the background a_3. Nevertheless, including the third parameter in the analysis increases $\hat{\sigma}_{a_1}$ and $\hat{\sigma}_{a_2}$ above the comparable values for Case I, Section 6.2.

Results determined by the direct approach (discussed in Section 4.4) are plotted in Figure 6.6 in terms of the dimensionless groups Λ_k:

$$\Lambda_1 \equiv \frac{\hat{\sigma}_{a_1} \cdot n^{1/2}}{K_{cy} \cdot (z_b - z_a)^{1/2}}, \tag{6.3.9}$$

$$\Lambda_2 \equiv \frac{\hat{\sigma}_{a_2} \cdot a_1 \cdot n^{1/2}}{K_{cy} \cdot a_2 \cdot (z_b - z_a)^{1/2}}, \tag{6.3.10}$$

and

$$\Lambda_3 \equiv \frac{\hat{\sigma}_{a_3} \cdot n^{1/2}}{K_{cy}}, \tag{6.3.11}$$

where z_a and z_b are defined by Equations 6.2.15 and 6.2.16. Although the curves are exact only for $n = 24$, they are good approximations for most values of n. For example, for $z_a = 0.0$, $z_b - z_a = 5.0$, $\omega_c = 0.0$, and $n = 4$, 14, 24, and infinity, $\Lambda_1 = 3.88$, 2.10, 2.04, and 2.00, respectively.

Also plotted in Figure 6.6 are curves for $z_b - z_a = 5.0$ and $\omega_c = 5.0$. As expected, the effect of ω_c is to increase the Λ_k's. However, as z_a increases, the curves approach the curves for $\omega_c = 0$.

Case II. *Constant fractional uncertainty for the variable y and constant uncertainty for the variable x.*

For this case the values of σ_{y_i} and σ_{x_i} are:

$$\sigma_{y_i} = K_{fy} \cdot y_i,$$
$$\sigma_{x_i} = K_{cx}. \tag{6.3.12}$$

The only difference between this case and Case II of Section 6.2 is that the function includes a third parameter (i.e., a_3). Equations 6.2.6, 6.2.7, 6.3.2, and 6.2.30 are valid for $\hat{F}_{a_1}{}^i$, $\hat{F}_{a_2}{}^i$, $\hat{F}_{a_3}{}^i$, and $\hat{L}_i$, respectively. Sub-

TABLE 6.3. Θ_1, Θ_2, and $a_3 \cdot \Theta_3$ as functions of $z_b - z_a$ for $a_3/a_1 = 0.0$, $z_a = 0.0$ and $\omega_f = 0$

$z_b - z_a$	Θ_1	Θ_2	$a_3 \cdot \Theta_3$
1.0	14.12	27.27	16.01 (units of a_3)
5.0	2.40	1.36	0.054
10.0	2.23	0.48	0.0003

stituting these equations and Equation 6.1.2 into Equation 4.4.1 gives:

$$\hat{C}_{11} = \frac{n}{K_{fy}{}^2 \cdot a_1{}^2} \cdot \left[\frac{\exp\,(-2a_2x)}{[\exp\,(-a_2x) + (a_3/a_1)]^2 + \omega_f{}^2 \exp\,(-2a_2x)} \right]_{av},$$

$$(6.3.13)$$

$$\hat{C}_{12} = \frac{n}{K_{fy}{}^2 \cdot a_1{}^2} \cdot \left[\frac{-a_1x \exp\,(-2a_2x)}{[\exp\,(-a_2x) + (a_3/a_1)]^2 + \omega_f{}^2 \exp\,(-2a_2x)} \right]_{v},$$

$$(6.3.14)$$

$$\hat{C}_{13} = \frac{n}{K_{fy}{}^2 \cdot a_1{}^2} \cdot \left[\frac{\exp\,(-a_2x)}{[\exp\,(-a_2x) + (a_3/a_1)]^2 + \omega_f{}^2 \exp\,(-2a_2x)} \right]_{av},$$

$$(6.3.15)$$

$$\hat{C}_{22} = \frac{n}{K_{fy}{}^2 \cdot a_1{}^2} \cdot \left[\frac{a_1{}^2x^2 \exp\,(-2a_2x)}{[\exp\,(-a_2x) + (a_3/a_1)]^2 + \omega_f{}^2 \exp\,(-2a_2x)} \right]_{av},$$

$$(6.3.16)$$

$$\hat{C}_{23} = \frac{n}{K_{fy}{}^2 \cdot a_1{}^2} \cdot \left[\frac{-a_1x \exp\,(-a_2x)}{[\exp\,(-a_2x) + (a_2/a_1)]^2 + \omega_f{}^2 \exp\,(-2a_2x)} \right]_{av},$$

$$(6.3.17)$$

$$\hat{C}_{33} = \frac{n}{K_{fy}{}^2 \cdot a_1{}^2} \cdot \left[\frac{1}{[\exp\,(-a_2x) + (a_3/a_1)]^2 + \omega_f{}^2 \exp\,(-2a_2x)} \right]_{av},$$

$$(6.3.18)$$

where ω_f is defined by Equation 6.2.34 to be $a_2 \cdot K_{cx}/K_{fy}$.

Equations for the $\hat{\sigma}$'s can be obtained by substituting Equations 6.3.13 through 6.3.18 into Equations 5.3.10 through 5.3.16. Even if ω_f is assumed to be zero, the resulting equations are extremely complicated. Although the derivation of the equations is not presented for this case, an examination of Equations 6.3.13 through 6.3.18 reveals that the $\hat{\sigma}$'s are dependent upon the background a_3.

Assuming that the x_i's are evenly spaced (i.e., Equation 4.4.3 is valid), results were determined using the direct approach (discussed in Section 4.4) and are plotted in terms of the dimensionless groups Θ_k:

$$\Theta_k \equiv \frac{\hat{\sigma}_{a_1} \cdot n^{1/2}}{K_{fy} \cdot a_k}. \qquad (6.3.19)$$

The curves are plotted in Figure 6.7 as functions of z_a and a_3/a_1 for $z_b - z_a = 5.0$, $n = 24$, and $\omega_f = 0.0$. For $a_3/a_1 = 0.0$, Θ_1 and Θ_2 are also plotted for $\omega_f = 1.0$. It can be seen that for this special case, the Θ's are proportional to $(1 + \omega_f{}^2)^{1/2}$.

TABLE 6.4. Comparison[a] of Θ_1 and Θ_2 for two- and three-parameter analyses as functions of z_a

z_a	Two-Parameter Analysis		Three-Parameter Analysis	
	Θ_1	Θ_2	Θ_1	Θ_2
0.0	2.000	0.693	2.403	1.356
1.0	2.625	0.693	3.621	1.356
2.0	3.277	0.693	4.913	1.356
3.0	3.943	0.693	6.231	1.356
4.0	4.617	0.693	7.562	1.356
5.0	5.296	0.693	8.901	1.356
6.0	5.978	0.693	10.244	1.356
7.0	6.663	0.693	11.591	1.356

[a] For both sets of results, $z_b - z_a = 5.0$, and $\omega_f = 0$. For the three-parameter analysis, $a_3/a_1 = 0$.

Although the curves are only exact for $n = 24$, they are good approximations for most values of n. For example, for $z_a = 0.0$, $z_b - z_a = 5.0$, $a_3/a_1 = 0.0$, $\omega_f = 0.0$, and $n = 4$, 24, and infinity, $\Theta_1 = 2.66$, 2.40, and 2.31, respectively.

Although not shown in Figure 6.7, the Θ's are functions of $z_b - z_a$. For $a_3/a_1 = 0.0$, $z_a = 0.0$, and $\omega_f = 0.0$, the Θ's are listed in Table 6.3 as functions of $z_b - z_a$.

The results from Case II of Section 6.2 and the results for this case are not the same even for $a_3/a_1 = 0$. The ϑ's of Case II, Section 6.2, reduce to the Θ's of this section for $\omega_f = 0$. Thus the ϑ's of Figure 6.3 can be compared to the Θ's of Figure 6.7. For $a_3/a_1 = 0.0$, $z_b - z_a = 5.0$, and $\omega_f = 0.0$, Θ_1 and Θ_2 are compared in Table 6.4.

The results included in Table 6.4 illustrate an important concept which is applicable to the analysis of functions of this type. Even when the term a_3 is negligible, if it is included as an unknown parameter in a least squares analysis of data exhibiting exponential decay, the resulting uncertainties $\hat{\sigma}_{a_1}$ and $\hat{\sigma}_{a_2}$ are increased significantly.

Case III. *Counting statistics for the variable y and constant uncertainty for the variable x.*

For this case the values of σ_{y_i} and σ_{x_i} are:

$$\sigma_{y_i} = K_{sy} \cdot (y_i)^{1/2},$$
$$\sigma_{x_i} = K_{cx}. \tag{6.3.20}$$

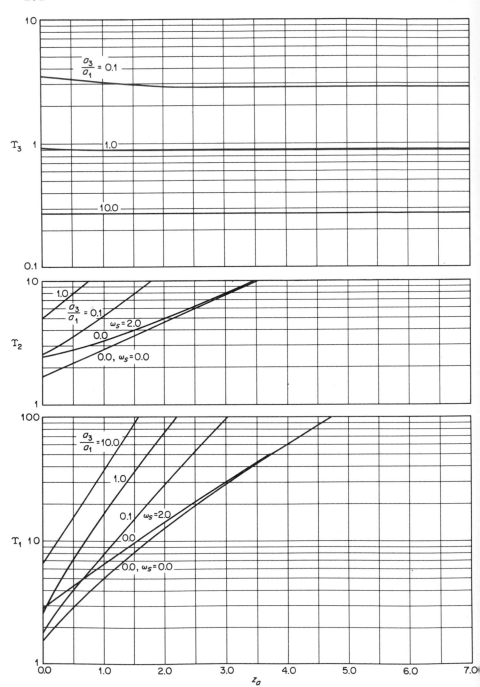

Figure 6.8. Υ_1, Υ_2, and Υ_3 as functions of z_a and a_3/a_1 for $z_b - z_a = 5.0$, $n = 24$, and $\omega_s = 0.0$ except as noted.

TABLE 6.5. Υ_1, Υ_2, and $a_3 \cdot \Upsilon_3$ as functions of $z_b - z_a$ for $a_3/a_1 = 0.0$, $z_a = 0.0$ and $\omega_s = 0.0$.

$z_b - z_a$	Υ_1	Υ_2	$a_3 \cdot \Upsilon_3$
1.0	19.00	34.63	21.00 (units of a_3)
5.0	1.58	1.72	0.141
10.0	1.45	1.05	0.0071

The only difference between this case and Case III of Section 6.2 is that the function includes a third parameter (i.e., a_3). Equations 6.2.6, 6.2.7, 6.3.2, and 6.2.54 are valid for $\hat{F}_{a_1}{}^i$, $\hat{F}_{a_2}{}^i$, $\hat{F}_{a_3}{}^i$, and $\hat{L}_i$, respectively. Substituting these equations and Equation 6.1.2 into Equation 4.4.1 gives:

$$\hat{C}_{11} = \frac{n}{K_{sy}^2 \cdot a_1} \cdot \left[\frac{\exp\left(-2a_2x\right)}{\exp\left(-a_2x\right) + (a_3/a_1) + \omega_s^2 \exp\left(-2a_2x\right)} \right]_{av}, \quad (6.3.21)$$

$$\hat{C}_{12} = \frac{n}{K_{sy}^2 \cdot a_1} \cdot \left[\frac{-a_1x \exp\left(-2a_2x\right)}{\exp\left(-a_2x\right) + (a_3/a_1) + \omega_s^2 \exp\left(-2a_2x\right)} \right]_{av}, \quad (6.3.22)$$

$$\hat{C}_{13} = \frac{n}{K_{sy}^2 \cdot a_1} \cdot \left[\frac{\exp\left(-a_2x\right)}{\exp\left(-a_2x\right) + (a_3/a_1) + \omega_s^2 \exp\left(-2a_2x\right)} \right]_{av}, \quad (6.3.23)$$

$$\hat{C}_{22} = \frac{n}{K_{sy}^2 \cdot a_1} \cdot \left[\frac{a_1^2x^2 \exp\left(-2a_2x\right)}{\exp\left(-a_2x\right) + (a_3/a_1) + \omega_s^2 \exp\left(-2a_2x\right)} \right]_{av}, \quad (6.3.24)$$

$$\hat{C}_{23} = \frac{n}{K_{sy}^2 \cdot a_1} \cdot \left[\frac{-a_1x \exp\left(-a_2x\right)}{\exp\left(-a_2x\right) + (a_3/a_1) + \omega_s^2 \exp\left(-2a_2x\right)} \right]_{av}, \quad (6.3.25)$$

$$\hat{C}_{33} = \frac{n}{K_{sy}^2 \cdot a_1} \cdot \left[\frac{1}{\exp\left(-a_2x\right) + (a_3/a_1) + \omega_s^2 \exp\left(-2a_2x\right)} \right]_{av}, \quad (6.3.26)$$

where ω_s is defined by Equation 6.2.58 to be $a_2 \cdot (a_1)^{1/2} \cdot K_{cx}/K_{sy}$.

Equations for the $\hat{\sigma}$'s can be obtained by substituting Equations 6.3.21 through 6.3.26 into Equations 5.3.10 through 5.3.16. Even if ω_s is assumed to be zero, the resulting equations are extremely complicated. Although the derivation of the equations is not presented for this case, an examination of Equations 6.3.21 through 6.3.26 reveals that the $\hat{\sigma}$'s are dependent upon the background a_3.

Assuming that the x_i's are evenly spaced (i.e., Equation 4.4.3 is valid), results were determined using the direct approach (discussed in Section 4.4) and are plotted in terms of the dimensionless groups Υ_k:

$$\Upsilon_k \equiv \frac{\hat{\sigma}_{a_k} \cdot (a_1)^{1/2} \cdot n^{1/2}}{K_{sy} \cdot a_k \cdot (z_b - z_a)^{1/2}}. \quad (6.3.27)$$

TABLE 6.6. Comparison[a] of Υ_1 and Υ_2 for two- and three-parameter analyses as functions of z_a

z_a	Two-Parameter Analysis		Three-Parameter Analysis	
	Υ_1	Υ_2	Υ_1	Υ_2
0.0	1.47	1.10	1.58	1.72
1.0	3.96	1.82	5.03	2.84
2.0	9.35	3.00	12.7	4.66
3.0	20.2	4.95	28.6	7.71
4.0	41.3	8.16	59.6	12.7
5.0	81.4	13.5	119.	20.9
6.0	156.	22.2	231.	34.6
7.0	294.	36.6	437.	57.0

[a] For both sets of results, $z_b - z_a = 5.0$, and $\omega_s = 0.0$. For the three-parameter analysis, $a_2/a_1 = 0.0$.

The results are plotted in Figure 6.8 as functions of z_a and a_3/a_1 for $z_b - z_a = 5.0$, $n = 24$, and $\omega_s = 0.0$. For $a_3/a_1 = 0.0$, Υ_1 and Υ_2 are also plotted for $\omega_s = 2$.

Although the curves are exact only for $n = 24$, they are good approximations for most values of n. For example, for $z_a = 0.0$, $z_b - z_a = 5.0$, $a_3/a_1 = 0.0$, and $\omega_s = 0.0$, for $n = 4, 14, 24$, and infinity, $\Upsilon_1 = 1.99, 1.60, 1.58$, and 1.57, respectively.

Although not shown in Figure 6.8, the Υ's are functions of $z_b - z_a$. For $a_3/a_1 = 0.0$, $z_a = 0.0$, and $\omega_s = 0.0$, the Υ's are listed in Table 6.5 as functions of $z_b - z_a$.

The results from Case III of Section 6.2 and the results for this case are not the same even for $a_3/a_1 = 0.0$. For $z_b - z_a = 5.0$, $a_3/a_1 = 0.0$, and $\omega_s = 0.0$, Υ_1 and Υ_2 are compared in Table 6.6.

The results included in Table 6.6 give further evidence to support the statement included at the end of the discussion related to Case II of this section. That is, even though the background a_3 is negligible, if it is included as an unknown parameter in a least squares analysis of data exhibiting exponential decay, the resulting uncertainties $\hat{\sigma}_{a_1}$ and $\hat{\sigma}_{a_2}$ are increased significantly.

6.4 Two Exponentials

For this class of experiments, the relationship between the dependent and independent variables is:

$$y = a_1 \exp(-a_2 x) + a_3 \exp(-a_4 x). \tag{6.4.1}$$

TABLE 6.7. Summary of cases analyzed for the *two exponentials* class of experiments

Case	σ_{y_i}	σ_{x_i}	Table
I	K_{cy}	K_{cx}	6.8
II	$K_{fy} \cdot y_i$	K_{cx}	6.9
III	$K_{sy} \cdot (y_i)^{1/2}$	K_{cx}	6.10

Equation 6.4.1 is a special case of Equation 6.1.3. The decay constants a_2 and a_4 are unknown parameters for experiments in this class.

The cases that are considered in this section are summarized in Table 6.7.

The derivation of equations for the C_{jk}'s is similar to the derivations included in Section 6.3. There are, however, several differences. The equation for $\hat{L}_i$ must be modified:

$$\hat{L}_i = \sigma_{y_i}{}^2 + [a_1 a_2 \exp{(-a_2 x)} + a_3 a_4 \exp{(-a_4 x)}] K_{cx}{}^2. \quad (6.4.2)$$

The equations for the $\hat{F}_{a_k}$'s are:

$$\hat{F}_{a_1}{}^i = -\exp{(-a_2 x_i)} \quad (6.4.3)$$

$$\hat{F}_{a_2}{}^i = a_1 x_i \exp{(-a_2 x_i)}, \quad (6.4.4)$$

$$\hat{F}_{a_3}{}^i = -\exp{(-a_4 x_i)}, \quad (6.4.5)$$

$$\hat{F}_{a_4}{}^i = a_3 x_i \exp{(-a_4 x_i)}. \quad (6.4.6)$$

For each case, the appropriate expressions for the $\hat{C}_{jk}$'s are derived by substituting the expression for σ_{y_i} from Table 6.7 into Equation 6.4.2 and then substituting Equations 6.4.2 through 6.4.6 into Equation 4.4.1. Equations for the $\hat{\sigma}$'s can then be derived by substituting the expressions

TABLE 6.8. Values[a] of the E_k's as functions of $z_b - z_a$, a_3/a_1, a_4/a_2, and ω_c

$z_b - z_a$	a_3/a_1	a_4/a_2	ω_c	E_1	E_2	E_3	E_4
5.0	0.1	2.0	0.0	200.	62.1	1957.	898.
5.0	1.0	2.0	0.0	199.	61.7	194.	89.2
5.0	10.0	2.0	0.0	201.	62.1	19.6	8.98
5.0	1.0	1.5	0.0	983.	185.	978.	226.
5.0	1.0	5.0	0.0	23.3	15.6	19.2	44.6
5.0	1.0	2.0	0.1	200.	62.	195.	89.9
5.0	1.0	2.0	1.0	242.	71.8	231.	118.
5.0	1.0	2.0	5.0	399.	107.	361.	239.
2.0	1.0	2.0	0.0	353.	125.	350.	132.
10.0	1.0	2.0	0.0	314.	93.9	301.	151.

[a] For each combination, $z_a = 0$ (Case I).

TABLE 6.9. Values[a] of the Θ_k's as functions of $z_b - z_a$, a_3/a_1, a_4/a_2 and ω_f

$z_b - z_a$	a_3/a_1	a_4/a_2	ω_f	Θ_1	Θ_2	Θ_3	Θ_4
5.0	0.1	2.0	0.0	15.5	3.40	134.	125.
5.0	1.0	2.0	0.0	17.9	3.86	15.2	15.8
5.0	10.0	2.0	0.0	32.8	6.64	3.46	3.89
5.0	1.0	1.5	0.0	97.6	14.7	92.6	32.0
5.0	1.0	5.0	0.0	3.27	1.01	17.2	18.5
5.0	1.0	2.0	0.1	18.0	3.88	15.3	15.9
5.0	1.0	2.0	1.0	27.1	5.78	22.8	24.8
5.0	1.0	2.0	5.0	102.	21.7	86.6	96.4
2.0	1.0	2.0	0.0	180.	58.9	175.	77.1
10.0	1.0	2.0	0.0	5.69	0.77	11.1	10.3

[a] For each combination, $z_a = 0$ (Case II).

for $\hat{C}_{jk}$ into the equations for the elements of the inverse matrix $\hat{C}^{-1}$. The complexity of the resulting equations is enormous.

An examination of Equations 6.4.4 and 6.4.6 reveals that if either a_1 or a_3 is zero, all values of $\hat{F}_{a_2}{}^i$ or $\hat{F}_{a_4}{}^i$ will be zero. From Equation 4.4.1 it can be concluded that all elements in one row and one column of the $\hat{C}$ matrix will be zero. If one row or one column of a matrix is zero, the determinant of the matrix is zero and the elements of the inverse matrix are infinite. For this reason, if a_1 or a_3 is equal to zero, the prediction analysis will not

TABLE 6.10. Values of the Ψ_k's as functions of $z_b - z_a$, a_3/a_1, a_4/a_2, and ω_s

$z_b - z_a$	a_3/a_1	a_4/a_2	ω_s	Ψ_1	Ψ_2	Ψ_3	Ψ_4
5.0	0.1	2.0	0.0	57.9	15.1	554.	333.
5.0	1.0	2.0	0.0	62.6	16.1	58.3	37.3
5.0	10.0	2.0	0.0	87.4	21.2	7.78	5.84
5.0	1.0	1.5	0.0	307.	51.3	302.	82.7
5.0	1.0	5.0	0.0	9.20	4.40	15.8	27.6
5.0	1.0	2.0	0.1	62.9	16.1	58.5	37.5
5.0	1.0	2.0	1.0	75.3	18.6	67.4	49.7
5.0	1.0	2.0	5.0	145.	32.7	121.	120.
2.0	1.0	2.0	0.0	255.	86.6	250.	102.
10.0	1.0	2.0	0.0	60.8	14.2	54.3	42.5

[a] For each combination, $z_a = 0$ (Case III).

yield meaningful results. In addition, if $a_2 = a_4$, $\hat{F}_{a_1}{}^i = \hat{F}_{a_2}{}^i$ and the analysis will also fail. (See the paragraph following Equation 4.8.14.)

Results determined using the direct approach are presented in Tables 6.8 through 6.10. Values of the appropriate dimensionless groups are presented as functions of $z_b - z_a$, a_3/a_1, a_4/a_2, and $\omega_{c,f,\text{or } s}$. For all cases included in the tables, $z_a = 0$.

For Case I the results are presented in terms of the dimensionless groups E_k:

$$E_k \equiv \frac{\hat{\sigma}_{a_k} \cdot a_1 \cdot n^{1/2}}{K_{cy} \cdot a_k},$$
(6.4.7)

for Case II, in terms of the dimensionless groups Θ_k:

$$\Theta_k \equiv \frac{\hat{\sigma}_{a_k} \cdot n^{1/2}}{K_{fy} \cdot a_k},$$
(6.4.8)

and for Case III, in terms of the dimensionless groups Ψ_k:

$$\Psi_k \equiv \frac{\hat{\sigma}_{a_k} \cdot (a_1)^{1/2} \cdot n^{1/2}}{K_{sy} \cdot a_k}.$$
(6.4.9)

The ω's are also dimensionless and are defined as follows. For Case I,

$$\omega_c \equiv \frac{a_1 \cdot a_2 \cdot K_{cx}}{K_{cy}},$$
(6.4.10)

for Case II,

$$\omega_f \equiv \frac{a_2 \cdot K_{cx}}{K_{fy}},$$
(6.4.11)

and for Case III,

$$\omega_s \equiv \frac{(a_1)^{1/2} \cdot a_2 \cdot K_{cx}}{K_{sy}}.$$
(6.4.12)

From Table 6.8 it is seen that E_1 and E_2 are insensitive to a_3/a_1 but E_3 and E_4 are approximately inversely proportional to this ratio. This result is best understood by examining the equation which defines E_k (i.e., Equation 6.4.7). From Tables 6.9 and 6.10 it is seen that Θ_1, Θ_2, Ψ_1, and Ψ_2 increase, while Θ_3, Θ_4, Ψ_3, and Ψ_4 decrease as the ratio a_3/a_1 increases.

For all cases the ratio a_4/a_2 has a strong influence upon the dimensionless groups. In fact, as a_4/a_2 approaches one, the values of all the groups approach infinity. Stated in practical terms, it is more difficult to determine the amplitudes and the decay constants the closer the ratio of the decay constants is to one. This behavior is explained by observing that as $a_4 \to a_2$, $\hat{F}_{a_3}{}^i \to \hat{F}_{a_1}{}^i$, the determinant of the $\hat{C}$ matrix approaches zero, and therefore all elements of the $\hat{C}^{-1}$ matrix approach infinity.

As expected, for all cases, an increase in the ω's causes increases in all the dimensionless groups. For $\omega_{c,f,or\ s} = 0.1$, however, this increase is observed to be small.

For Case I it is seen that as $z_b - z_a$ is increased from 2 to 5 and then to 10, all the E_k's first decrease and then increase. An optimum value of $z_b - z_a$ therefore exists in this region. For Case II the Θ_k's decrease monotonically in the region $2 < (z_b - z_a) < 10$. For Case III only, Ψ_4 increases as $z_b - z_a$ is increased from 5 to 10.

6.5 Exponential with Known Decay Constant

For this class of experiments, the relationship between the dependent and independent variables is:

$$y = a_1 \exp{(-\lambda_1 x)}. \tag{6.1.5}$$

The only unknown parameter is a_1. The decay constant λ_1 is assumed to be known.

The correct expressions for $\hat{F}_{a_1}{}^i$ and $\hat{L}_i$ are:

$$\hat{F}_{a_1}{}^i = -\exp{(-\lambda_1 x_i)}, \tag{6.5.1}$$

$$\hat{L}_i = \sigma_{y_i}{}^2 + [\lambda_1 a_1 \exp{(-\lambda_1 x_i)}]^2 \sigma_{x_i}{}^2. \tag{6.5.2}$$

Since there is only one unknown parameter, the matrices $\hat{C}$ and $\hat{C}^{-1}$ each consist of only one element:

$$\hat{C}_{11} = \sum_{i=1}^{n} \frac{(\hat{F}_{a_1}{}^i)^2}{L_i} = \sum_{i=1}^{n} \frac{\exp{(-2\lambda_1 x_i)}}{L_i}, \tag{6.5.3}$$

$$\hat{C}_{11}{}^{-1} = \frac{1}{\hat{C}_{11}} = \left(\sum_{i=1}^{n} \frac{\exp{(-2\lambda_1 x_i)}}{L_i} \right)^{-1}. \tag{6.5.4}$$

If σ_{x_i} is assumed to be negligible, Equation 6.5.4 reduces to a very simple equation:

$$\hat{C}_{11}{}^{-1} = \left(\sum_{i=1}^{n} \frac{\exp{(-2\lambda_1 x_i)}}{\sigma_{y_i}{}^2} \right)^{-1}. \tag{6.5.5}$$

If $\sigma_{y_i} = K_{cy}$, then

$$\hat{\sigma}_{a_1}^2 = \hat{C}_{11}^{-1} = \frac{K_{cy}^2}{n} \cdot \frac{1}{[\exp(-2\lambda_1 x)]_{av}} . \tag{6.5.6}$$

If the x_i's are evenly spaced (i.e., Equation 4.4.3 is valid), as n approaches infinity,

$$\hat{\sigma}_{a_1}^2 \rightarrow \frac{K_{cy}^2}{n} \cdot \frac{2(z_b - z_a)}{\exp(-2z_a) - \exp(-2z_b)} , \tag{6.5.7}$$

where z_b and z_a are defined as $\lambda_1 x_b$ and $\lambda_1 x_a$, respectively. Equations 4.4.7 and 4.4.8 define x_a and x_b.

If $\sigma_{y_i} = K_{fy} \cdot y_i$, then

$$\hat{\sigma}_{a_1}^2 = \hat{C}_{11}^{-1} = \frac{K_{fy}^2 \cdot a_1^2}{n} . \tag{6.5.8}$$

If $\sigma_{y_i} = K_{sy} \cdot (y_i)^{1/2}$, then

$$\hat{\sigma}_{a_1}^2 = \hat{C}_{11}^{-1} = \frac{K_{sy}^2 a_1}{n} \cdot \frac{1}{[\exp(-\lambda_1 x)]_{av}} . \tag{6.5.9}$$

If the x_i's are evenly spaced (i.e., Equation 4.4.3 is valid), as n approaches infinity,

$$\hat{\sigma}_{a_1}^2 \rightarrow \frac{K_{sy}^2 a_1}{n} \cdot \frac{z_b - z_a}{\exp(-z_a) - \exp(-z_b)} . \tag{6.5.10}$$

The results are summarized in Table 6.11.

TABLE 6.11. Summary of results from Section 6.5.[a]

Case	σ_{y_i}	σ_{x_i}	$\hat{\sigma}_{a_1}^2$
I	K_{cy}	0	$\dfrac{K_{cy}^2}{n} \cdot \dfrac{2(z_b - z_a)}{\exp(-2z_a) - \exp(-2z_b)}$
II	$K_{fy} \cdot y_i$	0	$\dfrac{K_{fy}^2}{n} \cdot a_1^2$
III	$K_{sy} \cdot (y_i)^{1/2}$	0	$\dfrac{K_{sy}^2}{n} \cdot \dfrac{a_1(z_b - z_a)}{\exp(-z_a) - \exp(-z_b)}$

[a] The equation for $\hat{\sigma}_{a_1}^2$ for Cases I and III are for experiments in which the values of the independent variable x are evenly spaced. For Case II this limitation is not necessary.

For Cases I and III, as z_a increases, $\hat{\sigma}_{a_1}$ also increases. For Case II, however, $\hat{\sigma}_{a_1}$ is independent of z_a. As an example, if $K_{fy} = 0.01$, and $n = 100$, from Equation 6.5.8 it is seen that $\hat{\sigma}_{a_1}/a_1$ is equal to $K_{fy}/n^{1/2} = 0.001$. That is, in the proposed experiment the unknown parameter a_1 should be determined to 0.1%. This conclusion is valid regardless of the choice of the limits z_a and z_b as long as systematic errors can be neglected.

6.6 Exponential-Plus-Background with Known Decay Constant

For this class of experiments, the relationship between the dependent and independent variables is:

$$y = a_1 \exp (-\lambda_1 x) + a_2. \tag{6.1.6}$$

The decay constant λ_1 is assumed to be known.

For this class of experiments, $\hat{F}_{a_1}{}^i$, $\hat{F}_{a_2}{}^i$, and $\hat{L}_i$ are:

$$\hat{F}_{a_1}{}^i = - \exp (-\lambda_1 x_i), \tag{6.6.1}$$

$$\hat{F}_{a_2}{}^i = -1, \tag{6.6.2}$$

$$\hat{L}_i = \sigma_{y_i}{}^2 + [\lambda_1 a_1 \exp (-\lambda_1 x_i)]^2 \sigma_{x_i}{}^2. \tag{6.6.3}$$

If σ_{x_i} is assumed to be negligible, simple expressions for $\hat{\sigma}_{a_1}$ and $\hat{\sigma}_{a_2}$ can be derived. Three cases are analyzed in this section and are summarized in Table 6.12.

For each case, the matrix elements $\hat{C}_{11}$, $\hat{C}_{12}$, and $\hat{C}_{22}$ are determined by substituting the expressions for σ_{y_i} and σ_{x_i} from Table 6.12 into Equation 6.6.3 and then substituting Equations 6.6.1 through 6.6.3 into Equation 4.4.1. The equations for the $\hat{\sigma}$'s are derived by substituting the expressions for the elements $\hat{C}_{11}$, $\hat{C}_{12}$, and $\hat{C}_{22}$ into Equations 6.2.20 through 6.2.22.

TABLE 6.12. Cases analyzed in Section 6.6

Case	a_2/a_1	σ_{a_1}	σ_{x_i}	Equations		
				$\hat{\sigma}_{a_1}$	$\hat{\sigma}_{a_2}$	$\hat{\sigma}_{12}$
I	Unrestricted	K_{cy}	0	6.6.7	6.6.8	6.6.9
II	0	$K_{fy} \cdot y_i$	0	6.6.13	6.6.14	6.6.15
III	0	$K_{sy} \cdot (y_i)^{1/2}$	0	6.6.19	6.6.20	6.6.21

If the x_i's are evenly spaced (i.e., Equation 4.4.3 is valid), as n approaches infinity,

Case I.

$$\hat{C}_{11} \rightarrow \frac{n}{K_{cy}{}^2} \cdot \frac{1}{2(z_b - z_a)} \cdot [\exp(-2z_a) - \exp(-2z_b)], \qquad (6.6.4)$$

$$\hat{C}_{12} \rightarrow \frac{n}{K_{cy}{}^2} \cdot \frac{1}{z_b - z_a} \cdot [\exp(-z_a) - \exp(-z_b)], \qquad (6.6.5)$$

$$\hat{C}_{22} \rightarrow \frac{n}{K_{cy}{}^2}, \qquad (6.6.6)$$

$$\hat{\sigma}_{a_1}{}^2 = \hat{C}_{11}{}^{-1} \rightarrow \frac{K_{cy}{}^2}{n}$$

$$\times \frac{2(z_b - z_a)^2}{(z_b - z_a)[\exp(-2z_a) - \exp(-2z_b)] - 2[\exp(-z_a) - \exp(-z_b)]^2},$$

$$(6.6.7)$$

$$\hat{\sigma}_{a_2}{}^2 = \hat{C}_{22}{}^{-1} \rightarrow \frac{K_{cy}{}^2}{n}$$

$$\times \frac{(z_b - z_a)[\exp(-2z_a) - \exp(-2z_b)]}{(z_b - z_a)[\exp(-2z_a) - \exp(-2z_b)] - 2[\exp(-z_a) - \exp(-z_b)]^2},$$

$$(6.6.8)$$

$$\hat{\sigma}_{12} = \hat{C}_{12}{}^{-1} \rightarrow \frac{K_{cy}{}^2}{n}$$

$$\times \frac{-2(z_b - z_a)[\exp(-z_a) - \exp(-z_b)]}{(z_b - z_a)[\exp(-2z_a) - \exp(-2z_b)] - 2[\exp(-z_a) - \exp(-z_b)]^2}.$$

$$(6.6.9)$$

Equations 6.6.7 through 6.6.9 are valid for any value of the ratio a_2/a_1.

For Cases II and III the equations for the $\hat{\sigma}$'s are complicated unless $a_2/a_1 = 0$. If it is assumed that a_2/a_1 is negligible (i.e., zero), as n approaches infinity:

Case II.

$$\hat{C}_{11} \to \frac{n}{K_{fy}{}^2 a_1{}^2}, \tag{6.6.10}$$

$$\hat{C}_{12} \to \frac{n}{K_{fy}{}^2 a_1{}^2} \cdot \frac{1}{(z_b - z_a)} \cdot [\exp(z_b) - \exp(z_a)], \tag{6.6.11}$$

$$\hat{C}_{22} \to \frac{n}{K_{fy}{}^2 a_1{}^2} \cdot \frac{1}{2(z_b - z_a)} \cdot [\exp(2z_b) - \exp(2z_a)], \tag{6.6.12}$$

$$\hat{\sigma}_{a_1}{}^2 = \hat{C}_{11}{}^{-1} \to \frac{K_{fy}{}^2 a_1{}^2}{n}$$

$$\times \frac{(z_b - z_a)[\exp(2z_b) - \exp(2z_a)]}{(z_b - z_a)[\exp(2z_a) - \exp(2z_a)] - 2[\exp(z_b)\exp(z_a)]^2}, \tag{6.6.13}$$

$$\hat{\sigma}_{a_2}{}^2 = \hat{C}_{22}{}^{-1} \to \frac{K_{fy}{}^2 a_1{}^2}{n}$$

$$\times \frac{2(z_b - z_a)^2}{(z_b - z_a)[\exp(2z_b) - \exp(2z_a)] - 2[\exp(z_b) - \exp(z_a)]^2}, \tag{6.6.14}$$

$$\hat{\sigma}_{12} = \hat{C}_{12}{}^{-1} \to \frac{K_{fy}{}^2 a_1{}^2}{n}$$

$$\times \frac{-2(z_b - z_a)[\exp(z_b) - \exp(z_a)]}{(z_b - z_a)[\exp(2z_b) - \exp(2z_a)] - 2[\exp(z_b) - \exp(z_a)]^2}. \tag{6.6.15}$$

Case III.

$$\hat{C}_{11} \to \frac{n}{K_{sy}{}^2 a_1} \cdot \frac{1}{z_b - z_a} \cdot [\exp(-z_a) - \exp(-z_b)], \tag{6.6.16}$$

$$\hat{C}_{12} \to \frac{n}{K_{sy}{}^2 a_1}, \tag{6.6.17}$$

$$\hat{C}_{22} \to \frac{n}{K_{sy}{}^2 a_1} \cdot \frac{1}{z_b - z_a} \cdot [\exp(+z_a) - \exp(+z_b)], \tag{6.6.18}$$

$$\hat{\sigma}_{a_1}^2 = \hat{C}_{11}^{-1} \rightarrow \frac{K_{sy}^2 a_1}{n}$$

$$\times \frac{(z_b - z_a)[\exp(z_b) - \exp(z_a)]}{[\exp(-z_a) - \exp(-z_b)][\exp(z_b) - \exp(z_a)] - (z_b - z_a)^2}, \quad (6.6.19)$$

$$\hat{\sigma}_{a_2}^2 = \hat{C}_{22}^{-1} \rightarrow \frac{K_{sy}^2 a_1}{n}$$

$$\times \frac{(z_b - z_a)[\exp(-z_a) - \exp(-z_b)]}{[\exp(-z_a) - \exp(-z_b)][\exp(z_b) - \exp(z_a)] - (z_b - z_a)^2}, \quad (6.6.20)$$

$$\hat{\sigma}_{12} = \hat{C}_{12} \rightarrow \frac{K_{sy}^2 a_1}{n}$$

$$\times \frac{-(z_b - z_a)^2}{[\exp(-z_a) - \exp(-z_b)][\exp(z_b) - \exp(z_a)] - (z_b - z_a)^2}. \quad (6.6.21)$$

The equations for Cases II and III are only valid as a_2/a_1 approaches zero. It is possible to derive equations for the more general problem (i.e., a_2/a_1 not equal to zero); however, the resulting equations are complicated.

It is interesting to compare the results of Section 6.5 and 6.6. Equations for the ratios of $\hat{\sigma}_{a_1}^2$ for the one and two parameter fits can be derived by dividing the appropriate equations from Section 6.6 by the equations from Section 6.5. The two sets of equations can only be compared if it is assumed that a_2 is negligible:

Case I.

$$\frac{(\hat{\sigma}_{a_1}^2)_{2\,par}}{(\hat{\sigma}_{a_1}^2)_{1\,par}} = \left(1 - \frac{2[\exp(-z_a) - \exp(-z_b)]^2}{(z_b - z_a) \cdot [\exp(-2z_a) - \exp(-2z_b)]}\right)^{-1}, \quad (6.6.22)$$

Case II.

$$\frac{(\hat{\sigma}_{a_1}^2)_{2\,par}}{(\hat{\sigma}_{a_1}^2)_{1\,par}} = \left(1 - \frac{2[\exp(z_b) - \exp(z_a)]^2}{(z_b - z_a)[\exp(2z_b) - \exp(2z_a)]^2}\right)^{-1}, \quad (6.6.23)$$

Case III.

$$\frac{(\hat{\sigma}_{a_1}^2)_{2\,par}}{(\hat{\sigma}_{a_1}^2)_{1\,par}} = \left(1 - \frac{(z_b - z_a)^2}{[\exp(-z_a) - \exp(-z_b)][\exp(z_b) - \exp(z_a)]}\right)^{-1}.$$

$$(6.6.24)$$

These ratios are clearly greater than one for all values of z_a and z_b. This is reasonable, because an increase in the number of parameters in an analysis results in a loss of accuracy if all other factors are the same.

6.7 Summary

In this chapter experiments for which the exponential function is used in the relationship between the dependent and independent variables were considered. The specific functions that were considered are listed in Table 6.13.

The difference between the functions discussed in Sections 6.2 and 6.3 as compared to the functions discussed in Section 6.5 and 6.6 is that a_2 is an unknown parameter and λ_1 is a known parameter. For example, analysis of data based upon the function considered in Section 6.3 will yield values for a_1, a_2, and a_3. Analysis of data based upon the function considered in Section 6.6 will yield values for only a_1 and a_2. The value of λ_1 is an input quantity.

For each function, three cases were considered. These were *constant error, constant fractional error, and counting statistics* for the dependent variable. For the independent variable, only *constant or zero error* was considered.

Several generalizations can be made about the results:

1. The predicted uncertainties are inversely proportional to the square root of the number of data points.
2. The predicted uncertainties are proportional to the constant K_{cy}, K_{fy}, or K_{sy}.
3. If all other factors are held constant, increasing the number of unknown parameters will increase the predicted uncertainties.
4. Increasing the range $(z_b - z_a)$ will decrease the predicted uncertainties if the decay constants are unknown parameters.

The analysis could have been extended to still other functions. It was felt, however, that the choice of functions provided enough scope to illustrate adequately the important points related to prediction analysis of these types of experiments. A word of caution with regard to experiments

TABLE 6.13. Summary of the functions considered in Chapter 6

Function	Section
$y = a_1 \exp(-a_2 x)$	6.2
$y = a_1 \exp(-a_2 x) + a_3$	6.3
$y = a_1 \exp(-a_2 x) + a_3 \exp(-a_4 x)$	6.4
$y = a_1 \exp(-\lambda_1 x)$	6.5
$y = a_1 \exp(-\lambda_1 x) + a_2$	6.6

of this type is worthwhile. The equations considered in Sections 6.2, 6.3, and 6.4 are nonlinear. Although no trouble is encountered in the prediction analysis of these functions, least squares analysis of experimental data will require iterative solutions. As the number of unknown parameters is increased, convergence to a solution becomes more difficult to obtain. The techniques suggested in Section 3.8 are often helpful in promoting convergence for these types of problems.

As an example of how the results in this chapter might be used, consider the following problem. The dependent and independent variables for a proposed experiment are to be related by Equation 6.1.2 (i.e., exponential-plus-background). "Counting statistics" applies for all values of the dependent variable y and the uncertainties for the measured values of the independent variable x are assumed to be zero. The range of values of x is 0.5 to 3.0, and the expected values of a_3/a_1 and a_2 are 0.1 and 2.0, respectively. If the number of data points is to be 256, and if the points are to be evenly spaced over the entire range, how large should a_1 be so that a_2 is determined to 1% accuracy? Spacing 256 points over the range 0.5 to 3.0 implies that $\Delta x \approx 0.01$, and therefore $x_a \approx 0.495$ and $x_b \approx 3.005$. With only a slight loss of accuracy, it can be assumed that $x_a = 0.5$ and $x_b = 3.0$ and then from Equations 6.2.15 and 6.2.16, $z_a = 1.0$ and $z_b = 6.0$. Case III of Section 6.3 is applicable and, from Figure 6.8, $\Upsilon_2 = 5.3$. For this problem, $K_{sy} = 1.0$, $\hat{\sigma}_{a_2}/a_2 = 0.01$, $n^{1/2} = 16$, $(z_b - z_a)^{1/2} = 2.236$, and therefore, from Equation 6.3.27, $(a_1)^{1/2} = 5.3 \times 1.0 \times 2.236/0.01 \times 16 = 74.0$, so $a_1 = 5476$. If y_i is the total number of counts in the ith channel of a multichannel analyzer, then the number of counts in the ith channel should be $a_1 \exp(-a_2 x_i)$. To satisfy the objective of the experiment, the experiment should be continued until the number of counts accumulated in the first channel is approximately equal to $5476 \exp(-2.0 \times 0.5) \approx 2000$.

It should be noted that Figure 6.8 is applicable to this problem only because $z_b - z_a = 5.0$. If data are required for combinations of z_a, $z_b - z_a$, a_3/a_1, and ω_s not included in this chapter, it might be possible to determine the required information by interpolation or extrapolation. An alternative procedure is to extend the results to the required combinations using the techniques and equations included in Chapter 4 and in this chapter.

Problems

1. The dependent and independent variables for a proposed experiment are to be related by the "simple exponential" function. The values of y_i are to be determined to 1% accuracy and the values of x_i are to be determined to ± 0.001. The expected values of a_1 and a_2 are 10000. and 2.0, respectively. The values of x_i are to be evenly spaced with the minimum value equal to 0.1, the spacing between points is to be equal to 0.01, and the total number of points is to be 128. Predict the values of σ_{a_1} and σ_{a_2} for the proposed experiment. *Ans.*: $\hat{\sigma}_{a_1} = 20.$, $\hat{\sigma}_{a_2} = 0.0024$.

2. If the spacing between points and the total number of points in Problem 1 is adjusted so that $n \cdot \Delta x$ remains constant, how many points are required for a_2 to be determined to about 1% accuracy?

3. The dependent and independent variables for a proposed experiment are to be related by the "exponential-plus-background" function. The uncertainties for the values of y_i are to be estimated by "counting statistics" and the uncertainties for the values of x_i are negligible. The range $z_b - z_a$ is to be equal to 5.0, z_a is to be equal to 0.5, the number of data points is to be 256, and the expected value of a_3/a_1 is 0.1. What should the value of a_1 be for a_2 to be determined to 0.1% accuracy?

4. For Problem 3, if $a_1 = 10000.$, how many data points are required for a_2 to be determined to 0.1% accuracy?

5. The dependent and independent variables for a proposed experiment are to be related by the "simple exponential" function. The values of y_i are to be determined to $\pm 100.$, and the uncertainties for the values of x_i are assumed to be negligible. If the expected values of a_1 and a_2 are 10000. and 1.0, $x_a = 0.0$, $x_b - x_a = 2.0$, and if the number of data points is to be 20, what is the predicted uncertainty for the value of y corresponding to $x = 3.0$?

Computer Projects

1. Recompute the results presented in Figure 6.7 for $z_b - z_a = 2.0$ and 10.0 as well as 5.0.

2. Recompute the results presented in Figure 6.8 for $z_b - z_a = 2.0$ and 10.0 as well as 5.0.

3. Predict the values of σ_{a1}, σ_{a2}, and σ_{12} as functions of the experimental variables for the following cases of the "simple exponential" class of experiments:

(A) $\sigma_{y_i} = K_{fy} \cdot y_i$ $\sigma_{x_i} = K_{fx} \cdot x_i$,
(B) $\sigma_{y_i}^2 = (K_{fy} \cdot y_i)^2 + K_{cy}^2$ $\sigma_{x_i} = 0$,
(C) $\sigma_{y_i 2} = K_{sy}^2 \cdot y_i + K_{cy}^2$ $\sigma_{x_i} = 0$.

For all cases, the values of x are to be evenly spaced.

4. Predict the values of σ_{a1}, σ_{a2}, and σ_{a3} as functions of the experimental variables for the same cases as in the previous problem but for the "exponential-plus-background" class of experiments.

5. For the following class of experiments:

$$y = a_1 \exp(-a_2 x) + a_3 x + a_4,$$

predict the values of σ_{a1}, σ_{a2}, σ_{a3}, and σ_{a4} as functions of the experimental variables for the following cases:

(A) $\sigma_{y_i} = K_{cy}$ $\sigma_{x_i} = K_{cx}$,
(B) $\sigma_{y_i} = K_{fy} \cdot y_i$ $\sigma_{x_i} = K_{cx}$,
(C) $\sigma_{y_i} = K_{sy} \cdot (y_i)^{\frac{1}{2}}$ $\sigma_{x_i} = K_{cx}$.

For all cases, the values of x are to be evenly spaced.

Chapter 7

The Sine Series

7.1 Introduction

Many experiments require a sine series to relate the dependent and independent variables:

$$y = \sum_{k=1}^{p} a_k \cdot \sin kx. \qquad (7.1.1)$$

The independent variable x is dimensionless and is equivalent to $\pi \cdot z/H$, where H is the range for the variable z. In all sections except 7.6, it is assumed that H is a known parameter. The objective of the experiment is to determine the unknown parameters a_k. A different analysis is required if H is also an unknown parameter.

In Section 7.2 the simple sine function is analyzed:

$$y = a_1 \sin x. \qquad (7.1.2)$$

In Section 7.3, the two-term series is considered:

$$y = a_1 \sin x + a_2 \sin 2x. \qquad (7.1.3)$$

In Section 7.4, the three-term series is considered:

$$y = a_1 \sin x + a_2 \sin 2x + a_3 \sin 3x. \qquad (7.1.4)$$

In Section 7.5, series with more than three terms are discussed. In Section 7.6, the simple sine function with the range as an unknown parameter is considered:

$$y = a_1 \sin \frac{x}{a_2}. \qquad (7.1.5)$$

The independent variable x is different from the x variable considered in

219

Equations 7.1.1 through 7.1.4. Since the ratio x/a_2 is dimensionless, x must have the same units as a_2. Equation 7.1.5 in nonlinear and therefore analysis of experimental data using this equation must employ iterative techniques.

It is assumed in this chapter that systematic errors are negligible. If systematic errors are considered as an important factor in a proposed experiment, the prediction analysis should be amended using the techniques discussed in Section 4.9.

7.2 The Simple Sine Function

Equation 7.1.2 is the relationship between the dependent and independent variables for experiments in this class:

$$y = a_1 \sin x. \tag{7.1.2}$$

The cases within this class that are considered in this section are summarized in Table 7.1.

TABLE 7.1. Summary of cases analyzed in Section 7.2

Case	σ_{y_i}	σ_{x_i}	Equation for $\hat{\sigma}_{a_1}$
I	K_{cy}	K_{cx}	7.2.14
II	$K_{fy} \cdot \lvert y_i \rvert$	K_{cx}	7.2.20
III	$K_{sy} \cdot (y_i)^{1/2}$	K_{cx}	7.2.25

For this class of experiments, $\hat{F}_{a_1}{}^i$ and $\hat{L}_i$ are:

$$\hat{F}_{a_1}{}^i = -\sin x_i, \tag{7.2.1}$$

$$\hat{L}_i = \sigma_{y_i}{}^2 + (a_1 \cos x_i)^2 \sigma_{x_i}{}^2. \tag{7.2.2}$$

Since there is only one unknown parameter, the matrices $\hat{C}$ and $\hat{C}^{-1}$ each consist of only one element:

$$\hat{C}_{11} = \sum_{i=1}^{n} \frac{(\hat{F}_{a_1}{}^i)^2}{\hat{L}_i} = \sum_{i=1}^{n} \frac{\sin^2 x_i}{\hat{L}_i}, \tag{7.2.3}$$

$$\hat{C}_{11}^{-1} = \frac{1}{\hat{C}_{11}} = \left(\sum_{i=1}^{n} \frac{\sin^2 x_i}{\hat{L}_i} \right)^{-1}. \tag{7.2.4}$$

From Equation 7.2.4:

$$\hat{\sigma}_{a_1}{}^2 = \hat{C}_{11}^{-1} = \left[n \cdot \left(\frac{\sin^2 x}{\hat{L}_i} \right)_{av} \right]^{-1}. \tag{7.2.5}$$

The average value of $\sin^2 x/\hat{L}_i$ must therefore be predicted for each case.

Case I. *Constant uncertainty for both the x and y variables.*

Substituting the expressions for σ_{y_i} and σ_{x_i} from Table 7.1 into Equation 7.2.2, we obtain:

$$\left(\frac{\sin^2 x}{\hat{L}_i}\right)_{av} = \left(\frac{\sin^2 x}{K_{cy}^2 + (a_1 \cos x)^2 K_{cx}^2}\right)_{av}. \tag{7.2.6}$$

If the values of x are evenly spaced (i.e., Equation 4.4.3 is valid), as n approaches infinity:

$$\left(\frac{\sin^2 x}{\hat{L}_i}\right)_{av} \rightarrow \frac{1}{K_{cy}^2(x_b - x_a)} \cdot \int_{x_a}^{x_b} \frac{\sin^2 x}{1 + \omega_c^2 \cos^2 x} \, dx, \tag{7.2.7}$$

where

$$\omega_c \equiv \frac{a_1 \cdot K_{cx}}{K_{cy}}, \tag{7.2.8}$$

and x_a and x_b are defined by Equations 4.4.7 and 4.4.8. Equation 7.2.7 can be evaluated using the following equality and integral:

$$\sin^2 x = 1 - \cos^2 x, \tag{7.2.9}$$

$$I_1 \equiv \int_{x_a}^{x_b} \frac{dx}{1 + \omega_c^2 \cos^2 x} = \frac{1}{(1 + \omega_c^2)^{1/2}} \cdot \left[\tan^{-1}\left(\frac{\tan x}{(1 + \omega_c^2)^{1/2}}\right)\right]_{x_a}^{x_b}. \tag{7.2.10}$$

Since

$$\int_{x_a}^{x_b} \frac{(1 + \omega_c^2 \cos^2 x)}{(1 + \omega_c^2 \cos^2 x)} \, dx = x_b - x_a, \tag{7.2.11}$$

subtracting Equation 7.2.10 from 7.2.11 and dividing by ω_c^2, we obtain:

$$\int \frac{\cos^2 x}{1 + \omega_c^2 \cos^2 x} \, dx = \frac{(x_b - x_a) - I_1}{\omega_c^2}. \tag{7.2.12}$$

If we substitute Equations 7.2.9, 7.2.10, and 7.2.12 into Equation 7.2.7:

$$\left(\frac{\sin^2 x}{\hat{L}_i}\right)_{av} \rightarrow \frac{1}{K_{cy}^2(x_b - x_a)} \cdot \left(I_1 - \frac{(x_b - x_a) - I_1}{\omega_c^2}\right). \tag{7.2.13}$$

Substituting Equation 7.2.13 into Equation 7.2.5 and rearranging terms gives:

$$A_1 \equiv \frac{\hat{\sigma}_{a_1} \cdot n^{1/2}}{K_{cy}} \rightarrow \left[\frac{1}{x_b - x_a} \cdot \left(I_1 - \frac{(x_b - x_a) - I_1}{\omega_c^2}\right)\right]^{-1/2}. \tag{7.2.14}$$

TABLE 7.2. Values of A_1 as a function of x_a, x_b, and ω_c

x_a	x_b	ω_c	A_1
0	π	0	1.414
		π	2.073
		5π	4.304
0.02π	0.98π	0	1.386
		π	2.031
		5π	4.191
0.1π	0.9π	0	1.273
		π	1.857
		5π	3.748
0.2π	0.8π	0	1.153
		π	1.625
		5π	3.197

The dimensionless group A_1 is seen to be a function of x_a, x_b, and ω_c. Values of A_1 are listed in Table 7.2 for several combinations of these variables. The most interesting point to notice is that the predicted uncertainty decreases as the values of x_a and x_b approach $\pi/2$. In fact, using Equation 7.2.7, it can be shown that as x_a and x_b approach $\pi/2$, $A_1 \rightarrow 1.0$ for all values of ω_c.

Case II. *Constant fractional uncertainty for the variable y and constant uncertainty for the variable x.*

If we substitute the expressions for σ_{y_i} and σ_{x_i} from Table 7.1 into Equation 7.2.2:

$$\left(\frac{\sin^2 x}{\hat{L}_i}\right)_{av} = \left(\frac{\sin^2 x}{K_{fy}{}^2(a_1 \sin x)^2 + K_{cx}{}^2(a_1 \cos x)^2}\right)_{av} . \qquad (7.2.15)$$

If the values of x are evenly spaced (i.e., Equation 4.4.3 is valid), as n approaches infinity:

$$\left(\frac{\sin^2 x}{\hat{L}_i}\right)_{av} \rightarrow \frac{1}{K_{fy}{}^2 a_1{}^2 (x_b - x_a)} \cdot \int_{x_a}^{x_b} \frac{\sin^2 x}{\sin^2 x + \omega_f{}^2 \cos^2 x} \, dx, \qquad (7.2.16)$$

where

$$\omega_f \equiv \frac{K_{cx}}{K_{fy}} . \qquad (7.2.17)$$

Equation 7.2.16 can be evaluated in a manner similar to that used for

TABLE 7.3. Values of B_1 as a function of x_a, x_b, and ω_f

x_a	x_b	ω_f	B_1
0	π	0	1.000
		π	2.035
		5π	4.298
0.02π	0.98π	0	1.000
		π	1.994
		5π	4.186
0.1π	0.9π	0	1.000
		π	1.823
		5π	3.743
0.2π	0.8π	0	1.000
		π	1.597
		5π	3.194

Equation 7.2.7:

$$I_2 \equiv \int_{x_a}^{x_b} \frac{dx}{\sin^2 x + \omega_f^2 \cos^2 x} = \frac{1}{\omega_f} \cdot \left[\tan^{-1}\left(\frac{\tan x}{\omega_f} \right) \right]_{x_a}^{x_b}, \qquad (7.2.18)$$

$$\left(\frac{\sin^2 x}{\hat{L}_i} \right)_{av} \rightarrow \frac{1}{K_{fy}{}^2 a_1{}^2 (x_b - x_a)} \cdot \left(I_2 - \frac{(x_b - x_a) - I_2}{\omega_f{}^2 - 1} \right), \qquad (7.2.19)$$

$$B_1 \equiv \frac{\hat{\sigma}_{a1} \cdot n^{1/2}}{a_1 \cdot K_{fy}} \rightarrow \left[\frac{1}{x_b - x_a} \cdot \left(I_2 - \frac{(x_b - x_a) - I_2}{\omega_f{}^2 - 1} \right) \right]^{-1/2}. \qquad (7.2.20)$$

The dimensionless group B_1 is a function of x_a, x_b, and ω_f and values of B_1 are listed in Table 7.3 for several combinations of these variables.

The value of B_1 equals 1.0 regardless of the values of x_a and x_b for $\omega_f = 0$. This observation can be explained using Equations 7.2.16 and 7.2.20. The equations can also be used to prove that as x_a and x_b approach $\pi/2$, B_1 approaches 1.0 for all values of ω_f.

Case III. *Counting statistics for the variable* y *and constant uncertainty for the variable* x.

Substituting the expressions for σ_{yi} and σ_{xi} from Table 7.1 into Equation 7.2.2 gives:

$$\left(\frac{\sin^2 x}{\hat{L}_i} \right)_{av} = \left(\frac{\sin^2 x}{K_{sy}{}^2 a_1 \sin x + (a_1 \cos x)^2 K_{cx}{}^2} \right)_{av}. \qquad (7.2.21)$$

If the values of x are evenly spaced (i.e. Equation 4.4.3 is valid), as n

TABLE 7.4. Values of Γ_1 as a function of x_a, x_b, and ω_s

x_a	x_b	ω_s	Γ_1
0	π	0	1.253
		π	2.053
		5π	4.301
0.02π	0.98π	0	1.229
		π	2.011
		5π	4.189
0.1π	0.9π	0	1.149
		π	1.839
		5π	3.745
0.2	0.8π	0	1.079
		π	1.610
		5π	3.195

approaches infinity:

$$\left(\frac{\sin^2 x}{\hat{L}_i}\right)_{av} \to \frac{1}{K_{sy}{}^2 a_1 (x_b - x_a)} \cdot \int_{x_a}^{x_b} \frac{\sin^2 x \, dx}{\sin x + \omega_s{}^2 \cos^2 x}, \quad (7.2.22)$$

where

$$\omega_s = \frac{K_{cx} \cdot (a_1)^{1/2}}{K_{sy}}. \quad (7.2.23)$$

If $\omega_s = 0$, Equation 7.2.22 reduces to a simple form:

$$\left(\frac{\sin^2 x}{\hat{L}_i}\right)_{av} \to \frac{1}{K_{sy}{}^2 a_1} \cdot \frac{\cos x_a - \cos x_b}{x_b - x_a}, \quad (7.2.24)$$

and

$$\Gamma_1 \equiv \frac{\hat{\sigma}_{a1} \cdot n^{1/2}}{K_{sy} \cdot (a_1)^{1/2}} \to \left(\frac{x_b - x_a}{\cos x_a - \cos x_b}\right)^{1/2}. \quad (7.2.25)$$

The more general equation for which ω_s is not equal to zero is complicated.

The dimensionless group Γ_1 is a function of x_a, x_b, and ω_s, and values of Γ_1 are listed in Table 7.4 for several combinations of these variables. The values for ω_s not equal to zero were determined using the direct approach (described in Section 4.4).

The values of Γ_1 approach 1.0 as x_a and x_b approach $\pi/2$. This statement is valid for all values of ω_s.

7.3 Two-Term Sine Series

Equation 7.1.3 is the relationship between the dependent and independent variables for experiments in this class:

$$y = a_1 \sin x + a_2 \sin 2x. \tag{7.1.3}$$

The cases within this class that are considered in this section are summarized in Table 7.5. For this class of experiments, $\hat{F}_{a_1}{}^i$, $\hat{F}_{a_2}{}^i$, and $\hat{L}_i$ are:

$$\hat{F}_{a_1}{}^i = -\sin x_i, \tag{7.3.1}$$

$$\hat{F}_{a_2}{}^i = -\sin 2x_i, \tag{7.3.2}$$

$$\hat{L}_i = \sigma_{y_i}{}^2 + (a_1 \cos x_i + 2a_2 \cos 2x_i)^2 \sigma_{x_i}{}^2. \tag{7.3.3}$$

The equations for the elements of the matrix $\hat{C}$ are determined by substituting Equations 7.3.1 through 7.3.3 into Equation 4.4.1:

$$\hat{C}_{11} = \sum_{i=1}^{n} \frac{\sin^2 x_i}{\hat{L}_i}, \tag{7.3.4}$$

$$\hat{C}_{12} = \sum_{i=1}^{n} \frac{\sin x_i \sin 2x_i}{\hat{L}_i}, \tag{7.3.5}$$

$$\hat{C}_{22} = \sum_{i=1}^{n} \frac{\sin^2 2x_i}{\hat{L}_i}. \tag{7.3.6}$$

Even if K_{cx} (i.e., σ_{x_i}) is set equal to zero, the resulting equations for the elements of $\hat{C}$ are extremely complicated for Cases II and III. For Case I, however, the equations are simple. If the values of x are evenly spaced (i.e., Equation 4.4.3 is valid), then as n approaches infinity,

$$\hat{C}_{11} = n \cdot \left(\frac{\sin^2 x}{K_{cy}{}^2} \right)_{av} \rightarrow$$

$$\frac{n}{K_{cy}{}^2} \cdot \frac{(x_b - x_a) - (\sin x_b \cos x_b - \sin x_a \cos x_a)}{2(x_b - x_a)}, \tag{7.3.7}$$

$$\hat{C}_{12} = n \cdot \left(\frac{\sin x \sin 2x}{K_{cy}{}^2} \right)_{av} \rightarrow \frac{n}{K_{cy}{}^2} \cdot \frac{2(\sin^3 x_b - \sin^3 x_a)}{3(x_b - x_a)}, \tag{7.3.8}$$

$$\hat{C}_{22} = n \cdot \left(\frac{\sin^2 2x}{K_{cy}{}^2} \right)_{av} \rightarrow$$

$$\frac{n}{K_{cy}{}^2} \cdot \frac{(x_b - x_a) - \frac{1}{2}(\sin 2x_b \cos 2x_b - \sin 2x_a \cos 2x_a)}{2(x_b - x_a)}. \tag{7.3.9}$$

TABLE 7.5. Summary of cases analyzed in Section 7.3

Case	σ_{y_i}	σ_{x_i}	Figure
I	K_{cy}	K_{cx}	7.1
II	$K_{fy} \cdot \lvert y_i \rvert$	K_{cx}	7.2
III	$K_{sy} \cdot (y_i)^{1/2}$	K_{cx}	7.3

If the values of x are symmetrical about $\pi/2$ (i.e., $x_b = \pi - x_a$), then $\hat{C}_{12}$ is equal to zero and the equations for $\hat{\sigma}_{a_1}^2$ and $\hat{\sigma}_{a_2}^2$ reduce to:

$$\hat{\sigma}_{a_1}^2 = \hat{C}_{11}^{-1} = \frac{1}{\hat{C}_{11}} \rightarrow \frac{K_{cy}^2}{n} \cdot \frac{2(\pi - 2x_a)}{(\pi - 2x_a) + 2 \sin x_a \cos x_a}, \quad (7.3.10)$$

$$\hat{\sigma}_{a_2}^2 = \hat{C}_{22}^{-1} = \frac{1}{\hat{C}_{22}} \rightarrow \frac{K_{cy}^2}{n} \cdot \frac{2(\pi - 2x_a)}{(\pi - 2x_a) + \sin 2x_a \cos 2x_a}. \quad (7.3.11)$$

For this special case (i.e., $x_b = \pi - x_a$) $\hat{\sigma}_{12}$ is equal to zero. The results for Case I are presented in Figure 7.1 in terms of A_k:

$$A_k \equiv \frac{\hat{\sigma}_{a_k} \cdot n^{1/2}}{K_{cy}}. \quad (7.3.12)$$

It should be noted that for Case I, A_1 and A_2 are independent of the ratio a_2/a_1 only for $\omega_c = 0$.

Although equations are not derived for Cases II and III, results are plotted in Figure 7.2 and 7.3. The results for Case II are presented in terms of B_k:

$$B_k \equiv \frac{\hat{\sigma}_{a_k} \cdot n^{1/2}}{K_{fy} \cdot a_1}, \quad (7.3.13)$$

and the results for Case III are presented in terms of Γ_k:

$$\Gamma_k \equiv \frac{\hat{\sigma}_{a_k} \cdot n^{1/2}}{K_{sy} \cdot (a_1)^{1/2}}. \quad (7.3.14)$$

The direct approach was used to compute the results. For Cases II and III the B_k's and Γ_k's are functions of a_2/a_1 even if K_{cx} (and therefore ω_f and ω_s) is equal to zero. The dimensionless groups ω_c, ω_f, and ω_s are defined by Equations 7.2.8, 7.2.17, and 7.2.23, respectively.

It should be noted that for Case III, the values of y_i must be equal to or greater than zero. There is no physical significance for negative values of y_i for this case. If x_a equals 0, and x_b equals π, and if a_1 and a_2 are posi-

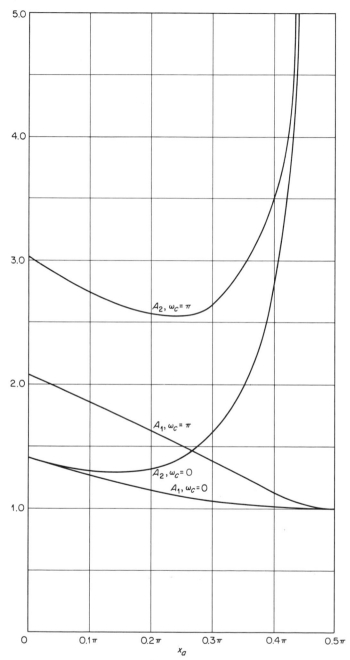

Figure 7.1. A_1 and A_2 as functions of x_a and ω_c for $x_b = \pi - x_a$. For $\omega_c = 0$ the results are valid for all values of a_2/a_1. For $\omega_c = \pi$, the results are exact only for $a_2/a_1 = 0.0$ (Case I).

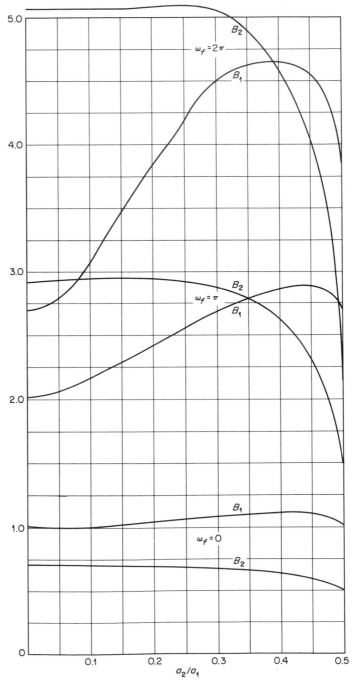

Figure 7.2. Values of B_1 and B_2 as functions of the ratio a_2/a_1 and ω_f for $n = 24$, $x_a = 0$, and $x_b = \pi$ (Case II).

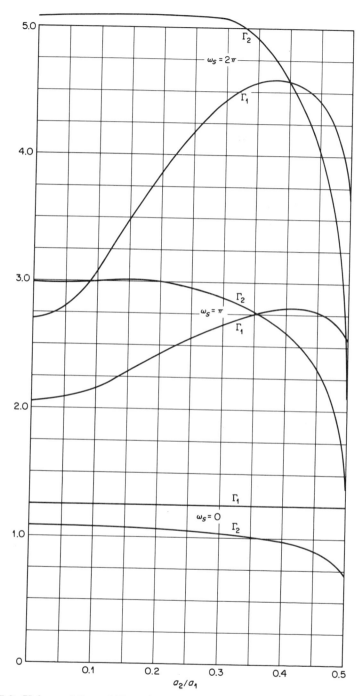

Figure 7.3. Values of Γ_1 and Γ_2 as functions of the ratio a_2/a_1 and ω_s for $n = 24$, $x_a = 0$, and $x_b = \pi$ (Case III.)

tive, then the criterion for ensuring that all the y_i's are greater than zero is:

$$\left(\frac{dy}{dx}\right)_{x=\pi} \leq 0. \qquad (7.3.15)$$

From Equations 7.1.3 and 7.3.15:

$$a_1 \geq 2a_2. \qquad (7.3.16)$$

That is, as long as the ratio a_2/a_1 is less than $\frac{1}{2}$, all the values of y_i are positive.

Although the curves in Figures 7.2 and 7.3 are exact only for $n = 24$, they are good approximations for all values of n. For example, for $a_2/a_1 = 0.2$, and $K_{cx} = 0.0$, the values of B_1 for $n = 3$, 24, and infinity are 1.0377, 1.0366, and 1.0366. The values of Γ_1 are 1.225, 1.253, and 1.254. The values of B_1 and Γ_1 for $n \rightarrow \infty$ were computed by increasing n until these quantities approached asymptotes.

Figures 7.2 and 7.3 are limited to the special condition $x_a = 0$ and $x_b = \pi$. The effect of changing x_a and x_b for Cases II and III is similar to the effect upon A_1 and A_2 for Case I (as shown in Figure 7.1).

7.4 Three-Term Sine Series

Equation 7.1.4 is the relationship between the dependent and independent variables for experiments in this class:

$$y = a_1 \sin x + a_2 \sin 2x + a_3 \sin 3x. \qquad (7.1.4)$$

The cases within this class that are considered in this section are summarized in Table 7.6. For this class of experiments, $\hat{F}_{a_1}{}^i$, $\hat{F}_{a_2}{}^i$, $\hat{F}_{a_3}{}^i$, and $\hat{L}_i$ are:

$$\hat{F}_{a_1}{}^i = -\sin x_i, \qquad (7.4.1)$$

$$\hat{F}_{a_2}{}^i = -\sin 2x_i, \qquad (7.4.2)$$

$$\hat{F}_{a_3}{}^i = -\sin 3x_i, \qquad (7.4.3)$$

$$\hat{L}_i = \sigma_{y_i}{}^2 + (a_1 \cos x_i + 2a_2 \cos 2x_i + 3a_3 \cos 3x_i)^2 \sigma_{x_i}{}^2. \qquad (7.4.4)$$

The equations for the elements of the matrix $\hat{C}$ are determined by sub-

TABLE 7.6. Summary of cases analyzed in Section 7.4

Case	σ_{y_i}	σ_{x_i}	x_a	x_b	Table
I	K_{cy}	K_{cx}	0	π	7.7
II	$K_{fy} \cdot \| y_i \|$	K_{cx}	0	π	7.8
III	$K_{sy} \cdot (y_i)^{1/2}$	K_{cx}	0	π	7.9

stituting Equations 7.4.1 through 7.4.4 into Equation 4.4.1:

$$\hat{C}_{11} = \sum_{i=1}^{n} \frac{\sin^2 x_i}{\hat{L}_i}, \tag{7.4.5}$$

$$\hat{C}_{12} = \sum_{i=1}^{n} \frac{\sin x_i \sin 2x_i}{\hat{L}_i}, \tag{7.4.6}$$

$$\hat{C}_{13} = \sum_{i=1}^{n} \frac{\sin x_i \sin 3x_i}{\hat{L}_i}, \tag{7.4.7}$$

$$\hat{C}_{22} = \sum_{i=1}^{n} \frac{\sin^2 2x_i}{L_i}, \tag{7.4.8}$$

$$\hat{C}_{23} = \sum_{i=1}^{n} \frac{\sin 2x_i \sin 3x_i}{L_i}, \tag{7.4.9}$$

$$\hat{C}_{33} = \sum_{i=1}^{n} \frac{\sin^2 3x_i}{\hat{L}_i}. \tag{7.4.10}$$

Developing analytical expressions for the $\hat{\sigma}$'s is not worthwhile because the resulting expressions are extremely complicated. Most of the results presented in this section were determined using the direct approach (which is discussed in Section 4.4).

For Case I, the A_k's are independent of a_2/a_1 and a_3/a_1 if $\omega_c = 0.0$. If, in addition, $x_a = 0$ and $x_b = \pi$, as n approaches infinity,

$$A_1 = A_2 = A_3 \rightarrow \sqrt{2}. \tag{7.4.11}$$

This equation can be derived using Equations 7.4.5 through 7.4.10. For $\omega_c = 0$, $x_a = 0$ and $x_b = \pi$,

$$\hat{C}_{11} = \hat{C}_{22} = \hat{C}_{33} \rightarrow \frac{n}{K_{cy}^2} \cdot \frac{1}{2}, \tag{7.4.12}$$

$$\hat{C}_{12} = \hat{C}_{13} = \hat{C}_{23} = 0, \tag{7.4.13}$$

$$\hat{\sigma}_{a_k} = \hat{C}_{kk}^{-1} = \frac{1}{\hat{C}_{kk}} \rightarrow \frac{K_{cy}^2}{n} \cdot 2. \tag{7.4.14}$$

Equation 7.4.11 follows from Equation 7.4.14 and the definition of the A_k's (i.e., Equation 7.3.12).

Values of the A_k's computed using the direct approach for values of ω_o other than zero are listed in Table 7.7.

TABLE 7.7. Values of A_1, A_2, and A_3 as functions of ω_c for $a_2/a_1 = 0$, $a_3/a_1 = 0$, $x_a = 0$, and $x_b = \pi$

ω_c	A_1	A_2	A_3
0	1.414	1.414	1.414
π	2.633	3.038	3.038
2π	4.663	5.205	5.208
3π	6.813	7.397	7.424
4π	8.998	9.582	9.669
5π	11.197	11.760	11.935

For Cases II and III, the analysis is limited to only those problems for which $x_a = 0$, $x_b = \pi$, and all values of y_i are equal to or greater than zero. If the coordinate system is chosen so that the values of a_1 and a_2 are positive, two conditions must be satisfied to ensure that all values of y_i are equal to or greater than zero:

$$\left(\frac{dy}{dx}\right)_{x=\pi} \le 0, \tag{7.4.15}$$

$$\eta \equiv \frac{y}{a_1 \sin x} \ge 0 \quad \text{for all values of } x. \tag{7.4.16}$$

The first condition is satisfied if

$$\frac{a_2}{a_1} \le \frac{1}{2} + \frac{3}{2} \cdot \frac{a_3}{a_1}. \tag{7.4.17}$$

Equation 7.4.17 can be derived by evaluating the derivative of 7.1.4 at $x = \pi$. To satisfy the second condition, the value of x corresponding to the minimum value of η must be located. Using the following equalities:

$$\sin 2x = 2 \sin x \cos x, \tag{7.4.18}$$

$$\sin 3x = 3 \sin x - 4 \sin^3 x, \tag{7.4.19}$$

and substituting them into Equation 7.1.4 and then into Equation 7.4.16, we obtain:

$$\eta = 1 + \frac{2a_2}{a_1} \cos x + \frac{3a_3}{a_1} - \frac{4a_3}{a_1} \sin^2 x \ge 0, \tag{7.4.20}$$

$$\eta = 1 - \frac{a_3}{a_1} + 2 \frac{a_2}{a_1} \cos x + \frac{4a_3}{a_1} \cos^2 x \ge 0. \tag{7.4.21}$$

The minimum value of η is located by taking the derivative of η with

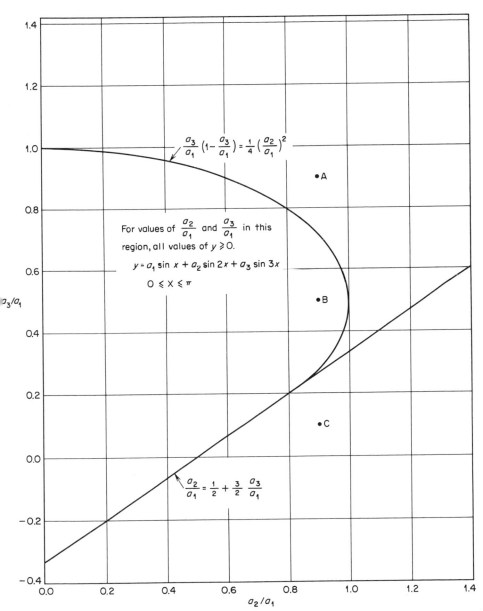

Figure 7.4. Region in which $y \geq 0$.

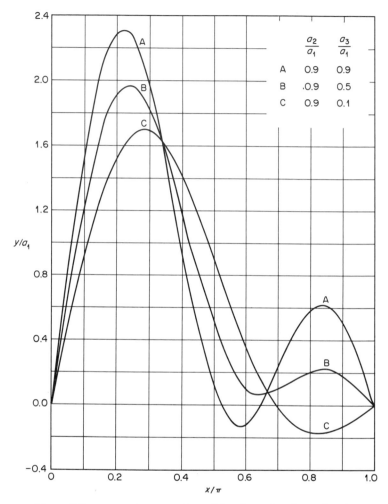

Figure 7.5. y vs. x for several combinations of a_2/a_1 and a_3/a_1.

respect to $\cos x$ and setting it equal to zero:

$$\cos x = -\frac{a_2}{4a_3} \quad \text{at } \eta_{\min}. \qquad (7.4.22)$$

Since the absolute value of $\cos x$ cannot exceed one, $\eta_{\min}$ exists only if $|a_2/a_3| \leq 4$. Substituting Equation 7.4.22 into Equation 7.4.21 and re-arranging terms, we obtain:

$$\frac{a_3}{a_1} \cdot \left(1 - \frac{a_3}{a_1}\right) \geq \frac{1}{4}\left(\frac{a_2}{a_1}\right)^2. \qquad (7.4.23)$$

TABLE 7.8. Values[a] of B_1, B_2, and B_3 for several combinations of a_2/a_1, a_3/a_1, and ω_f

a_2/a_1	a_3/a_1	ω_f	B_1	B_2	B_3
0.0	−0.333	0	1.067	0.010	0.358
		π	3.281	1.292	1.307
		2π	4.958	2.580	2.137
0.0	−0.2	0	1.269	0.372	0.546
		π	3.358	4.032	3.093
		2π	5.965	7.835	5.765
0.0	0.0	0	1.225	0.707	0.707
		π	2.537	2.928	2.928
		2π	4.609	5.149	5.151
0.2	0.0	0	1.208	0.748	0.678
		π	2.573	3.005	3.071
		2π	4.446	5.304	5.536
0.4	0.0	0	1.149	0.849	0.566
		π	2.924	3.133	3.158
		2π	4.816	5.433	5.851
0.5	0.0	0	1.056	0.861	0.376
		π	3.284	3.226	2.632
		2π	5.532	5.491	4.892
0.0	0.5	0	1.020	1.182	0.991
		π	2.401	2.787	2.361
		2π	3.589	3.631	3.332
0.8	0.5	0	1.188	1.300	0.757
		π	3.450	3.498	1.924
		2π	4.579	4.552	2.723
1.0	0.5	0	1.073	1.081	0.535
		π	4.104	3.878	1.890
		2π	5.989	5.163	2.921
0.0	0.8	0	1.013	1.184	1.087
		π	3.689	4.334	3.366
		2π	6.659	7.207	5.749
0.0	1.0	0	1.048	0.408	1.069
		π	4.435	4.950	3.823
		2π	7.822	7.144	6.009

[a] The combinations included in this table satisfy the condition that all values of y_i are equal to or greater than zero. The values of x_a and x_b are 0 and π for all the combinations (Case II).

TABLE 7.9. Values[a] of Γ_1, Γ_2, and Γ_3 for several combinations of a_2/a_1, a_3/a_1, and ω_s

a_2/a_1	a_3/a_1	ω_s	Γ_1	Γ_2	Γ_3
0.0	−0.333	0	1.333	0.379	0.538
		π	3.077	1.361	1.266
		2π	4.777	2.615	2.098
0.0	−0.2	0	1.328	0.881	0.892
		π	3.318	4.063	3.093
		2π	5.931	7.839	5.754
0.0	0.0	0	1.301	1.084	1.049
		π	2.575	2.975	2.969
		2π	4.630	5.173	5.173
0.2	0.0	0	1.299	1.084	1.026
		π	2.587	3.028	3.096
		2π	4.449	5.306	5.541
0.4	0.0	0	1.291	1.084	0.934
		π	2.864	3.091	3.162
		2π	4.735	5.380	5.837
0.5	0.0	0	1.276	1.084	0.715
		π	3.183	3.134	2.626
		2π	5.428	5.393	4.864
0.0	0.5	0	1.214	1.326	1.219
		π	2.482	2.742	2.416
		2π	3.641	3.552	3.353
0.8	0.5	0	1.195	1.290	1.041
		π	3.124	2.976	2.026
		2π	4.110	3.840	2.955
1.0	0.5	0	1.173	1.194	0.782
		π	3.494	3.166	2.039
		2π	5.308	4.274	3.019
0.0	0.8	0	1.148	1.370	1.230
		π	3.654	4.129	3.263
		2π	6.632	7.050	5.670
0.0	1.0	0	1.090	1.253	1.131
		π	4.229	4.439	3.505
		2π	7.667	6.624	5.737

[a] The combinations included in this table satisfy the condition that all values of y_i are equal to or greater than zero. The values of x_a and x_b are 0 and π for all the combinations (Case III).

Equations 7.4.17 and 7.4.23 are plotted in Figure 7.4. In Figure 7.5, values of y are plotted as a function of x for several combinations of a_2/a_1 and a_3/a_1.

Values of B_k (defined by Equation 7.3.13) are listed in Table 7.8 for combinations of a_2/a_1, a_3/a_1 and ω_f within the region shown in Figure 7.4.

Values of Γ_k (defined by Equation 7.3.14) are listed in Table 7.9 for combinations of a_2/a_1, a_3/a_1 and ω_s within the region shown in Figure 7.4.

7.5 Higher-Order Sine Series

Equation 7.1.1 is the general relationship for experiments in this class:

$$y = \sum_{k=1}^{p} a_k \sin kx. \tag{7.1.1}$$

The integer p is defined as the order of the series. As p increases, the complexity of the analysis increases.

The analysis is limited to the cases listed in Table 7.10.

TABLE 7.10. Summary of cases analyzed in Section 7.5

Case	σ_{yi}	σ_{xi}	x_a	x_b	a_k/a_1	p
I	K_{cy}	K_{cx}	0	π	0.0	Up to 7
II	$K_{cy} \cdot y_i$	K_{cx}	0	π	0.0	Up to 7
III	$K_{sy} \cdot (y_i)^{1/2}$	K_{cx}	0	π	0.0	Up to 7

The limitation on the ratios a_k/a_1 is severe. This limitation implies that the fundamental mode (i.e., $a_1 \sin x$) predominates over the higher harmonics (i.e., $a_k \sin kx$, $k \geq 2$). For many experiments, this situation is approximately correct and for these experiments the following results yield good estimates of the $\hat{\sigma}_{a_k}$'s. To extend the analysis to cases in which the ratios a_k/a_1 are not small or to cases in which x_a and x_b are not 0 and π, the direct approach described in Section 4.4 should be used.

Results for Cases I through III are listed in Tables 7.11 through 7.13, respectively. The results for Case I are presented in terms of the dimensionless groups A_k:

$$A_k \equiv \frac{\hat{\sigma}_{a_k} \cdot n^{1/2}}{K_{cy}}, \qquad k = 1, 2, \cdots, p. \tag{7.5.1}$$

the results for Case II are presented in terms of B_k:

$$B_k \equiv \frac{\hat{\sigma}_{a_k} \cdot n^{1/2}}{K_{fy} \cdot a_1}, \qquad k = 1, 2, \cdots, p, \tag{7.5.2}$$

TABLE 7.11. Values[a] of A_k for several combinations of p and ω_c
(where $\omega_c \equiv K_{cy} \cdot a_1/K_{cx}$)

p	ω_c	A_1	A_2	A_3	A_4	A_5	A_6	A_7
1	0	1.414						
2	0	1.414	1.414					
3	0	1.414	1.414	1.414				
4	0	1.414	1.414	1.414	1.414			
5	0	1.414	1.414	1.414	1.414	1.414		
6	0	1.414	1.414	1.414	1.414	1.414	1.414	
7	0	1.414	1.414	1.414	1.414	1.414	1.414	1.414
1	π	2.073						
2	π	2.073	3.038					
3	π	2.633	3.038	3.038				
4	π	2.633	3.445	3.038	3.308			
5	π	2.633	3.445	3.445	3.038	3.038		
6	π	2.633	3.445	3.445	3.445	3.038	3.038	
7	π	2.633	3.445	3.445	3.445	3.445	3.038	3.038

[a] The ratios $a_k/a_1 = 0$, $x_a = 0$, and $x_b = \pi$ (Case I).

TABLE 7.12. Values[a] of B_k for several combinations of p and ω_f
(where $\omega_f \equiv K_{cx}/K_{cy}$)

p	ω_f	B_1	B_2	B_3	B_4	B_5	B_6	B_7
1	0	1.000						
2	0	1.000	0.707					
3	0	1.225	0.707	0.707				
4	0	1.225	1.000	0.707	0.707			
5	0	1.225	1.000	1.000	0.707	0.707		
6	0	1.225	1.000	1.000	1.000	0.707	0.707	
7	0	1.225	1.000	1.000	1.000	1.000	0.707	0.707
1	π	2.035						
2	π	2.035	2.928					
3	π	2.537	2.928	2.928				
4	π	2.537	3.297	2.928	2.928			
5	π	2.537	3.297	3.297	2.928	2.928		
6	π	2.537	3.297	3.297	3.297	2.928	2.928	
7	π	2.537	3.297	3.297	3.297	3.297	2.928	2.928

[a] The ratios $a_k/a_1 = 0$, $x_a = 0$ and $x_b = \pi$ (Case II).

TABLE 7.13. Values of Γ_k for several combinations of p and ω_s
(where $\omega_s \equiv K_{cx} \cdot (a_1)^{1/2} / K_{sy}$)

p	ω_s	Γ_1	Γ_2	Γ_3	Γ_4	Γ_5	Γ_6	Γ_7
1	0	1.253						
2	0	1.253	1.084					
3	0	1.301	1.084	1.049				
4	0	1.301	1.164	1.049	1.032			
5	0	1.301	1.164	1.142	1.032	1.023		
6	0	1.301	1.164	1.142	1.132	1.023	1.020	
7	0	1.301	1.164	1.142	1.132	1.127	1.020	1.014
1	π	2.053						
2	π	2.053	2.975					
3	π	2.575	2.975	2.969				
4	π	2.575	3.350	2.969	2.968			
5	π	2.575	3.350	3.347	2.968	2.967		
6	π	2.575	3.350	3.347	3.343	2.967	2.966	
7	π	2.575	3.350	3.347	3.343	3.340	2.966	2.965

[a] The ratios $a_k/a_1 = 0$, $x_a = 0$ and $x_b = \pi$ (Case III).

and the results for Case III are presented in terms of Γ_k:

$$\Gamma_k \equiv \frac{\hat{\sigma}_{a_k} \cdot n^{1/2}}{K_{sy} \cdot (a_1)^{1/2}}, \qquad k = 1, 2, \cdots, p. \qquad (7.5.3)$$

For all cases, the values of x are assumed to be evenly spaced (i.e., Equation 4.4.3 is valid). The results were determined using the direct approach with $n = 24$.

The most interesting point to notice is that, for all the cases considered, the values of the A_k's, B_k's, and Γ_k's are not strongly dependent upon the order p. For the range $x_a = 0$ to $x_b = \pi$, the sine terms are orthogonal. That is:

$$\int_0^\pi \sin jx \sin kx \, dx = 0, \qquad \text{if } j \neq k. \qquad (7.5.4)$$

The orthogonal nature of the sine function for the given range explains the weak dependence of the A_k's, B_k's, and Γ_k's upon the order p.

It is interesting to compare the results obtained in this section with the results from Section 5.4. For the polynomial function (i.e., Equation 5.1.1) it was shown that an increase in the number of unknown parameters beyond the correct number had a marked effect upon the predicted uncertainties $\hat{\sigma}_{a_k}$. For the sine series the effect is seen to be small. It should be mentioned, however, that orthogonal polynomials can also be used for the analysis of data.[1]

7.6 The Simple Sine Function with Unknown Range

Equation 7.1.5 is the relationship for experiments in this class:

$$y = a_1 \sin \frac{x}{a_2}.$$ (7.1.5)

Defining H as the range, the relationship between H and a_2 is simply:

$$a_2 = \frac{H}{\pi}.$$ (7.6.1)

Equation 7.1.5 is plotted in Figure 7.6.

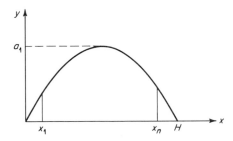

Figure 7.6. The simple sine function with unknown range.

The purpose of the experiment is often to determine a_2 (and thus H) by fitting Equation 7.1.5 to a set of data points. The data points fall within the region $0 < x_i < H$. Equation 7.1.5 is nonlinear, and therefore iterative techniques are required to determine the least squares values of a_1 and a_2. Equation 7.6.1 can then be used to determine H.

For this class of experiments, $\hat{F}_{a_1}{}^i$, $\hat{F}_{a_2}{}^i$, and $\hat{L}_i$ are:

$$\hat{F}_{a_1}{}^i = - \sin \frac{x_i}{a_2},$$ (7.6.2)

$$\hat{F}_{a_2}{}^i = \frac{a_1 \cdot x_i}{a_2{}^2} \cos \frac{x_i}{a_2},$$ (7.6.3)

$$\hat{L}_i = \sigma_{y_i}{}^2 + \left(\frac{a_1}{a_2} \cos \frac{x_i}{a_2}\right)^2 \sigma_{x_i}{}^2.$$ (7.6.4)

[1] A discussion of orthogonal polynomials is included in F. B. Hildebrand's book "Introduction to Numerical Analysis," McGraw-Hill Book Co., New York, 1956.

The elements of the matrix $\hat{C}$ are:

$$\hat{C}_{11} = \sum_{i=1}^{n} \frac{\sin^2 (x_i/a_2)}{\hat{L}_i}, \tag{7.6.5}$$

$$\hat{C}_{12} = \sum_{i=1}^{n} - \frac{a_1}{a_2^2} \cdot \frac{x_i \sin (x_i/a_2) \cos (x_i/a_2)}{\hat{L}_i}, \tag{7.6.6}$$

$$\hat{C}_{22} = \sum_{i=1}^{n} \frac{a_1^2}{a_2^4} \cdot \frac{x_i^2 \cos^2 (x_i/a_2)}{\hat{L}_i}. \tag{7.6.7}$$

From Equations 7.6.5 through 7.6.7:

$$\hat{\sigma}_{a_1}^2 = \hat{C}_{11}^{-1} = \left(\frac{x^2 \cos^2 (x/a_2)}{\hat{L}_i} \right)_{av} \Big/ (n \cdot D), \tag{7.6.8}$$

$$\hat{\sigma}_{a_2}^2 = \hat{C}_{22}^{-1} = \frac{a_2^4}{a_1^2} \cdot \left(\frac{\sin^2 (x/a_2)}{\hat{L}_i} \right)_{av} \Big/ (n \cdot D), \tag{7.6.9}$$

$$\hat{\sigma}_{12} = \hat{C}_{12}^{-1} = \frac{a_2^2}{a_1} \cdot \left(\frac{x \sin (x/a_2) \cos (x/a_2)}{\hat{L}_i} \right)_{av} \Big/ (n \cdot D), \tag{7.6.10}$$

where

$$D \equiv \left(\frac{\sin^2 (x/a_2)}{\hat{L}_i} \right)_{av} \cdot \left(\frac{x^2 \cos^2 (x/a_2)}{\hat{L}_i} \right)_{av} - \left(\frac{x \sin (x/a_2) \cos (x/a_2)}{\hat{L}_i} \right)_{av}^2. \tag{7.6.11}$$

If it is assumed that the values of x_i are evenly spaced (i.e., Equation 4.4.3 is valid), analytical expression for the $\hat{\sigma}$'s can be obtained. However, the resulting expressions are extremely complicated, even for the simple case $\sigma_{y_i} = K_{cy}$ and $\sigma_{x_i} = K_{cx}$.

The cases considered in this section are listed in Table 7.14.

For all cases, the direct approach was used to compute the results. The results are exact only for $n = 24$, but are good approximations for all possible values of n. For Case I, the results are presented in Figure 7.7 in

TABLE 7.14. Summary of cases analyzed in Section 7.6

Case	σ_{y_i}	σ_{x_i}	x_a/a_2	x_b/a_2	K_{cx}/a_2
I	K_{cy}	K_{cx}	0 to $\frac{1}{2}\pi$	$\pi - x_a/a_2$	$0.00, 0.02, 0.05$
II	$K_{fy} \cdot y_i$	K_{cx}	0 to $\frac{1}{2}\pi$	$\pi - x_a/a_2$	$0.00, 0.02, 0.05$
III	$K_{sy} \cdot (y_i)^{1/2}$	K_{cx}	0 to $\frac{1}{2}\pi$	$\pi - x_a/a_2$	$0.00, 0.02, 0.05$

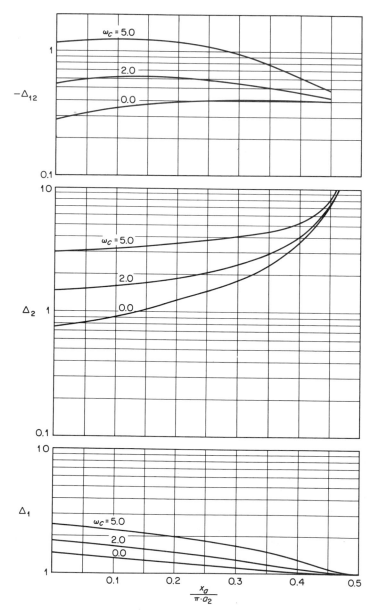

Figure 7.7. Δ_1, Δ_2, and $-\Delta_{12}$ as functions of $x_a/\pi \cdot a_2$ and ω_c for $x_b/a_2 = \pi - x_a/a_2$ and $n = 24$ (Case I).

terms of Δ_1, Δ_2, Δ_{12}, and ω_c:

$$\Delta_1 \equiv \frac{\hat{\sigma}_{a1} \cdot n^{1/2}}{K_{cy}}, \qquad (7.6.12)$$

$$\Delta_2 \equiv \frac{\hat{\sigma}_{a2} \cdot a_1 \cdot n^{1/2}}{a_2 \cdot K_{cy}}, \qquad (7.6.13)$$

$$\Delta_{12} \equiv \frac{\hat{\sigma}_{12} \cdot a_1 \cdot n}{a_2 \cdot K_{cy}^2}, \qquad (7.6.14)$$

$$\omega_c \equiv \frac{a_1}{a_2} \cdot \frac{K_{cx}}{K_{cy}}. \qquad (7.6.15)$$

For Case II, the results are presented in Figure 7.8 in terms of Θ_1, Θ_2, Θ_{12}, and ω_f:

$$\Theta_1 \equiv \frac{\hat{\sigma}_{a1} \cdot n^{1/2}}{a_1 \cdot K_{fy}}, \qquad (7.6.16)$$

$$\Theta_2 \equiv \frac{\hat{\sigma}_{a2} \cdot n^{1/2}}{a_2 \cdot K_{fy}}, \qquad (7.6.17)$$

$$\Theta_{12} \equiv \frac{\hat{\sigma}_{12} \cdot n}{a_1 \cdot a_2 \cdot K_{fy}^2}, \qquad (7.6.18)$$

$$\omega_f \equiv \frac{K_{cx}}{a_2 \cdot K_{fy}}. \qquad (7.6.19)$$

For Case III, the results are presented in Figure 7.9 in terms of Ψ_1, Ψ_2, Ψ_{12}, and ω_s:

$$\Psi_1 \equiv \frac{\hat{\sigma}_{a1} \cdot n^{1/2}}{(a_1)^{1/2} \cdot K_{sy}}, \qquad (7.6.20)$$

$$\Psi_2 \equiv \frac{\hat{\sigma}_{a2} \cdot (a_1)^{1/2} \cdot n^{1/2}}{a_2 \cdot K_{sy}}, \qquad (7.6.21)$$

$$\Psi_{12} \equiv \frac{\hat{\sigma}_{12} \cdot n}{a_2 \cdot K_{sy}^2}, \qquad (7.6.22)$$

$$\omega_s \equiv \frac{(a_1)^{1/2}}{a_2} \cdot \frac{K_{cx}}{K_{sy}}. \qquad (7.6.23)$$

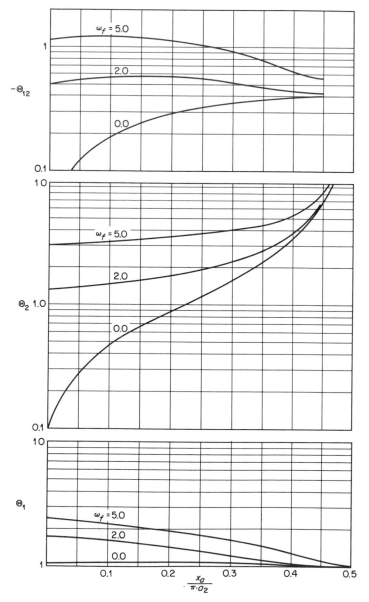

Figure 7.8. Θ_1, Θ_2, and $-\Theta_{12}$ as functions of $x_a/\pi \cdot a_2$ and ω_f for $x_b/a_2 = \pi - x_a/a_2$ and $n = 24$ (Case II).

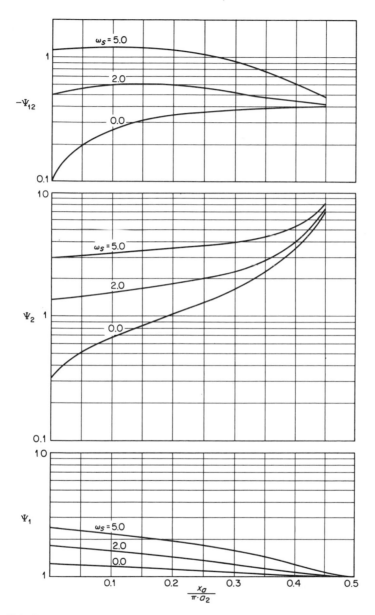

Figure 7.9. Ψ_1, Ψ_2, and $-\Psi_{12}$ as functions of $x_a/\pi \cdot a_2$ and ω_s for $x_b/a_2 = \pi - x_a/a_2$ and $n = 24$ (Case III).

The Δ's, Θ's, Ψ's, and ω's are all dimensionless. The curves are plotted as functions of $x_a/(\pi \cdot a_2)$. The variable x_a is defined by Equation 4.4.7. For all cases, as $x_a/(\pi \cdot a_2) = x_a/H$ approaches 0.5 (and therefore $x_b/(\pi \cdot a_2) = x_b/H$ approaches 0.5), $\hat{\sigma}_{a_2}$ approaches infinity. As one would expect, to design an experiment for determining H, the range of values of x should be as close as possible to the limits 0 and H.

7.7 Summary

In this chapter, experiments for which the sine function is used in the relationship between the dependent and independent variables were considered. The specific functions discussed in this chapter are listed in Table 7.15.

TABLE 7.15. Summary of the functions discussed in Chapter 7

Function	Section
$y = a_1 \sin x$	7.2
$y = a_1 \sin x + a_2 \sin 2x$	7.3
$y = a_1 \sin x + a_2 \sin 2x + a_3 \sin 3x$	7.4
$y = \sum_{k=1}^{p} a_k \sin kx$	7.5
$y = a_1 \sin (x/a_2)$	7.6

The functions considered in Sections 7.2 through 7.5 are linear and the function considered in Section 7.6 is nonlinear. The nonlinear equation is required for experiments in which the range for the values of x is an unknown parameter.

For each function, three cases were considered. These were *constant error, constant fractional error, and counting statistics* for the dependent variable. For the independent variable, only *constant and zero error* were considered.

Several generalizations can be made about the results:

1. The predicted uncertainties are inversely proportional to the square root of the number of data points.
2. The predicted uncertainties are proportional to the constant K_{cy}, K_{fy}, or K_{sy}.
3. As the data points are bunched closer and closer to the center of the range, the predicted uncertainty $\hat{\sigma}_{a_1}$ decreases while the $\hat{\sigma}_{a_k}$'s for $k > 1$ approach infinity.

4. If all other factors are held constant, increasing the number of un-known parameters only increases the predicted uncertainties slightly.

As an example of how the results in this chapter might be used, con-sider the following problem. The dependent and independent variables are to be related by Equation 7.1.3 (i.e., two-term sine series). The values of y_i are to be measured to ± 0.01, the values of x_i are to be measured to $\pm 1\%$ of the range $(x_b - x_a)$, $x_a = 0.0$ and $x_b = \pi$ (i.e. $\sigma_{x_i} = 0.01\pi$). The expected values of a_1 and a_2 are 1.0 and 0.05. What are the predicted un-certainties $\hat{\sigma}_{a_1}$ and $\hat{\sigma}_{a_2}$ if the total number of data points is 100?

Case I of Section 7.3 is applicable to this problem. From Equation 7.2.7, $\omega_c = 1.0 \times 0.01\pi/0.01 = \pi$. For $\omega_c = \pi$, Figure 7.1 is exact only if a_2/a_1 is equal to zero. Since the expected value of a_2/a_1 is small, with only a slight loss of accuracy, it can be assumed that a_2/a_1 is equal to zero. From Figure 7.1, $A_1 = 2.07$ and $A_2 = 3.03$, and from the definition of A_k, $\hat{\sigma}_{a_1} = 2.07 \times 0.01/(100)^{1/2} = 0.00207$, and $\hat{\sigma}_{a_2} = 3.03 \times 0.01/(100)^{1/2} = 0.00303$.

It should be noted that if data are required for combinations of the dimensionless groups not included in this chapter, it might be possible to determine the required information by interpolation or extrapolation. An alternative procedure is to extend the results to the required combinations using the techniques and equations included in Chapter 4 and in this chapter.

Problems

1. The dependent and independent variables for a proposed experiment are to be related by a "two-term sine series." The values of y_i are to be determined to 0.5% accuracy, and the uncertainties for the values of x_i are negligible. The expected values of a_1 and a_2 are 100. and 25. The values of x_i are to be evenly spaced, x_a is to be 0, and x_b is to be π. If the number of data points is 25, what are the predicted values of σ_{a_1} and σ_{a_2} for the proposed experiment?

2. Adjust the number of data points in Problem 1 so that the ratio a_2/a_1 is deter-mined to 0.2% accuracy. (Note—Equation 2.6.8 is required for solution of this problem.)

3. The dependent and independent variables for a proposed experiment are to be related by the "simple sine function with unknown range." The uncertainties for the values of y_i are to be estimated by "counting statistics," and the values of x_i are to be determined to ± 0.04. If the expected value of a_1 and a_2 are 10000. and 2.0, and the values x_a/a_2 and x_b/a_2 are to be 0.1 π and 0.9 π, how many data points are required so that a_2 will be determined to 1% accuracy.

4. For Problem 3, if the values of x_i are determined by an alternative method such that the uncertainties will be negligible, how many data points are required so that a_2 will be determined to 1% accuracy?

Computer Projects

1. Recompute the results presented in Table 7.7 for $a_2/a_1 = 0.1$ and 0.5 as well as 0.0.

2. Recompute the results presented in Table 7.7 for $a_3/a_1 = 0.1$ and 0.5, as well as 0.0.

3. Predict the values of σ_{a_1}, σ_{a_2}, and σ_{12} as functions of the experimental variables for the following cases of the "two-term sine series" class of experiments:

   (A) $\sigma_{y_i} = K_{fy} \cdot y_i$ $\sigma_{x_i} = K_{fy} \cdot x_i$

   (B) $\sigma_{y_i}^2 = (K_{fy} \cdot y_i)^2 + K_{cy}^2$ $\sigma_{y_i} = 0.$

   (C) $\sigma_{y_i}^2 = K_{sy}^2 \cdot y_i + K_{cy}^2$ $\sigma_{x_i} = 0.$

   For all cases, the values of x are to be evenly spaced.

4. Repeat the previous problem for the "simple sine function with unknown range" class of experiments.

5. For the following class of experiments:

   $$y = a_1 \sin x + a_2 \sin 2x + a_3,$$

   predict the values of σ_{a_1}, σ_{a_2}, and σ_{a_3} as functions of the experimental variables for the following cases:

   (A) $\sigma_{y_i} = K_{cy}$ $\sigma_{x_i} = K_{cx}$

   (B) $\sigma_{y_i} = K_{fy} \cdot |y_i|$ $\sigma_{x_i} = K_{cx}$

   (C) $\sigma_{y_i} = K_{sy} \cdot (y_i)^{1/2}$ $\sigma_{x_i} = K_{cx}$

   For all cases, the values of x are to be evenly spaced.

Chapter 8

The Gaussian Function

8.1 Introduction

Results for a wide variety of experiments are similar to the forms shown in Figures 8.1 and 8.2.

The equation used to relate y and x for experiments of the type shown in Figure 8.1 is:

$$y = a_1 \exp\left[-\left(\frac{x - x_0}{s}\right)^2\right]. \tag{8.1.1}$$

Equation 8.1.1 is often referred to as the Gaussian function. If x_0 and s are also unknown parameters, Equation 8.1.1 must be modified to conform to the notation used throughout the book:

$$y = a_1 \exp\left[-\left(\frac{x - a_3}{a_2}\right)^2\right]. \tag{8.1.2}$$

The equation used to relate y and x for experiments of the type shown in Figure 8.2 is:

$$y = a_1 \exp\left[-\left(\frac{x - x_{01}}{s_1}\right)^2\right] + a_2 \exp\left[-\left(\frac{x - x_{02}}{s_2}\right)^2\right]. \tag{8.1.3}$$

If the x_0's and s's are unknown parameters, the modified equation is:

$$y = a_1 \exp\left[-\left(\frac{x - a_3}{a_2}\right)^2\right] + a_4 \exp\left[-\left(\frac{x - a_6}{a_5}\right)^2\right]. \tag{8.1.4}$$

Equations 8.1.1 and 8.1.3 are linear, and Equations 8.1.2 and 8.1.4 are nonlinear.

Prediction analyses for experiments of the types expressed by Equations 8.1.1 through 8.1.4 are included in the following sections. Other variations

249

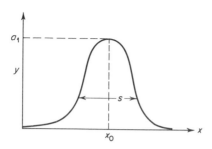

Figure 8.1. Single Gaussian.

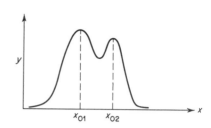

Figure 8.2. Sum of two Gaussians.

could also have been included; however, it was felt that these four types of experiments constitute a large percentage of all experiments requiring the Gaussian function.

It is assumed in this chapter that systematic errors are negligible. If systematic errors are considered as an important factor in a proposed experiment, the prediction analysis should be amended using the techniques discussed in Section 4.9.

8.2 Single Gaussian with Unknown Amplitude

Equation 8.1.1 is the relationship between the dependent and independent variables for experiments in this class:

$$y = a_1 \exp\left[-\left(\frac{x - x_0}{s}\right)^2\right]. \tag{8.1.1}$$

Only the amplitude a_1 is unknown. The constants x_0 and s are known and are treated as input quantities for an analysis. The cases that are analyzed in this section are summarized in Table 8.1.

For this class of experiments, $\hat{F}_{a_1}{}^i$ and $\hat{L}_i$ are:

$$\hat{F}_{a_1}{}^i = -\exp\left[-\left(\frac{x_i - x_0}{s}\right)^2\right], \tag{8.2.1}$$

$$\hat{L}_i = \sigma_{y_i}{}^2 + \left\{a_1 \cdot \frac{2(x_i - x_0)}{s^2} \cdot \exp\left[-\left(\frac{x_i - x_0}{s}\right)^2\right]\right\}^2 \cdot \sigma_{x_i}{}^2. \tag{8.2.2}$$

Since there is only one unknown parameter, the matrices $\hat{C}$ and $\hat{C}^{-1}$ both have only one element:

$$\hat{C}_{11} = \sum_{i=1}^{n} \frac{\exp\{-2[(x_i - x_0)/s]^2\}}{\hat{L}_i}, \tag{8.2.3}$$

$$\hat{\sigma}_{a_1}{}^2 = \hat{C}_{11}{}^{-1} = \frac{1}{\hat{C}_{11}} = \left(\sum_{i=1}^{n} \frac{\exp\{-2[(x_i - x_0)/s]^2\}}{\hat{L}_i}\right)^{-1}. \tag{8.2.4}$$

TABLE 8.1. Summary of cases analyzed in Section 8.2

Case	σ_{y_i}	σ_{x_i}	Figure
I	K_{cy}	K_{cx}	8.3
II	$K_{fy} \cdot y_i$	K_{cx}	8.3
III	$K_{sy} \cdot (y_i)^{1/2}$	K_{cx}	8.3

If it is assumed that $\sigma_{x_i} = 0$, then simple analytical solutions can be obtained for all the cases. Substituting the relationships for σ_{y_i} from Table 8.1 and Equation 8.2.2 into Equation 8.2.4 gives:

Case I

$$\hat{\sigma}_{a_1}^2 = \left(K_{cy}^2 \Big/ n \cdot \left\{ \exp\left[-2\left(\frac{x - x_0}{s}\right)^2 \right] \right\}_{av} \right), \qquad (8.2.5)$$

Case II

$$\hat{\sigma}_{a_1}^2 = \frac{a_1^2 \cdot K_{fy}^2}{n}, \qquad (8.2.6)$$

Case III

$$\hat{\sigma}_{a_1}^2 = \left(a_1 \cdot K_{sy}^2 \Big/ n \cdot \left\{ \exp\left[-\left(\frac{x - x_0}{s}\right)^2 \right] \right\}_{av} \right). \qquad (8.2.7)$$

If the values of x are evenly spaced (i.e., Equation 4.4.3 is valid), and if $x_b - x_0$ equals $x_0 - x_a$, then as n approaches infinity:

$$\left\{ \exp\left[-\left(\frac{x - x_0}{s}\right)^2 \right] \right\}_{av} \rightarrow \left\{ \int_{x_a}^{x_b} \exp\left[-\left(\frac{x - x_0}{s}\right)^2 \right] dx \Big/ (x_b - x_a) \right\}$$

$$= \frac{\text{erf } (t)}{t} \cdot \frac{\sqrt{\pi}}{2}, \qquad (8.2.8)$$

where

$$\text{erf } (t) \equiv \frac{2}{\sqrt{\pi}} \int_0^t \exp\left(-u^2 \right) du, \qquad (8.2.9)$$

and

$$t \equiv \frac{x_b - x_0}{s}. \qquad (8.2.10)$$

Similarly,

$$\left\{ \exp\left[-2\left(\frac{x-x_0}{s}\right)^2 \right] \right\}_{av} \rightarrow \frac{\text{erf}\,(\sqrt{2}\cdot t)}{\sqrt{2}\cdot t}\cdot\frac{\sqrt{\pi}}{2}. \qquad (8.2.11)$$

Substituting Equations 8.2.8 and 8.2.11 into 8.2.5 and 8.2.7, and then rearranging terms, we obtain:

Case I

$$E_1 \equiv \frac{\hat{\sigma}_{a1}\cdot n^{1/2}}{K_{cy}} \rightarrow \left(\frac{\sqrt{2}\cdot t}{\text{erf}\,(\sqrt{2}\cdot t)}\cdot\frac{2}{\sqrt{\pi}}\right)^{1/2}, \qquad (8.2.12)$$

Case II

$$\Theta_1 \equiv \frac{\hat{\sigma}_{a1}\cdot n^{1/2}}{a_1\cdot K_{fy}} \rightarrow 1, \qquad (8.2.13)$$

Case III

$$\Psi_1 \equiv \frac{\hat{\sigma}_{a1}\cdot n^{1/2}}{(a_1)^{1/2}\cdot K_{sy}} \rightarrow \left(\frac{t}{\text{erf}\,(t)}\cdot\frac{2}{\sqrt{\pi}}\right)^{1/2}. \qquad (8.2.14)$$

To evaluate the error function [i.e., $\text{erf}\,(t)$], use can be made of the relationship between this function and the normal probability integral:

$$\text{Normal probability integral} \equiv \frac{1}{(2\pi)^{1/2}}\int_{-t}^{t} \exp\left(-\frac{u^2}{2}\right) du = \text{erf}\left(\frac{t}{\sqrt{2}}\right).$$

$$(8.2.15)$$

The normal probability integral and the error function are tabulated in many reference and textbooks.[1]

If σ_{x_i} is not equal to zero, Equations 8.2.12 through 8.2.14 are not valid. Results for such cases were determined using the direct approach and are presented for Case I in terms of E_1 and ω_c:

$$\omega_c \equiv \frac{a_1\cdot K_{cx}}{(x_b - x_0)\cdot K_{cy}}, \qquad (8.2.16)$$

---

[1] "Handbook of Chemistry and Physics," edited by Charles D. Hodgman et al., Chemical Rubber Co., Cleveland, published annually; P. G. Hoel, "Introduction to Mathematical Statistics," John Wiley & Sons, Inc., New York, 1954, "Handbook of Mathematical Functions," NBS Applied Mathematics Series No. 55, U.S. Govt. Printing Office, Washington, D.C., etc,

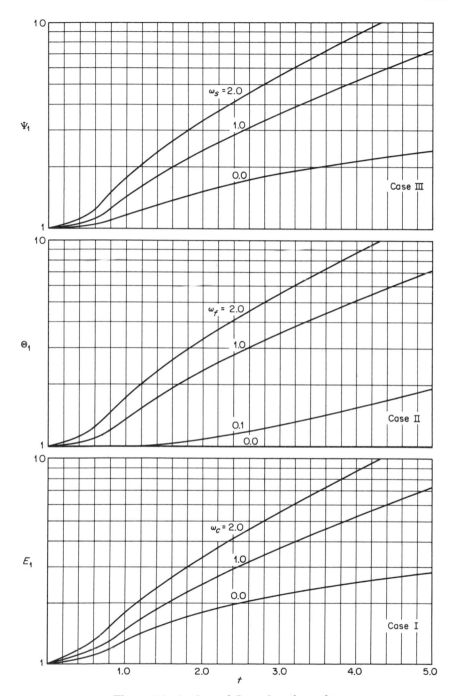

Figure 8.3. Δ_1, Θ_1, and Ψ_1 as functions of t.

for Case II in terms of Θ_1 and ω_f:

$$\omega_f \equiv \frac{K_{cx}}{(x_b - x_0) \cdot K_{fy}}, \qquad (8.2.17)$$

and for Case III in terms of Ψ_1 and ω_s:

$$\omega_s \equiv \frac{(a_1)^{1/2} \cdot K_{cx}}{(x_b - x_0) \cdot K_{sy}}. \qquad (8.2.18)$$

The results are plotted in Figure 8.3.

For all cases, Figure 8.3 shows that E_1, Θ_1, and Ψ_1 approach one as t approaches zero. This observation can be explained by examining the expansion of the error function:

$$\text{erf}\,(t) = \frac{2t}{\sqrt{\pi}} \cdot \left[1 - \frac{t^2}{3.1!} + \frac{t^4}{5.2!} - \frac{t^6}{7.3!} + \cdots \right]. \qquad (8.2.19)$$

As $t \to 0$, $\text{erf}\,(t) \to 2t/\sqrt{\pi}$. Substituting this result into Equations 8.2.12 and 8.2.14, it is seen that E_1 and Ψ_1, as well as Θ_1, approach one as t approaches zero. These equations are, however, only valid for $K_{cx} = 0$. Examining Equation 8.2.2, it is seen that all values of $\hat{L}_i$ approach σ_{y_i} as t (i.e., $x_b - x_0/s$) approaches zero, and therefore the limit of one is also valid for values of K_{cx} other than zero.

For large values of t, Equation 8.2.19 is not a useful series for evaluating $\text{erf}(t)$. An alternative expansion of the error function which can be used for large values of t is:

$$\text{erf}(t) = 1 - \frac{e^{-t^2}}{t\sqrt{\pi}} \cdot \left[1 - \frac{1}{2t^2} + \frac{1 \cdot 3}{(2t^2)^2} - \frac{1 \cdot 3 \cdot 5}{(2t^2)^3} + \cdots \right]. \qquad (8.2.20)$$

8.3 Single Gaussian with Three Unknown Parameters

Equation 8.1.2 is the relationship between the dependent and independent variables for experiments in this class:

$$y = a_1 \exp\left[-\left(\frac{x - a_3}{a_2} \right)^2 \right]. \qquad (8.1.2)$$

The parameter a_1 is often referred to as the *amplitude* of the peak, a_2 as the *width* of the peak, and a_3 as the *location* of the peak. For this class of experiments, the purpose is to determine a_2 and a_3, as well as a_1. The cases that are analyzed in this section are summarized in Table 8.2.

TABLE 8.2. Summary of cases analyzed in Section 8.3

Case	σ_{y_i}	σ_{x_i}	Figure
I	K_{cy}	K_{cx}	8.4
II	$K_{fy} \cdot y_i$	K_{cx}	8.5
III	$K_{sy} \cdot (y_i)^{1/2}$	K_{cx}	8.6

For this class of experiments, $\hat{F}_{a_1}{}^i$, $\hat{F}_{a_2}{}^i$, $\hat{F}_{a_3}{}^i$, and $\hat{L}_i$ are:

$$\hat{F}_{a_1}{}^i = - \exp\left[-\left(\frac{x_i - a_3}{a_2}\right)^2\right], \tag{8.3.1}$$

$$\hat{F}_{a_2}{}^i = -a_1 \cdot \frac{2}{a_2} \cdot \left(\frac{x_i - a_3}{a_2}\right)^2 \cdot \exp\left[-\left(\frac{x_i - a_3}{a_2}\right)^2\right], \tag{8.3.2}$$

$$\hat{F}_{a_3}{}^i = -a_1 \cdot \frac{2(x_i - a_3)}{a_2{}^2} \cdot \exp\left[-\left(\frac{x_i - a_3}{a_2}\right)^2\right], \tag{8.3.3}$$

$$\hat{L}_i = \sigma_{y_i}{}^2 + \left\{a_1 \cdot \frac{2(x_i - a_3)}{a_2{}^2} \cdot \exp\left[-\left(\frac{x_i - a_3}{a_2}\right)^2\right]\right\}^2 \cdot \sigma_{x_i}{}^2. \tag{8.3.4}$$

Equations 8.3.1 through 8.3.4 can be simplified by using the following substitution:

$$t_i \equiv \frac{x_i - a_3}{a_2}. \tag{8.3.5}$$

The resulting equations are:

$$\hat{F}_{a_1}{}^i = - \exp(-t_i{}^2), \tag{8.3.6}$$

$$\hat{F}_{a_2}{}^i = - \frac{2a_1}{a_2} t_i{}^2 \exp(-t_i{}^2), \tag{8.3.7}$$

$$\hat{F}_{a_3}{}^i = - \frac{2a_1}{a_2} t_i \exp(-t_i{}^2), \tag{8.3.8}$$

$$\hat{L}_i = \sigma_{y_i}{}^2 + \left[\frac{2a_1}{a_2} t_i \exp(-t_i{}^2)\right] \sigma_{x_i}{}^2. \tag{8.3.9}$$

Attempts to develop analytical expressions for the $\hat{\sigma}_{a_k}$'s are complicated not only by the fact that there are three unknown parameters, but also by the complexity of the individual elements of the matrix $\hat{C}$. Even if the σ_{x_i}'s are assumed to be zero, the complexity of the resulting equations is

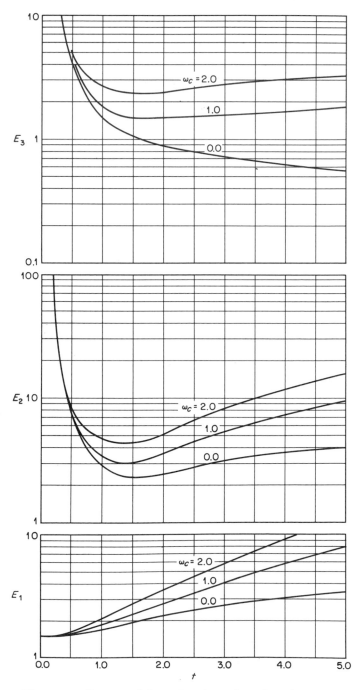

Figure 8.4. E_1, E_2, and E_3 as functions of t and ω_c (Case I).

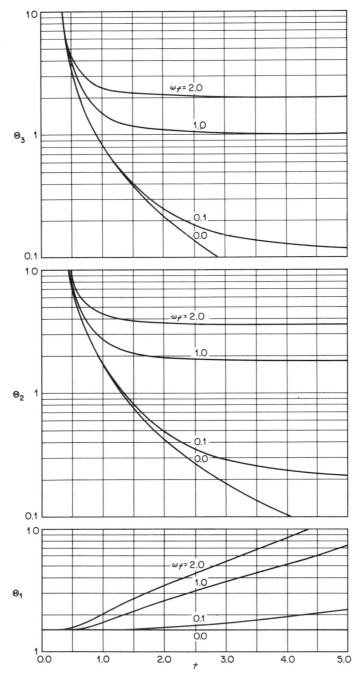

Figure 8.5. Θ_1, Θ_2, Θ_3 as functions of t and ω_f (Case II).

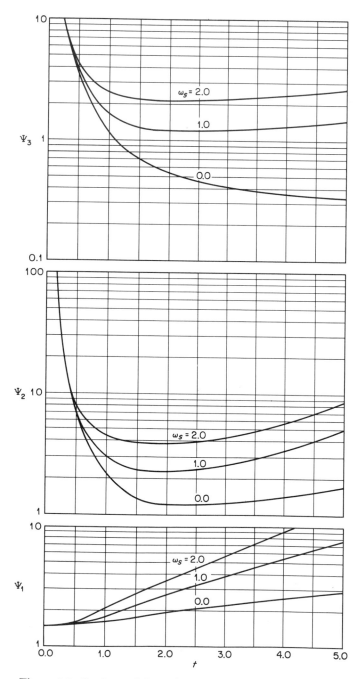

Figure 8.6. Ψ_1, Ψ_2, and Ψ_3 as functions of t and ω_s (Case III).

enormous. Using the direct approach, results can easily be obtained and are presented in Figures 8.4 through 8.6. For Case I the results are presented in terms of E_k and ω_c:[2]

$$E_k \equiv \frac{\hat{\sigma}_{a_k} \cdot n^{1/2}}{a_k} \cdot \frac{a_1}{K_{cy}}, \tag{8.3.10}$$

$$\omega_c \equiv \frac{a_1 \cdot K_{cx}}{(x_b - a_3) \cdot K_{cy}}; \tag{8.3.11}$$

for Case II the results are presented in terms of Θ_k and ω_f:[2]

$$\Theta_k \equiv \frac{\hat{\sigma}_{a_k} \cdot n^{1/2}}{a_k \cdot K_{fy}}, \tag{8.3.12}$$

$$\omega_f \equiv \frac{K_{cx}}{(x_b - a_3) \cdot K_{fy}}; \tag{8.3.13}$$

and for Case III the results are presented in terms of Ψ_k and ω_s:[2]

$$\Psi_k \equiv \frac{\hat{\sigma}_{a_k} \cdot n^{1/2} \cdot (a_1)^{1/2}}{a_k \cdot K_{sy}}, \tag{8.3.14}$$

$$\omega_s \equiv \frac{(a_1)^{1/2} \cdot K_{cx}}{(x_b - a_3) \cdot K_{sy}}. \tag{8.3.15}$$

For each case, it is assumed that the values of x_i are evenly spaced (i.e., Equation 4.4.3 is valid), and are symmetrical about a_3. The dimensionless groups are plotted for a range of t values from 0.0 to 5.0.

As t approaches zero it is observed that E_1, Θ_1, and Ψ_1 approach 1.5, while the other E_k's, Θ_k's and Ψ_k's approach infinity. These limits can be explained by first considering the limits of Equations 8.3.6 through 8.3.9. As t approaches zero:

$$\hat{F}_{a_1}{}^i \rightarrow -1, \tag{8.3.16}$$

$$\hat{F}_{a_2}{}^i \rightarrow -\frac{2a_1}{a_2} t_i^2, \tag{8.3.17}$$

$$\hat{F}_{a_3}{}^i \rightarrow -\frac{2a_1}{a_2} t_i, \tag{8.3.18}$$

and

$$\hat{L}_i \rightarrow K_{cy}{}^2 \qquad \text{for Case I,} \tag{8.3.19}$$

$$\hat{L}_i \rightarrow a_1{}^2 K_{fy}{}^2 \qquad \text{for Case II,} \tag{8.3.20}$$

$$\hat{L}_i \rightarrow a_1 K_{sy}{}^2 \qquad \text{for Case III.} \tag{8.3.21}$$

[2] If x_a is unequal to zero, the equations for E_3, Θ_3, and Ψ_3 should be altered by replacing a_3 in the denominator by $a_3 - x_a$.

For Case I, from Equations 8.3.16 through 8.3.21 it can be seen that if the values of x_i are evenly spaced and are symmetric about a_3, then as t approaches zero, substitution into Equation 4.4.5 yields:

$$\hat{C}_{11} \rightarrow \frac{n}{K_{cy}{}^2}, \tag{8.3.22}$$

$$\hat{C}_{12} \rightarrow \frac{n}{K_{cy}{}^2} \cdot (t_i{}^2)_{\text{av}} \cdot \frac{2a_1}{a_2} = \frac{n}{K_{cy}{}^2} \cdot \frac{t^2}{3} \cdot \frac{2a_1}{a_2}, \tag{8.3.23}$$

$$\hat{C}_{13} \rightarrow \frac{n}{K_{cy}{}^2} \cdot (t_i)_{\text{av}} \cdot \frac{2a_1}{a_2} = 0, \tag{8.3.24}$$

$$\hat{C}_{22} \rightarrow \frac{n}{K_{cy}{}^2} \cdot (t_i{}^4)_{\text{av}} \cdot \frac{4a_1{}^2}{a_2{}^2} = \frac{n}{K_{cy}{}^2} \cdot \frac{t^4}{5} \cdot \frac{4a_1{}^2}{a_2{}^2}, \tag{8.3.25}$$

$$\hat{C}_{23} \rightarrow \frac{n}{K_{cy}{}^2} \cdot (t_i{}^3)_{\text{av}} \cdot \frac{4a_1{}^2}{a_2{}^2} = 0, \tag{8.3.26}$$

$$\hat{C}_{33} \rightarrow \frac{n}{K_{cy}{}^2} \cdot (t_i{}^2)_{\text{av}} \cdot \frac{4a_1{}^2}{a_2{}^2} = \frac{n}{K_{cy}{}^2} \cdot \frac{t^2}{3} \cdot \frac{4a_1{}^2}{a_2{}^2}. \tag{8.3.27}$$

Inverting the matrix, it is seen that:

$$\hat{C}_{11}{}^{-1} \rightarrow \frac{K_{cy}{}^2}{n} \cdot \frac{9}{4}, \tag{8.3.28}$$

$$\hat{C}_{22}{}^{-1} \quad \text{and} \quad \hat{C}_{33}{}^{-1} \rightarrow \infty, \tag{8.3.29}$$

and therefore $E_1 \rightarrow 1.5$, $E_2 \rightarrow \infty$, and $E_3 \rightarrow \infty$. Using the same approach, these limits can also be derived for the Θ_k's and the Ψ_k's.

If the purpose of the experiment is to measure a_2, for Case I it is observed that the optimum value of t is about 1.5. This optimum is insensitive to the value of ω_c. For Case II the value of Θ_2 approaches an asymptote as t becomes large. For Case III the optimum value of t is about 2.0 and is insensitive to ω_c.

Comparing the results from Sections 8.2 and 8.3, it is seen that for all values of t, ω_c, ω_f, and ω_s, the values of E_1, Θ_1, and Ψ_1 are greater in Section 8.3. For small values of t, the values in Section 8.3 are 50% higher than the values in Section 8.2. This increase represents the loss of accuracy associated with increasing the number of unknown parameters from one to three.

8.4 Two Gaussians with Unknown Amplitudes

Equation 8.1.3 is the relationship between the dependent and independent variables for experiments in this class:

$$y = a_1 \exp\left[-\left(\frac{x - x_{01}}{s_1}\right)^2\right] + a_2 \exp\left[-\left(\frac{x - x_{02}}{s_2}\right)^2\right]. \quad (8.1.3)$$

Only the amplitudes a_1 and a_2 are unknown. The constants x_{01}, x_{02}, s_1, and s_2 are known and are treated as input quantities for the analysis. The cases that are analyzed in this section are summarized in Table 8.3.

For this class of experiments, $\hat{F}_{a_1}{}^i$, $\hat{F}_{a_2}{}^i$, and $\hat{L}_i$ are:

$$\hat{F}_{a_1}{}^i = -\exp\left[-\left(\frac{x_i - x_{01}}{s_1}\right)^2\right], \quad (8.4.1)$$

$$\hat{F}_{a_2}{}^i = -\exp\left[-\left(\frac{x_i - x_{02}}{s_2}\right)^2\right], \quad (8.4.2)$$

$$\hat{L}_i = \sigma_{y_i}{}^2 + \left\{a_1 \cdot \frac{2(x_i - x_{01})}{s_1{}^2} \cdot \exp\left[-\left(\frac{x_i - x_{01}}{s_1}\right)^2\right] + a_2 \cdot \frac{2(x_i - x_{02})}{s_2{}^2}\right.$$

$$\left. \times \exp\left[-\left(\frac{x_i - x_{02}}{s_2}\right)^2\right]\right\}^2 \cdot \sigma_{x_i}{}^2. \quad (8.4.3)$$

The equations can be simplified by using the following substitutions:

$$t_{1i} \equiv \frac{x_i - x_{01}}{s_1}, \quad (8.4.4)$$

$$t_{2i} \equiv \frac{x_i - x_{02}}{s_2}. \quad (8.4.5)$$

Substituting Equations 8.4.4 and 8.4.5 into Equations 8.4.1 through 8.4.3 gives:

$$\hat{F}_{a_1}{}^i = -\exp\left(-t_{1i}{}^2\right), \quad (8.4.6)$$

$$\hat{F}_{a_2}{}^i = -\exp\left(-t_{2i}{}^2\right), \quad (8.4.7)$$

$$\hat{L}_i = \sigma_{y_i}{}^2 + \left(\frac{2a_1}{s_1} \cdot t_{1i} \exp\left(-t_{1i}{}^2\right) + \frac{2a_2}{s_2} \cdot t_{2i} \exp\left(-t_{2i}{}^2\right)\right)^2 \cdot \sigma_{x_i}{}^2. \quad (8.4.8)$$

Equations for the elements of the matrix $\hat{C}$ can be obtained by substituting Equations 8.4.6 through 8.4.8 into Equation 4.4.1:

$$\hat{C}_{11} = n \cdot \left(\frac{\exp\left(-2t_{1i}^2\right)}{\hat{L}_i} \right)_{av},$$
(8.4.9)

$$\hat{C}_{12} = n \cdot \left(\frac{\exp\left[-(t_{1i}^2 + t_{2i}^2)\right]}{\hat{L}_i} \right)_{av},$$
(8.4.10)

$$\hat{C}_{22} = n \cdot \left(\frac{\exp\left(-2t_{2i}^2\right)}{\hat{L}_i} \right)_{av}.$$
(8.4.11)

If the "peaks" are widely separated (i.e., $|x_{02} - x_{01}| \gg s_1$ and s_2), then $\hat{C}_{12} \ll \hat{C}_{11}$ and $\hat{C}_{22}$ and the expressions for the $\hat{\sigma}$'s reduce to:

$$\hat{\sigma}_{a_1}^2 = \hat{C}_{11}^{-1} = \frac{\hat{C}_{22}}{\hat{C}_{11} \cdot \hat{C}_{22} - \hat{C}_{12}^2} \approx \frac{1}{\hat{C}_{11}},$$
(8.4.12)

$$\hat{\sigma}_{a_2}^2 = \hat{C}_{22}^{-1} = \frac{\hat{C}_{11}}{\hat{C}_{11} \cdot C_{22} - \hat{C}_{12}^2} \approx \frac{1}{\hat{C}_{22}},$$
(8.4.13)

$$\hat{\sigma}_{12} = \hat{C}_{12}^{-1} = \frac{-\hat{C}_{12}}{\hat{C}_{11} \cdot \hat{C}_{22} - C_{12}^2} \approx 0.$$
(8.4.14)

Comparing Equations 8.4.12 and 8.4.13 with Equation 8.2.4, it is seen that the resulting expressions for $\hat{\sigma}_{a_1}$ and $\hat{\sigma}_{a_2}$ for this class of experiments reduce to the expressions for the *single Gaussian with one unknown parameter* if the peaks are widely separated.

When the peaks are not widely separated, the solution is more complicated. In fact, even if σ_{x_i} is assumed to be equal to zero, the complexity of the solution is enormous. It is instructive to simplify the analysis still further. If a_2 and σ_{x_i} are both assumed to be zero, then for Case III:

$$\hat{L}_i = \hat{\sigma}_{y_i}^2 = K_{sy}^2 \cdot a_1 \exp\left(-t_{1i}^2\right),$$
(8.4.15)

and

$$\hat{C}_{11} = \frac{n}{a_1 K_{sy}^2} \cdot \left[\exp\left(-t_{1i}^2\right)\right]_{av},$$
(8.4.16)

$$\hat{C}_{12} = \frac{n}{a_1 K_{sy}^2} \cdot \left[\exp\left(-t_{2i}^2\right)\right]_{av},$$
(8.4.17)

$$\hat{C}_{22} = \frac{n}{a_1 K_{sy}^2} \cdot \left[\exp\left(-2t_{2i}^2 + t_{1i}^2\right)\right]_{av}.$$
(8.4.18)

TABLE 8.3. Summary of cases analyzed in Section 8.4

Case	σ_{yi}	σ_{xi}	Table
I	K_{cy}	K_{cx}	8.4
II	$K_{fy} \cdot y_i$	K_{cx}	8.5
III	$K_{sy} \cdot (y_i)^{1/2}$	K_{cx}	8.6

If it is assumed that the values of x_i are evenly spaced, (i.e., Equation 4.4.3 is valid), then Equation 8.2.8 can be used to derive expressions for $\hat{C}_{11}$ and $\hat{C}_{12}$. The derivation of an expression for $\hat{C}_{22}$ is, however, more complicated. Considering only the exponent of the exponential, and assuming that $s_1 = s_2 = s$, then,

$$-2t_{2i}^2 + t_{1i}^2 = -2\left(\frac{x - x_{01}}{s}\right)^2 + \left(\frac{x - x_{02}}{s}\right)^2 = -\left(\frac{x - x_0}{s}\right)^2 + \frac{Q}{s^2},$$

$$(8.4.19)$$

where x_0 and Q are two unknown constants defined by Equation 8.4.19. Solving for these constants, we obtain:

$$-2(x^2 - 2xx_{01} + x_{01}^2) + x^2 - 2xx_{02} + x_{02}^2 = -x^2 + 2xx_0 - x_0^2 + Q,$$

$$(8.4.20)$$

$$4xx_{01} - 2xx_{02} = 2xx_0, \qquad (8.4.21)$$

$$-2x_{01}^2 + x_{02}^2 = Q - x_0^2, \qquad (8.4.22)$$

$$x_0 = 2x_{01} - x_{02}, \qquad (8.4.23)$$

$$Q = -2x_{01}^2 + x_{02}^2 + 4x_{01}^2 - 4x_{01}x_{02} + x_{02}^2 = 2(x_{02} - x_{01})^2. \qquad (8.4.24)$$

Substituting Equations 8.4.19, 8.4.23, and 8.4.24 into Equation 8.4.18 gives:

$$\hat{C}_{22} = \frac{n}{a_1 K_{sy}^2} \cdot \exp\left[\frac{2(x_{02} - x_{01})^2}{s^2}\right] \left\{\exp\left[-\left(\frac{x - x_0}{s}\right)^2\right]\right\}_{av}. \qquad (8.4.25)$$

Equation 8.2.8 is only valid if $x_b - x_0 = x_0 - x_a$. The equation can be modified:

$$\left\{\exp\left[-\left(\frac{x - x_0}{s}\right)^2\right]\right\}_{av} \rightarrow \left\{\int_{x_a}^{x_b} \exp\left[-\left(\frac{x - x_0}{s}\right)^2\right] dx \Big/ (x_b - x_a)\right\}$$

$$= \frac{s \cdot \sqrt{\pi}}{2(x_b - x_a)} \cdot \left[\text{erf}\left(\frac{x_b - x_0}{s}\right) - \text{erf}\left(\frac{x_a - x_0}{s}\right)\right]. \qquad (8.4.26)$$

TABLE 8.4. Values[a] of E_1 for various combinations of a_2/a_1, ω_c, and $|x_{01} - x_{02}|/s$

a_2/a_1	ω_c	$|x_{01} - x_{02}|/s$				
		0.25	0.75	1.25	1.75	2.25
0.1	0.0	11.48	4.307	3.177	2.893	2.834
1.0	0.0	11.48	4.307	3.177	2.893	2.834
10.0	0.0	11.48	4.307	3.177	2.893	2.834
0.1	1.0	38.58	11.42	7.492	7.240	7.311
1.0	1.0	68.70	19.66	8.388	4.813	6.497
10.0	1.0	321.4	82.65	38.43	20.22	6.592
0.1	5.0	162.6	45.95	32.43	32.48	32.90
1.0	5.0	308.7	91.26	35.05	14.22	26.93
10.0	5.0	1455.	357.6	175.4	96.89	26.71

[a] For all combinations, $s_1 = s_2 = s$ and $x_{01} - x_a = x_b - x_{01} = 5s$. The results were determined using the direct approach with $n = 24$ (Case I).

By substituting Equation 8.4.26 into Equations 8.4.16, 8.4.17 and 8.4.25, expressions for the elements of the matrix $\hat{C}$ can be derived. As x_{02} approaches x_{01}, the constant Q approaches zero, and therefore $\exp(Q^2/s^2)$ approaches one. In addition, the constant x_0 also approaches x_{01}. The elements $\hat{C}_{22}$ and $\hat{C}_{12}$ therefore approach $\hat{C}_{11}$ and the $\hat{s}$'s all approach in-

TABLE 8.5. Values[a] of Θ_1 for various combinations of a_2/a_1, ω_f, and $|x_{01} - x_{02}|/s$

a_2/a_1	ω_f	$|x_{01} - x_{02}|/s$				
		0.25	0.75	1.25	1.75	2.25
0.1	0.0	1.412	1.284	1.281	1.268	1.250
1.0	0.0	1.996	1.475	1.389	1.340	1.303
10.0	0.0	5.256	1.797	1.530	1.427	1.363
0.1	1.0	19.37	8.111	7.045	7.048	7.084
1.0	1.0	46.29	18.20	8.800	4.880	6.117
10.0	1.0	174.0	49.60	30.52	18.65	6.763
0.1	5.0	95.00	37.02	31.75	32.24	32.56
1.0	5.0	229.9	85.28	34.92	14.27	26.66
10.0	5.0	867.4	247.4	152.0	92.27	26.69

[a] For all combinations, $s_1 = s_2 = s$ and $x_{01} - x_a = x_b - x_{01} = 5s$. The results were determined using the direct approach with $n = 24$ (Case II).

TABLE 8.6. Values[a] of Ψ_1 for various combinations of a_2/a_1, ω_s, and $|x_{01} - x_{02}|/s$

a_2/a_1	ω_s	$|x_{01} - x_{02}|/s$				
		0.25	0.75	1.25	1.75	2.25
0.1	0.0	7.301	3.123	2.595	2.459	2.410
1.0	0.0	9.791	3.921	2.962	2.632	2.490
10.0	0.0	21.96	6.837	4.069	3.112	2.706
0.1	1.0	30.23	9.203	7.165	7.125	7.193
1.0	1.0	58.59	18.87	8.575	4.847	6.298
10.0	1.0	267.7	65.44	34.06	19.39	6.666
0.1	5.0	128.2	39.39	31.91	32.33	32.67
1.0	5.0	265.9	87.59	34.91	14.24	26.77
10.0	5.0	1179.	290.1	161.2	94.22	26.69

[a] For all combinations, $s_1 = s_2 = s$ and $x_{01} - x_a = x_b - x_{01} = 5s$. The results were determined using the direct approach with $n = 24$ (Case III).

finity. It should be remembered that a_2 was assumed to be zero. In conclusion, even if a_2 is zero, if an experiment is performed in which it is hoped to determine a_1 and a_2 by a least squares analysis, the resulting estimates of the uncertainties σ_{a_1} and σ_{a_2} will be large if $|x_{01} - x_{02}|$ is small compared to s_1 or s_2.

Results determined by the direct approach are listed in Tables 8.4 through 8.6. For all the results $s_1 = s_2 = s$, and $x_{01} - x_a = x_b - x_{01} = 5s$. For Case I the results are expressed in terms of E_1 (defined by Equation 8.3.10), for Case II in terms of Θ_1 (defined by Equation 8.2.12), and for Case III in terms of Ψ_1 (defined by Equation 8.2.14). The results are presented for various combinations a_2/a_1, $\omega_{c,f,\text{or } s}$ (defined by Equations 8.2.16 through 8.2.18), and $|x_{02} - x_{01}|/s$.

It should be emphasized that if $s_1 \neq s_2$ the results in Tables 8.4 through 8.6 are invalid. In addition, all the results are for the special case $x_{01} - x_a = x_b - x_{01} = 5s$. The results are exact only for $n = 24$, but are good approximations for most values of n. For the cases and combinations studied, the differences between the values of E_1, Θ_1, and Ψ_1 determined at $n = 15$ and $n \to \infty$ were less than 1%.

From Table 8.4 it is observed that E_1 is not a function of a_2/a_1 if $\omega_c = 0$. However, for values of ω_c other than zero, E_1 is a function of a_2/a_1. For all cases (except Case I, $\omega_c = 0$) the effect of increasing a_2/a_1 is dependent upon the "dimensionless peak separation" $|x_{01} - x_{02}|/s$. For $|x_{01} - x_{02}|/s = 1.75$ and for ω_c, ω_f, and $\omega_s = 1.0$ and 5.0, the values of E_1, Θ_1, and Ψ_1 are seen to first decrease and then increase as a_2/a_1 is increased. This behavior

TABLE 8.7. Summary of cases analyzed in Section 8.5

Case	σ_{y_i}	σ_{x_i}	Table
I	K_{cy}	K_{cx}	8.8
II	$K_{fy} \cdot y_i$	K_{cx}	8.9
III	$K_{sy} \cdot (y_i)^{1/2}$	K_{cx}	8.10

can only be understood by examining Equation 8.4.3. In the region between x_{01} and x_{02}, the two terms which constitute the derivative $F_x{}^i$ have opposite signs. As a_2/a_1 increases from zero, in this region the absolute value of the derivative, and therefore $\hat{L}_i$, decreases, passes through a minimum, and then increases.

Comparing the results from Sections 8.2 and 8.4, it is observed that as $|x_{01} - x_{02}|/s$ increases and as a_2/a_1 decreases, the values of E_1, Θ_1, and Ψ_1 for the *two Gaussians class of experiments* approach the values for the *single Gaussian class*. The comparison must be made with the values in Figure 8.3 corresponding to $t = 5.0$, because all the results in this section are only valid for $t = 5.0$. This observation is reasonable because as $|x_{01} - x_{02}|/s$ becomes large, the "interaction" between the two peaks becomes small. In fact, $\hat{\sigma}_{12}$ approaches zero as $|x_{01} - x_{02}|/s$ becomes large.

TABLE 8.8. Values[a] of E_1, E_2, and E_3 for various combinations of a_4/a_1, ω_c, and $|a_3 - a_6|/a_2$

a_4/a_1	ω_c	$|a_3 - a_6|/a_2$	E_1	E_2	E_3
All values	0.0	2.25	3.503	5.710	0.790
		1.75	5.329	9.070	1.800
		1.25	40.87	19.97	7.110
		0.75	654.1	75.74	54.34
0.1	1.0	2.25	8.037	13.00	2.212
		1.75	10.03	18.78	3.876
		1.25	71.74	37.28	13.43
		0.75	1361.	156.9	114.8
1.0	1.0	2.25	7.635	12.63	2.110
		1.75	8.117	17.88	3.604
		1.25	89.01	43.37	16.44
		0.75	1611.	179.3	134.0
10.0	1.0	2.25	10.31	21.65	4.427
		1.75	67.75	47.93	16.59
		1.25	405.8	117.1	59.32
		0.75	5185.	497.2	413.0

[a] For all combinations $a_5 = a_2$ and $a_3 - x_a = x_b - a_3 = 5a_2$. The results were determined using the direct approach with $n = 24$ (Case I).

8.5 Two Gaussians, Each with Three Unknown Parameters

Equation 8.1.4 is the relationship between the dependent and independent variables for experiments in this class:

$$y = a_1 \exp\left[-\left(\frac{x - a_3}{a_2}\right)^2\right] + a_4 \exp\left[-\left(\frac{x - a_6}{a_5}\right)^2\right]. \quad (8.1.4)$$

For this class of experiment each of the six a_k's is determined by least squares analysis of the data. The cases that are analyzed in this section are summarized in Table 8.7.

For this class of experiments the $\hat{F}_{a_k}i$'s and $\hat{L}_i$ are:

$$\hat{F}_{a_1}i = -\exp\left[-\left(\frac{x_i - a_3}{a_2}\right)^2\right], \quad (8.5.1)$$

$$\hat{F}_{a_2}i = -a_1 \cdot \frac{2}{a_2} \cdot \left(\frac{x_i - a_3}{a_2}\right)^2 \cdot \exp\left[-\left(\frac{x_i - a_3}{a_2}\right)^2\right], \quad (8.5.2)$$

$$\hat{F}_{a_3}i = -a_1 \cdot \frac{2(x_i - a_3)}{a_2^2} \cdot \exp\left[-\left(\frac{x_i - a_3}{a_2}\right)^2\right], \quad (8.5.3)$$

$$\hat{F}_{a_4}i = -\exp\left[-\left(\frac{x_i - a_6}{a_5}\right)^2\right], \quad (8.5.4)$$

$$\hat{F}_{a_5}i = -a_4 \cdot \frac{2}{a_5} \cdot \left(\frac{x_i - a_6}{a_5}\right)^2 \cdot \exp\left[-\left(\frac{x_i - a_6}{a_5}\right)^2\right], \quad (8.5.5)$$

$$\hat{F}_{a_6}i = -a_4 \cdot \frac{2(x_i - a_6)}{a_5^2} \cdot \exp\left[-\left(\frac{x_i - a_6}{a_5}\right)^2\right], \quad (8.5.6)$$

$$\hat{L}_i = \sigma_{y_i}^2 + \left\{a_1 \cdot \frac{2(x_i - a_3)}{a_2^2} \cdot \exp\left[-\left(\frac{x_i - a_3}{a_2}\right)^2\right] + a_4 \cdot \frac{2(x_i - a_6)}{a_5^2}\right.$$

$$\left. \times \exp\left[-\left(\frac{x_i - a_6}{a_5}\right)\right]\right\}^2 \cdot \sigma_{x_i}^2. \quad (8.5.7)$$

For this class of experiments, computation of the $\hat{\sigma}$'s requires inversion of a 6×6 matrix. The complexity of such an inversion is feasible only with the aid of a digital computer, and therefore the direct approach is mandatory.

For Case I the results are presented in Table 8.8 in terms of E_1, E_2, and E_3 (defined by Equation 8.3.10), as functions of a_4/a_1, ω_c (defined by

TABLE 8.9. Values[a] of Θ_1, Θ_2, and Θ_3 for various combination of a_4/a_1, ω_f | $a_3 - a_6$ |/a_2

a_4/a_1	ω_f	| $a_3 - a_6$ |/a_2	Θ_1	Θ_2	Θ_3
0.1	0.0	2.25	1.834	0.215	0.176
		1.75	2.063	0.240	0.206
		1.25	2.607	0.280	0.259
		0.75	5.561	0.405	0.444
1.0		2.25	2.307	0.265	0.235
		1.75	3.036	0.310	0.298
		1.25	4.826	0.395	0.420
		0.75	13.25	0.670	0.860
10.0		2.25	3.281	0.325	0.319
		1.75	5.101	0.415	0.444
		1.25	10.08	0.610	0.733
		0.75	37.72	1.370	2.034
0.1	1.0	2.25	7.381	3.350	1.699
		1.75	7.543	3.880	2.096
		1.25	12.81	4.955	3.127
		0.75	75.01	8.525	7.650
1.0		2.25	7.118	3.525	1.792
		1.75	6.609	4.190	2.309
		1.25	28.42	6.650	4.897
		0.75	194.1	14.06	15.17
10.0		2.25	10.24	4.925	2.963
		1.75	44.08	8.070	6.441
		1.25	142.0	13.35	13.33
		0.75	717.7	32.42	43.27

[a] For all combinations $a_5 = a_2$ and $a_3 - x_a = x_b - a_3 = 5a_2$. The results were determined using the direct approach with $n = 24$ (Case II).

Equation 8.3.11), and | $a_3 - a_6$ |/a_2. For all combinations $a_3 - x_a = x_b - a_3 = 5a_2$, and $a_5 = a_2$.

For Case II the results are presented in Table 8.9 in terms of Θ_1, Θ_2, and Θ_3 (defined by Equation 8.3.12), as functions of a_4/a_1, ω_f, (defined by Equation 8.3.13) and | $a_3 - a_6$ |/a_2. For all combinations $a_3 - x_a = x_b - a_3 = 5a_2$ and $a_5 = a_2$.

For Case III the results are presented in Table 8.10 in terms of Ψ_1, Ψ_2, and Ψ_3 (defined by Equation 8.3.14) as functions of a_4/a_1, ω_s defined by Equation 8.3.15) and | $a_3 - a_6$ |/a_2. For all combinations $a_3 - x_a = x_b - a_3 = 5a_2$ and $a_5 = a_2$.

From Table 8.8 it is noted that the E_k's are not functions of a_4/a_1 if $\omega_c = 0$. However, for values of ω_c other than zero the E_k's are functions of a_4/a_1. For all cases (except Case I, $\omega_c = 0$) the effect of increasing a_4/a_1 is dependent upon the "dimensionless peak separation" | $a_3 - a_6$ |/a_2.

TABLE 8.10. Values[a] of Ψ_1, Ψ_2, and Ψ_3 for various combinations of a_4/a_1, ω_s, and $|a_3 - a_6|/a_2$

a_4/a_1	ω_s	$|a_3 - a_6|/a_2$	Ψ_1	Ψ_2	Ψ_3
0.1	0.0	2.25	3.004	2.480	0.471
		1.75	3.227	3.400	0.803
		1.25	10.95	5.885	2.162
		0.75	136.6	17.10	12.04
1.0		2.25	3.052	2.935	0.608
		1.75	4.238	4.260	1.175
		1.25	19.50	7.855	3.339
		0.75	222.6	24.05	18.50
10.0		2.25	3.402	3.690	0.892
		1.75	7.670	5.765	1.917
		1.25	39.22	11.67	5.806
		0.75	445.5	41.21	34.88
0.1	1.0	2.25	7.718	7.265	1.834
		1.75	7.941	9.270	2.572
		1.25	24.43	14.51	5.684
		0.75	311.4	38.99	28.46
1.0		2.25	7.399	7.695	1.915
		1.75	6.886	10.04	2.861
		1.25	47.42	19.69	9.075
		0.75	588.7	60.18	49.17
		2.25	10.12	12.25	3.800
10.0		1.75	54.56	23.61	11.11
		1.25	244.8	47.96	30.91
		0.75	2113.	160.6	155.8

[a] For all combinations $a_5 = a_2$ and $a_3 - x_a = x_b - a_3 = 5a_2$. The results were determined using the direct approach with $n = 24$ (Case III).

The functional dependence is complicated as a result of the complexity of the equation for $\hat{L}_i$ (i.e., Equation 8.5.7).

Comparing the results from Sections 8.3 and 8.5, it is observed that as $|a_3 - a_6|/a_2$ increases and as a_4/a_1 decreases, the values of E_k, Θ_k, and Ψ_k for the *two Gaussians class of experiments* approach the values for the *single Gaussian* class. (The comparison must be made with the values in Figures 8.4 through 8.6 corresponding to $t = 5.0$ because all the results in this section are only valid for $t = 5.0$.) This observation is reasonable because as $|a_3 - a_6|/a_2$ becomes large, the "interaction" between the two peaks becomes small. In fact, all the covariances between the two peaks (i.e., $\hat{\sigma}_{14}$, $\hat{\sigma}_{15}$, $\hat{\sigma}_{16}$, $\hat{\sigma}_{24}$, $\hat{\sigma}_{25}$, $\hat{\sigma}_{26}$, $\hat{\sigma}_{34}$, $\hat{\sigma}_{35}$, and $\hat{\sigma}_{36}$) approach zero as $|a_3 - a_6|/a_2$ becomes large.

8.6 Summary

In this chapter, experiments for which the Gaussian function is used to relate the dependent and independent variables were considered. The specific functions discussed are listed in Table 8.11.

The functions considered in Sections 8.2 and 8.4 are linear, and the functions considered in Sections 8.3 and 8.5 are nonlinear. The nonlinear equations are required for experiments in which the *peak locations* and the *peak widths* are unknown parameters.

For each function, three cases were considered. These were *constant error, constant fractional error, and counting statistics* for the dependent variable. For the independent variable, only *constant or zero error* was considered.

Several generalizations can be made about the results:

1. The predicted uncertainties are inversely proportional to the square root of the number of data points.

2. The predicted uncertainties are proportional to the constant K_{cy}, K_{fy}, or K_{sy}.

3. If the peaks are widely separated, the results for Sections 8.4 and 8.5 reduce to the comparable results for Sections 8.2 and 8.3.

4. The $\hat{\sigma}$'s for Sections 8.3 and 8.5 are larger than the comparable $\hat{\sigma}$'s for Sections 8.2 and 8.4 because of the increased number of unknown parameters.

5. For Sections 8.4 and 8.5, as the *location* of the second peak approaches the location of the first peak, all the $\hat{\sigma}$'s approach infinity. This observation is valid regardless of the ratio of *amplitudes* of the two peaks.

As an example of how the results in this chapter might be used, consider the following problem. The dependent and independent variables are to

TABLE 8.11. Summary of the functions discussed in Chapter 8

Function	Section
$y = a_1 \exp\left[-\left(\dfrac{x - x_0}{s} \right)^2 \right]$	8.2
$y = a_1 \exp\left[-\left(\dfrac{x - a_3}{a_2} \right)^2 \right]$	8.3
$y = a_1 \exp -\left(\dfrac{x - x_{01}}{s_1} \right)^2 + a_2 \exp\left[-\left(\dfrac{x - x_{02}}{s_2} \right)^2 \right]$	8.4
$y = a_1 \exp\left[-\left(\dfrac{x - a_3}{a_2} \right)^2 \right] + a_4 \exp\left[-\left(\dfrac{x - a_6}{a_5} \right)^2 \right]$	8.5

be related by Equation 8.1.2. "Counting statistics" applies for all values of the dependent variable y_i and the uncertainties for the measured values of x_i are assumed to be zero. The values of x_i are to be evenly spaced and the range is to extend from $x_a \approx a_3 - 3a_2$ to $x_b \approx a_3 + 3a_2$. The number of data points is to be 256. If a_3 is to be measured to 0.1% accuracy, what should be the value of a_1? For this problem, Case III of Section 8.3 is applicable. The dimensionless groups t and ω_s are determined from Equations 8.2.10 and 8.3.15 and are 3.0 and 0.0, respectively. From Figure 8.6, $\psi_3 = 0.42$, and from Equation 8.3.14, $(a_1)^{1/2} = 0.42 \times 1.0/0.001 \times (256)^{1/2} = 26.2$; thus $a_1 = 686$ (i.e., a_1 should be approximately equal to 700).

It should be noted that if data are required for combinations of the dimensionless groups not included in this chapter, it might be possible to determine the required information by interpolation or extrapolation. An alternative procedure is to extend the results to the required combinations using the techniques and equations included in Chapter 4 and in this chapter.

Problems

1. The dependent and independent variables for a proposed experiment are to be related by a "single Gaussian with three unknown parameters." The values of y_i are to be determined to 2.0% accuracy and the uncertainties for the values of x_i are negligible. The expected values of a_1, a_2, and a_3 are 100., 2.0, and 20.0 respectively. The values of x_i are to be evenly spaced, and the range x_a to x_b is to be from 15. to 25. If the number of data points is 50, what are the predicted values of σ_{a_1}, σ_{a_2}, and σ_{a_3}? $Ans.: \hat{\sigma}_{a_1} = 0.42$, $\hat{\sigma}_{a_2} = 0.0015$, $\hat{\sigma}_{a_3} = 0.0020$.

2. Adjust the number of data points in Problem 1 so that a_3 is determined to 1% accuracy.

3. The dependent and independent variables for a proposed experiment are to be related by "two Gaussians with unknown amplitudes." The uncertainties for the values of y_i are to be estimated by "counting statistics," and the uncertainties for the values of x_i are negligible. The peak separation $x_{01} - x_{02}$ is equal to $2s$ (where $s_1 = s_2 = s$), the expected values of a_1 and a_2 are both equal to 10000., and the values of x_i are evenly spaced. The range is symmetrical about x_{01} and $x_b - x_a = 10s$. How many data points are required for a_1 to be determined to 1% accuracy?

4. If a_2 in Problem 3 is reduced to 100., how many data points are required for a_1 to be determined to 1% accuracy?

Computer Projects

1. Recompute the results presented in Tables 8.4, 8.5, and 8.6 for $x_{01} - x_a = x_b - x_{01} = 2s$ and 10s, as well as 5s.

2. Recompute the results presented in Tables 8.8, 8.9 and 8.10 for $a_3 - x_a = x_b - a_3 = 2a_2$, and $10a_2$, as well as $5a_2$.

3. Predict the values of σ_{a_1}, σ_{a_2}, and σ_{a_3} as functions of the experimental variables for the following cases of the "single Gaussian with three unknown parameters" class of experiments:

    (A) $\sigma_{y_i} = K_{fy} \cdot y_i$ $\qquad\qquad$ $\sigma_{x_i} = K_{fx} \cdot x_i$,

    (B) $\sigma_{y_i}^2 = (K_{fy} \cdot y_i)^2 + K_{cy}^2$ $\qquad$ $\sigma_{x_i} = 0$,

    (C) $\sigma_{y_i}^2 = (K_{sy}^2 \cdot y_i) + K_{cy}^2$ $\qquad$ $\sigma_{x_i} = 0$.

    For all cases, the values of x_i are to be evenly spaced.

4. Repeat the previous problem for the "two Gaussians with unknown amplitudes" class of experiments.

5. For the following class of experiments:

$$y = a_1 \exp\left[-\left(\frac{x - x_0}{a_2}\right)^2\right],$$

    predict the values of σ_{a_1} and σ_{a_2} as functions of the experimental variables for the following cases:

    (A) $\sigma_{y_i} = K_{cy}$ $\qquad\qquad$ $\sigma_{x_i} = K_{cx}$,

    (B) $\sigma_{y_i} = K_{fy} \cdot y_i$ $\qquad\quad$ $\sigma_{x_i} = K_{cx}$,

    (C) $\sigma_{y_i} = K_{sy} \cdot (y_i)^{1/2}$ $\qquad$ $\sigma_{x_i} = K_{cx}$.

    For all cases the values of x_i are to be evenly spaced. Compare the results with the results of Sections 8.2 and 8.3.

Chapter 9

Applying Prediction Analysis to a Complicated Experiment

9.1 Introduction

The prediction analyses developed in Chapters 5 through 8 have several important similarities:

1. The functions considered are common mathematical models used for experiments in many diverse areas of science and technology.

2. For all the cases considered, the dependent variable y is a function of a single independent variable x.

3. The uncertainties associated with the dependent and independent variables (i.e., σ_{y_i} and σ_{x_i}) are assumed to be related to the variables by one of three simple models—*constant error, constant fractional error, or counting statistics* (i.e., Equation 4.5.2, 4.5.3, or 4.5.4).

4. Systematic errors are assumed to be negligible.

In this chapter, the prediction analysis of a proposed experiment is developed.[1] There are several basic differences between the following analysis and the analyses of the preceding chapters. For the following analysis:

1. The dependent variable y is a function of three independent variables.

2. The uncertainties associated with the variables are related to the variables in a complicated manner.

3. Systematic errors are not assumed to be negligible.

[1] In Sections 3.7 and 3.8, problems associated with least squares analysis of a similar experiment are discussed. In Section 3.8 it is shown that the least squares analyses of experiments of the type considered in this chapter are plagued by convergence problems. This situation is typical of many complicated experiments. The techniques discussed in Section 3.8 are often useful for promoting convergence.

9.2 Statement of the Problem

The dependent variable y is related to the variables x_1 and x_2 according to Equation 9.2.1:

$$y = \frac{(1 + a_1^2 x_1)(1 + a_2^2 x_1)}{(1 + a_1^2 x_2)(1 + a_2^2 x_2)}.$$

(9.2.1)

For the proposed experiment, the variables x_1 and x_2 are not independent. They are related to three independent variables v_1, v_2, and v_3:

$$x_1 \equiv \left(\frac{\pi}{v_1}\right)^2 + \left(\frac{\pi}{v_2}\right)^2 + \left(\frac{\pi}{v_3}\right)^2,$$

(9.2.2)

$$x_2 \equiv \left(\frac{k\pi}{v_1}\right)^2 + \left(\frac{\pi}{v_2}\right)^2 + \left(\frac{\pi}{v_3}\right)^2.$$

(9.2.3)

The integer k may be equal to or greater than two and varies from one data point to the next. All values from $k = 2$ to the maximum value of k are included as data points. The maximum value is an experimental variable.

Physically, the variables v_j are the *extrapolated dimensions* of a rectangular container in which the experiment is performed. The container is shown in Figure 9.1. The variables v_1, v_2, and v_3 are related to the dimensions u_1, u_2, and u_3 as follows;

$$v_1 = u_1 + 2d,$$

(9.2.4)

$$v_2 = u_2 + 2d,$$

(9.2.5)

$$v_3 = u_3 + 2d,$$

(9.2.6)

where d is called the *extrapolation length*.

The dependent variable y is actually the ratio of two quantities which

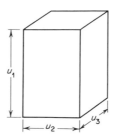

Figure 9.1. Experimental container.

are measured as a part of the experiment:

$$y = \frac{z_2}{z_1}.$$ (9.2.7)

The uncertainties in the values of y can be related to the uncertainties in z_1 and z_2. For this experiment z_1 and z_2 are uncorrelated and therefore Equation 2.6.8 may be used.

$$\left(\frac{\sigma_y}{y}\right)^2 = \left(\frac{\sigma_{z_1}}{z_1}\right)^2 + \left(\frac{\sigma_{z_2}}{z_2}\right)^2.$$ (9.2.8)

For the proposed experiment, the relative uncertainty in z_2 can be related to the relative uncertainty in z_1:

$$\frac{\sigma_{z_2}}{z_2} = \frac{1 + Px_2}{1 + Px_1} \cdot \frac{\sigma_{z_1}}{z_1},$$ (9.2.9)

where x_1 and x_2 are defined by Equations 9.2.2 and 9.2.3, and P is a property of the material within the experimental container.

The parameter P is a potential source of a systematic error. That is, an error in P will result in errors in all the values of σ_{z_2}/z_2 and therefore in all the values of σ_y. The effect on a_1 and a_2 will, however, be small. In addition, for the proposed experiment P is known to high precision, so it is assumed that P is not a source of a systematic error.

The uncertainties in the variables x_1 and x_2 are almost entirely due to the uncertainty in the extrapolation length d. The uncertainties in the dimensions u_1, u_2, and u_3 are negligibly small compared to the uncertainty in d. The uncertainty in d leads to uncertainties in x_1 and x_2 which are entirely of a systematic nature. In Section 4.9, ϵ is used to denote a systematic error. The proper notation for the systematic uncertainty of the extrapolation length d is therefore ϵ_d. The systematic uncertainties for the variables x_1 and x_2 are denoted as ϵ_{x_1} and ϵ_{x_2} and can be related to ϵ_d. From Equation 2.6.6:

$$\epsilon_{x_1} = \frac{\partial x_1}{\partial d} \cdot \epsilon_d = \left(\frac{\partial x_1}{\partial v_1} \cdot \frac{\partial v_1}{\partial d} + \frac{\partial x_1}{\partial v_2} \cdot \frac{\partial v_2}{\partial d} + \frac{\partial x_1}{\partial v_3} \cdot \frac{\partial v_3}{\partial d}\right) \cdot \epsilon_d.$$ (9.2.10)

This equation is an approximation and is exact only as ϵ_d approaches zero. From Equations 9.2.4 through 9.2.6:

$$\frac{\partial v_1}{\partial d} = \frac{\partial v_2}{\partial d} = \frac{\partial v_3}{\partial d} = 2.$$ (9.2.11)

Substituting Equation 9.2.11 into Equation 9.2.10, and then determining the other partial derivatives using Equation 9.2.2, we obtain:

$$\epsilon_{x_1} = -4\epsilon_d \cdot \left(\frac{\pi^2}{v_1^{\,3}} + \frac{\pi^2}{v_2^{\,3}} + \frac{\pi^2}{v_3^{\,3}} \right). \tag{9.2.12}$$

Similarly,

$$\epsilon_{x_2} = -4\epsilon_d \cdot \left(\frac{k^2\pi^2}{v_1^{\,3}} + \frac{\pi^2}{v_2^{\,3}} + \frac{\pi^2}{v_3^{\,3}} \right). \tag{9.2.13}$$

These equations can be checked in a simple manner. For example, if it is assumed that $v_1 = 10.2\pi$ and $v_2 = v_3 = 5.2\pi$, then from Equation 9.2.2:

$$x_1 = \left(\frac{1}{10.2} \right)^2 + 2\left(\frac{1}{5.2} \right)^2 = 0.0836. \tag{9.2.14}$$

If the error in d (i.e., ϵ_d) is $+0.1\pi$, then the true value of v_1 is 10.0π, the true values of v_2 and v_3 are 5.0π, the true value of x_1 is 0.0900, and ϵ_{x_1} is -0.0064. Computing ϵ_{x_1} from Equation 9.2.12:

$$\epsilon_{x_1} = -4 \times 0.1\pi \left(\frac{\pi^2}{(10\pi)^3} + \frac{2\pi^2}{(5\pi)^3} \right) = -0.0068. \tag{9.2.15}$$

The true value of ϵ_{x_1} and the value computed by Equation 9.2.12 are in close agreement. They are not exactly the same because Equation 9.2.10 is actually an approximation. Equation 9.2.13 can be checked in a similar manner, and close agreement will also be observed.

Since the uncertainties in the variables x_1 and x_2 are assumed to be systematic in nature, the statistical uncertainties σ_{x_1} and σ_{x_2} are assumed to be zero. This assumption simplifies the computation of the predicted statistical uncertainties $\hat{\sigma}_{a_1}$ and $\hat{\sigma}_{a_2}$.

The purpose of the experiment is to determine the values of the parameters a_1, a_2, and $a_1^2 + a_2^2$. The purpose of the prediction analysis is to predict how the experimental variables and parameters will affect the accuracies of the values of a_1, a_2, and $a_1^2 + a_2^2$ as determined by least squares analysis of the experimental data.

9.3 Design of the Experiment

The experiment is not necessarily limited to measurement in one container. In fact, the greater the number of containers, the more accurate will be the resulting values of a_1, a_2, and $a_1^2 + a_2^2$. The limit on the number of containers is basically an economic consideration.

Each data point corresponds to a particular value of k for a given container. The minimum value of k is two, and the maximum value is an experimental variable. Defining K_j as the maximum value of k for the jth container, and J as the total number of containers, the total number of data points n is:

$$n = \sum_{j=1}^{J} (K_j - 1). \tag{9.3.1}$$

For each container there are five experimental variables. These are u_{1j}, u_{2j}, u_{3j}, $(\sigma_{z_1}/z_1)_j$, and K_j. The total number of experimental variables is therefore $5J$. Theoretically, the number of experimental variables is large; however, the number can be reduced considerably if practical aspects of the problem are taken into consideration. The following set of assumptions reduce the number to $J + 3$:

1. The value of K_j is the same for all containers and is denoted as K.
2. The value of σ_{z_1}/z_1 is the same for all containers.
3. The dimensions u_{2j} and u_{3j} are equal.
4. The shape factor R (defined as the ratio $u_{2j}/u_{1j} = u_{3j}/u_{1j}$) is the same for all containers.

The $J + 3$ variables are therefore u_{1j} ($j = 1, 2, \cdots, J$), K, σ_{z_1}/z_1 and R. The number of data points (from Equation 9.3.1) is $J \cdot (K - 1)$. Three input parameters affect the analysis. These are d, ϵ_d, and P. The unknown parameters are a_1 and a_2.

The parameters d, P, a_1, and a_2 are all functions of the physical properties of the material within the containers. For a given material these parameters cannot be changed. The parameter ϵ_d is the uncertainty associated with the parameter d. The value of K can be increased, but an increase in K can only be made by changes in the design that increase the cost and time requirements of the experiment. Similarly, the value of σ_{z_1}/z_1 can be reduced by changes which require additional investments in time and money. The dimensions and shape of the containers will also affect the design and cost of the experiment. However, for each combination of the variables and parameters, there is a set of dimensions u_{1j} which minimize the predicted uncertainties of the parameters a_1 and a_2.

The experiment is to be designed for a material in which the parameters are:

(1) $a_1 = 10$ cm,
(2) $a_2 = 4$ cm,
(3) $P = 3530$ cm^2,
(4) $d = 2.5$ cm,
(5) $\epsilon_d = 0.5$ cm.

The listed values of a_1 and a_2 can be considered as results of a theoretical analysis of the problem. The purpose of the experiment is to verify the theoretical calculation of these parameters. The values of P, d, and ϵ_d are known from other experiments.

9.4 Prediction Analysis of the Proposed Experiment

From Equation 4.9.5:

$$\text{Predicted value of } \sigma_{a_k}{}^2 = \hat{\sigma}_{a_k}{}^2 + \hat{\epsilon}_{a_k}{}^2, \tag{9.4.1}$$

where $\hat{\sigma}_{a_k}$ is the predicted statistical uncertainty for a_k, and $\hat{\epsilon}_{a_k}$ is the systematic uncertainty for a_k resulting from the uncertainty in the extrapolation length d. Other sources of systematic uncertainty are assumed to be negligible.

To determine $\hat{\sigma}_{a_k}$, the direct approach (discussed in Section 4.4) must be employed. The experiment is too complicated to expect a simple analytical solution to the problem.

The equations for the derivatives $\hat{F}_{a_1}{}^i$ and $\hat{F}_{a_2}{}^i$ are determined by taking the derivatives of $\hat{F}^i$ with respect to the parameters a_1 and a_2. From Equation 3.6.1:

$$\hat{F}^i = y_i - \frac{(1 + a_1^2 x_{1i})(1 + a_2^2 x_{1i})}{(1 + a_1^2 x_{2i})(1 + a_2^2 x_{2i})}, \tag{9.4.2}$$

and therefore:

$$\hat{F}_{a_1}{}^i \equiv \frac{\partial \hat{F}^i}{\partial a_1} = \frac{(1 + a_1^2 x_{1i})(1 + a_2^2 x_{1i})}{(1 + a_1^2 x_{2i})(1 + a_2^2 x_{2i})} \cdot \frac{2a_1(x_{2i} - x_{1i})}{(1 + a_1^2 x_{1i})(1 + a_1^2 x_{2i})}. \tag{9.4.3}$$

Substituting Equation 9.2.1 into Equation 9.4.3 gives:

$$\hat{F}_{a_1}{}^i = y_i \cdot \frac{2a_1(x_{2i} - x_{1i})}{(1 + a_1^2 x_{1i})(1 + a_1^2 x_{2i})}. \tag{9.4.4}$$

Similarly:

$$\hat{F}_{a_2}{}^i = y_i \cdot \frac{2a_2(x_{2i} - x_{1i})}{(1 + a_2^2 x_{1i})(1 + a_2^2 x_{2i})}. \tag{9.4.5}$$

From Equation 3.6.16:

$$\hat{L}_i = \sigma_{y_i}{}^2 + (\hat{F}_{x_1}{}^i \cdot \sigma_{x_{1i}})^2 + (\hat{F}_{x_2}{}^i \cdot \sigma_{x_{2i}})^2. \tag{9.4.6}$$

Since the values of σ_{x1_i} and σ_{x2_i} are assumed to be zero, Equation 9.4.6 reduces to a simple form:

$$\hat{L}_i = \sigma_{y_i}^2. \tag{9.4.7}$$

In the preceding equations, the index i refers to the ith data point. This index can be related to the index k of the jth container:

$$i = (k - 1) + (j - 1) \cdot (K - 1). \tag{9.4.8}$$

The elements of the matrix $\hat{C}$ are:

$$\hat{C}_{11} = \sum_{i=1}^{n} \frac{\hat{F}_{a1}{}^i \cdot \hat{F}_{a1}{}^i}{\hat{L}_i}, \tag{9.4.9}$$

$$\hat{C}_{12} = \sum_{i=1}^{n} \frac{\hat{F}_{a1}{}^i \cdot \hat{F}_{a2}{}^i}{\hat{I}_{i}}, \tag{9.4.10}$$

$$\hat{C}_{22} = \sum_{i=1}^{n} \frac{\hat{F}_{a2}{}^i \cdot \hat{F}_{a2}{}^i}{\hat{L}_i}. \tag{9.4.11}$$

The summation of data points over the index i is actually a double summation over the indices j and k:

$$\sum_{i=1}^{n} = \sum_{j=1}^{J} \sum_{k=2}^{K}. \tag{9.4.12}$$

Substituting Equations 9.2.8, 9.2.9, 9.4.4, 9.4.5, and 9.4.12 into Equations 9.4.9 through 9.4.11, and dropping the index i,

$$\hat{C}_{11} = \sum_{j=1}^{J} \sum_{k=2}^{K} \left[\frac{2a_1(x_2 - x_1)}{(1 + a_1^2 x_1)(1 + a_1^2 x_2)} \right]^2 \Big/ D, \tag{9.4.13}$$

$$\hat{C}_{12} = \sum_{j=1}^{J} \sum_{k=2}^{K} \left(\frac{4a_1 a_2(x_2 - x_1)^2}{(1 + a_1^2 x_1)(1 + a_1^2 x_2)(1 + a_2^2 x_1)(1 + a_2^2 x_2)} \Big/ D \right), \tag{9.4.14}$$

$$\hat{C}_{22} = \sum_{j=1}^{J} \sum_{k=2}^{K} \left[\frac{2a_2(x_2 - x_1)}{(1 + a_2^2 x_1)(1 + a_2^2 x_2)} \right]^2 \Big/ D, \tag{9.4.15}$$

where

$$D \equiv \left(\frac{\sigma_y}{y} \right)^2 = \left(\frac{\sigma_{z1}}{z_1} \right)^2 \cdot \left[\left(\frac{1 + P x_2}{1 + P x_1} \right)^2 + 1 \right]. \tag{9.4.16}$$

The procedure for computing the values of $\hat{\sigma}_{a_1}$ and $\hat{\sigma}_{a_2}$ and $\hat{\sigma}_{(a_1^2+a_2^2)}$ is as follows:

1. Compute the values of x_1 and x_2 for each combination of k and j using Equations 9.2.2 and 9.2.3.
2. Compute the elements of the matrix $\hat{C}$ using Equations 9.4.13 through 9.4.16.
3. Invert the matrix and then compute the values of $\hat{\sigma}_{a_1}$, $\hat{\sigma}_{a_2}$, and $\hat{\sigma}_{12}$ in the usual manner (i.e., according to Equations 4.3.3 and 4.3.4).
4. To predict the variance of the sum $a_1^2 + a_2^2$, an equation must be derived which relates this variance to $\hat{\sigma}_{a_1}$, $\hat{\sigma}_{a_2}$, and $\hat{\sigma}_{12}$. For this experiment there is no reason to assume that a_1 and a_2 are uncorrelated, so Equation 2.6.18 must be used. From Equations 2.6.20 and 2.6.21:

$$[\hat{\sigma}_{(a_1^2+a_2^2)}]^2 = 4a_1^2\hat{\sigma}_{a_1}^2 + 4a_2^2\hat{\sigma}_{a_2}^2 + 8a_1a_2\hat{\sigma}_{12}. \qquad (9.4.17)$$

The complexity of Equations 9.2.12 and 9.2.13 as well as Equation 9.2.1 is too great to expect simple analytical solutions for $\hat{\epsilon}_{a_1}$, $\hat{\epsilon}_{a_2}$, and $\hat{\epsilon}_{(a_1^2+a_2^2)}$. The direct approach must therefore be used. To accomplish this, the following procedure is suggested:

1. For the given values of u_j, R, and d, for each combination of k and j compute the values of x_1 and x_2 using Equations 9.2.4 through 9.2.6, 9.2.2, and 9.2.3.
2. Compute the values of y for each combination of k and j using Equation 9.2.1.
3. Recompute the values of x_1 and x_2 after changing the value of d by an amount equal to ϵ_d. With a slight loss of accuracy, this step could also be performed using Equations 9.2.12 and 9.2.13.
4. Using the values of y computed in Step 2 and the values of x_1 and x_2 computed in Step 3, perform a least squares analysis to determine the values of a_1 and a_2 which best fit the data. Denoting these values as a_1' and a_2', the values of $\hat{\epsilon}_{a_1}^2$, $\hat{\epsilon}_{a_2}^2$, and $\hat{\epsilon}_{(a_1^2+a_2^2)}$ are determined using Equations 9.4.18 through 9.4.20:

$$\hat{\epsilon}_{a_1}^2 = (a_1' - a_1)^2, \qquad (9.4.18)$$

$$\hat{\epsilon}_{a_2}^2 = (a_2' - a_2)^2, \qquad (9.4.19)$$

$$[\hat{\epsilon}_{(a_1^2+a_2^2)}]^2 = (a_1'^2 + a_2'^2 - a_1^2 - a_2^2)^2. \qquad (9.4.20)$$

The final step in the analysis is to determine the predicted values of σ_{a_k} and $\sigma_{(a_1^2+a_2^2)}$ (i.e., the total predicted uncertainties) using Equation 9.4.1. The analysis should be repeated for enough combinations of the experimental variables to ensure logical decisions regarding the design and planning of the experiment.

9.5 Results of the Analysis

The *total predicted uncertainties* are denoted as σ_{a_1}, σ_{a_2}, and $\sigma_{(a_1{}^2+a_2{}^2)}$ and are presented in terms of the following dimensionless groups:

$$\Gamma_1{}^t \equiv \frac{\sigma_{a_1}/a_1}{\sigma_{z_1}/z_1}, \tag{9.5.1}$$

$$\Gamma_2{}^t \equiv \frac{\sigma_{a_2}/a_2}{\sigma_{z_1}/z_1}, \tag{9.5.2}$$

$$\Gamma_{12}{}^t \equiv \frac{\sigma_{(a_1{}^2+a_2{}^2)}/(a_1{}^2 + a_2{}^2)}{\sigma_{z_1}/z_1}. \tag{9.5.3}$$

These total predicted uncertainties include both the *statistical* and *systematic* uncertainties (see Equation 9.4.1). The dimensionless groups containing the statistical and systematic uncertainties are denoted as:

$$\Gamma_1{}^{st} \equiv \frac{\hat{\sigma}_{a_1}/a_1}{\sigma_{z_1}/z_1}, \tag{9.5.4}$$

$$\Gamma_2{}^{st} \equiv \frac{\hat{\sigma}_{a_2}/a_2}{\sigma_{z_1}/z_1}, \tag{9.5.5}$$

$$\Gamma_{12}{}^{st} \equiv \frac{\hat{\sigma}_{(a_1{}^2+a_2{}^2)}/(a_1{}^2 + a_2{}^2)}{\sigma_{z_1}/z_1}, \tag{9.5.6}$$

$$\Gamma_1{}^{sy} \equiv \frac{\hat{\epsilon}_{a_1}/a_1}{\sigma_{z_1}/z_1}, \tag{9.5.7}$$

$$\Gamma_2{}^{sy} \equiv \frac{\hat{\epsilon}_{a_2}/a_2}{\sigma_{z_1}/z_1}, \tag{9.5.8}$$

$$\Gamma_{12}{}^{sy} \equiv \frac{\hat{\epsilon}_{(a_1{}^2+a_2{}^2)}/(a_1{}^2 + a_2{}^2)}{\sigma_{z_1}/z_1}. \tag{9.5.9}$$

The Γ's can be related using Equation 9.4.1:

$$(\Gamma_1{}^t)^2 = (\Gamma_1{}^{st})^2 + (\Gamma_1{}^{sy})^2, \tag{9.5.10}$$

$$(\Gamma_2{}^t)^2 = (\Gamma_2{}^{st})^2 + (\Gamma_2{}^{sy})^2, \tag{9.5.11}$$

$$(\Gamma_{12}{}^t)^2 = (\Gamma_{12}{}^{st})^2 + (\Gamma_{12}{}^{sy})^2. \tag{9.5.12}$$

The dimensionless groups $\Gamma_1{}^{st}$, $\Gamma_1{}^{sy}$, $\Gamma_2{}^{st}$, $\Gamma_2{}^{sy}$, and $\Gamma_{12}{}^t$ are plotted in Figure 9.2 as functions of v_1 for an experiment performed using only one container. The values of the variables K, R, and σ_{z_1}/z_1 were assumed to be 8, 1.0, and 0.01, respectively. The values of the parameters a_1, a_2, P, d, and ϵ_d are listed in Section 9.3. Several interesting points should be noted:

1. The values of $\Gamma_1{}^{st}$ and $\Gamma_2{}^{st}$ are large compared to $\Gamma_1{}^{sy}$ and $\Gamma_2{}^{sy}$ and therefore from Equations 9.5.9 and 9.5.10, $\Gamma_1{}^t$ and $\Gamma_2{}^t$ are approximately equal to $\Gamma_1{}^{st}$ and $\Gamma_2{}^{st}$, respectively. In other words, for the combination of parameters and variables under discussion, the systematic uncertainties resulting from an uncertainty in the extrapolation length d are small compared to the statistical uncertainties. From the results there appears to be very little incentive for reducing the uncertainty ϵ_d.

2. The optimum values of v_1 for measuring a_1, a_2, and $a_1{}^2 + a_2{}^2$ are different. The optimum value for measuring $a_1{}^2 + a_2{}^2$ is larger than the values for a_1 and a_2 because of the influence of the covariance σ_{12}. For this experiment the covariance is negative.

3. The quantities $\Gamma_1{}^{sy}$ and $\Gamma_2{}^{sy}$ decrease monotonically as v_1 increases. This result could have been anticipated by examining Equations 9.2.12 and 9.2.13. Both ϵ_{x_1} and ϵ_{x_2} decrease as v_1 increases.

4. An examination of Equations 9.4.13 through 9.4.16 reveals that all the values of C are inversely proportional to $(\sigma_{z_1}/z_1)^2$ and therefore the predicted statistical uncertainties $\hat{\sigma}_{a_1}$, $\hat{\sigma}_{a_2}$, and $\hat{\sigma}_{(a_1{}^2+a_2{}^2)}$ are proportional to σ_{z_1}/z_1. That is, the curves for $\Gamma_1{}^{st}$, $\Gamma_2{}^{st}$, and $\Gamma_{12}{}^{st}$ are not dependent upon the value of σ_{z_1}/z_1. The predicted systematic errors $\hat{\epsilon}_{a_1}$, $\hat{\epsilon}_{a_2}$, and $\hat{\epsilon}_{(a_1{}^2+a_2{}^2)}$ are independent of this ratio, and therefore $\Gamma_1{}^{sy}$, $\Gamma_2{}^{sy}$ and $\Gamma_{12}{}^{sy}$ are inversely proportional to σ_{z_1}/z_1. The groups $\Gamma_1{}^t$, $\Gamma_2{}^t$, and $\Gamma_{12}{}^t$ therefore must also be functions of σ_{z_1}/z_1.

As an example of how the results might be used, assume that the primary objective of the experiment is to measure a_1 to at least 5% accuracy. Assume that a cubic container 95 cm on each side is available for the experiment. The extrapolated dimensions are therefore equal to $95 + 2 \times 2.5 = 100$ cm and from Figure 9.2 it is seen that the container is close to the optimum size for $\sigma_{z_1}/z_1 = 0.01$. Assume that $K = 8$ and σ_{z_1}/z_1 can be varied to satisfy the experimental objective.

From Equations 9.5.1, 9.5.4, 9.5.6, and 9.5.10, and from Figure 9.2,

$$(\Gamma_1{}^t)^2 = \left(\frac{0.05}{\sigma_{z_1}/z_1}\right)^2 = 11.6^2 + \left(1.9 \cdot \frac{0.01}{\sigma_{z_1}/z_1}\right)^2. \qquad (9.5.13)$$

Solving for σ_{z_1}/z_1, a value of 0.004 is obtained. That is, z_1 must be measured to an accuracy of 0.4% if a_1 is to be determined to 5% accuracy. It should

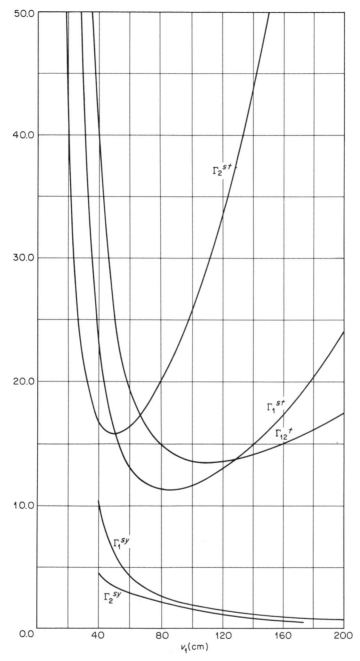

Figure 9.2. Γ_1^{st}, Γ_1^{sy}, Γ_2^{st}, Γ_2^{sy}, and Γ_{12}^{t} as functions of v_1 for $K = 8$, $R = 1.0$, and $\sigma_{z_1}/z_1 = 0.01$.

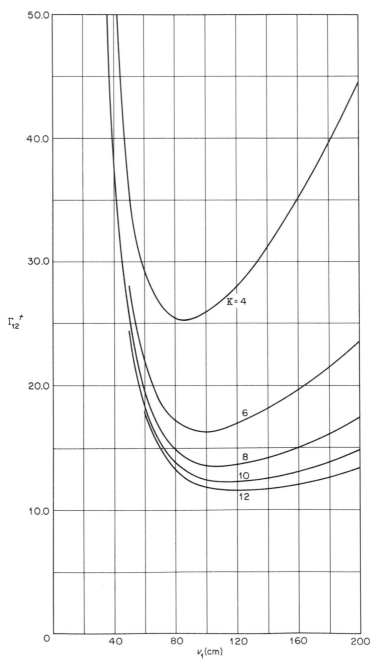

Figure 9.3. $\Gamma_{12}{}^t$ as a function of v_1 and K for $R = 1.0$ and $\sigma_{z_1}/z_1 = 0.01$.

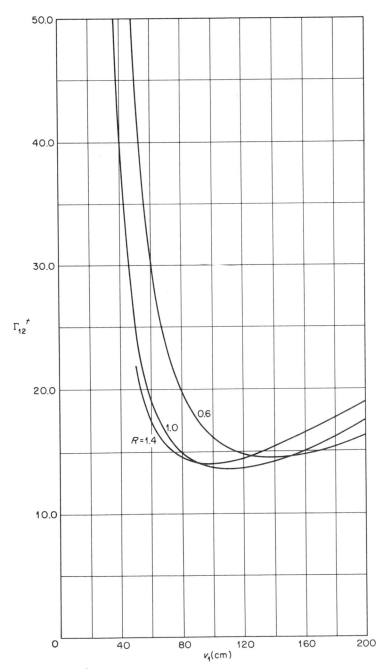

Figure 9.4. $\Gamma_{12}{}^{t}$ as a function of v_1 and R for $K = 8$ and $\sigma_{z_1}/z_1 = 0.01$.

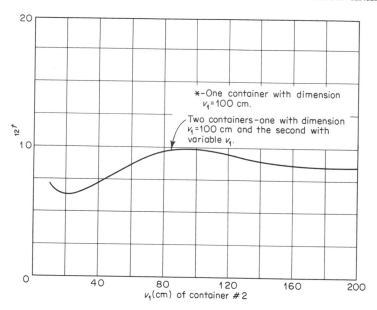

Figure 9.5. $\Gamma_{12}{}^t$ as a function of v_1 of container No. 2 for $K = 8$, $R = 1.0$, and $\sigma_{z_1}/z_1 = 0.01$.

be noted that the systematic error provides a limit for the accuracies to which a_1, a_2 and $a_1{}^2 + a_2{}^2$ can be measured. For example, even if σ_{z_1}/z_1 is reduced to 0.0, for the case under consideration the best accuracy to which a_1 can be measured is 1.9%.

By extending the value of K, the accuracy of the experiment can be improved. To evaluate the improvement in accuracy that can be obtained by increasing K, the calculations were repeated for other values of K and the results (for only $\Gamma_{12}{}^t$) are plotted in Figure 9.3. It is observed that the optimum value of v_1 increases and the minimum value of $\Gamma_{12}{}^t$ decreases as K is increased. The fractional improvement in accuracy decreases as K is increased. For example, there is a much greater incentive to increase K from 4 to 8 than from 8 to 12.

The shape factor R is an experimental variable. Values of $\Gamma_{12}{}^t$ are plotted in Figure 9.4 as a function of v_1 for three values of R—0.6, 1.0, and 1.4. It can be seen that the optimum value of v_1 increases as R decreases; however, the minimum value of $\Gamma_{12}{}^t$ is obtained for a shape factor equal to one.

The values of the Γ's can be reduced significantly if data from two containers rather than just one are analyzed together. In Figure 9.5, $\Gamma_{12}{}^t$ is plotted as a function of v_1 of the second container. The value of v_1 for the first container is 100 cm. Values of K, R, and σ_{z_1}/z_1 are 8, 1.0, and 0.01, respectively. As the dimensions of the second container become increasingly different from those of the first, $\Gamma_{12}{}^t$ decreases and eventually reaches a minimum. The optimum value of v_1 for the second container is observed to be 20 cm. For the combination of two containers with dimensions 100 and 20 cm, the value of $\Gamma_{12}{}^t$ is observed to be 6.35 as compared to 13.67 for a single container with a dimension of 100 cm. To avoid building a second container, a shortcut is to use only the first container but perform the experiment twice. From Fig. 9.5 it is observed that a value for $\Gamma_{12}{}^t$ of 9.87 is predicted. This value of $\Gamma_{12}{}^t$ is lower than the value of 13.67 that is predicted for the experiment when it is performed only once, but higher than the minimum predicted value of 6.35. The experimenter must decide if the potential 35% improvement in accuracy [i.e., $100 \times (9.87 - 6.35)/9.87$] is worth the additional cost of building a second container. If a second container is worthwhile, the possibility of a third or even fourth container should also be explored.

It is interesting to note that performing the experiment twice does *not* reduce $\Gamma_{12}{}^t$ by exactly a factor of $\sqrt{2}$. The value of $\Gamma_{12}{}^{st}$ is reduced by this factor; however, $\Gamma_{12}{}^{sy}$ remains unchanged. The net result (using Equation 9.5.12) is that $\Gamma_{12}{}^t$ is reduced by a factor which is less than $\sqrt{2}$. For the case considered in Figure 9.5, however, $\Gamma_{12}{}^{sy}$ is small compared to $\Gamma_{12}{}^{st}$, and thus the reduction is nearly equal to a factor of $\sqrt{2}$ (i.e., $13.67/\sqrt{2} = 9.65 \approx 9.87$).

9.6 Conclusions

In this chapter prediction analysis was applied to the problem of planning a complicated experiment. The results of the analysis provide a *quantitative* basis for designing the experiment. For example:

1. For a value of $K = 8$, a shape factor $R = 1.0$, and a value of $\sigma_{z_1}/z_1 = 0.01$, in Figure 9.2 it is seen that the optimum value of v_1 for measuring $a_1^2 + a_2^2$ is 110 cm.

2. From Figure 9.3 it is seen that the accuracy of the results is improved by increasing the value of K. However, the fractional improvement in accuracy becomes smaller as K is increased.

3. From Figure 9.4 it is seen that the best accuracy can be achieved using a shape factor R equal to 1.

4. From Figure 9.5 it is seen that the accuracy of the results can be improved by increasing the number of containers.

The important point to recognize in this chapter is that prediction analysis can be applied to a complicated experiment. Extremely useful results can be obtained regardless of whether or not the analysis is completed analytically. The experiment discussed in this chapter could be analyzed only by using the direct approach. Nevertheless, the results yield a useful picture of the experiment which permits planning based upon quantitative predictions of the uncertainties of the experimental results.

Once the techniques of *performing* and *using* prediction analyses have been mastered, the experimentalist has a powerful tool at his disposal. This is self-evident because the probability of a successful experiment is increased considerably if the experiment is adequately analyzed in the planning phase.

Index

289